UNDERSTANDING ECOLOGY

By

Dr. Veena

Dept. of Zoology

M.M.H. College

Ghaziabad (U.P.)

(India)

DISCOVERY PUBLISHING HOUSE PVT. LTD.

NEW DELHI-110 002

First Published-2009

ISBN 978-81-8356-456-4

Published by:

DISCOVERY PUBLISHING HOUSE PVT. LTD.
4831/24, Ansari Road, Prahlad Street,
Darya Ganj, New Delhi-110002 (India)
Phone: 23279245 • Fax: 91-11-23253475
E-mail: dphbooks@rediffmail.com
dphtemp@indiatimes.com
Website: www.discoverypublishinghouse.com

Printed at:

Sachin Printers
Delhi

Preface

The present title "Understanding Ecology" has been written for those students interested in careers in diverse fields of biological sciences. It provides a structured approach to learning by covering all the important topics in a uniform, systematic format. The book has been comprehensively designed incorporating recent advances in this fast moving field. It also provides accessible information on ecology in compact form for undergraduate students in biology and related life sciences. It is intelligible to the educated layman, though it deals with some complex ideas. It is an adequate text for all the requirements of students in this area. In addition, busy lecturers who require a quick reference compendium will find it useful, particularly for tutional planning. Simple, yet hopefully clear figures and tables are provided throughout the book.

The over-riding goal of this book, and indeed of the whole *Understanding series*, is to present the essential information concering ecology in a compact, readily accessible form which leads itself to student learning and revision. The convergence of various approaches has generated a rich panorama of detail, the significance of which we are still attempting to unraval. The present text has been written as an introduction to this rapidly growing field.

To make the work more comprehensive and informative, the author has consulted many authoritative books, research journals, abstracts, monographs etc., so there can be no claim to originality except in the manner of treatment.

The author expresses his thanks to his friends and colleagues whose continue inspirations have initiated him to bring out this book.

The author expresses his gratitude to Mr. Wasan and staff of M/s Discovery Publishing House Pvt. Ltd. for their whole hearted co-operation in the publication of this book.

In the mean time, the author will remain sincerely responsible for any shortcomings of the book and be grateful to the readers for their suggestions and constructive criticism for the continuous betterment of the book. He takes this opportunity to appeal to the readers to send their suggestions straightaway to his Publisher.

Author

Preface

The present book on Environmental Ecology has been written for [illegible] students and [illegible] in the field of [illegible]. It provides a structured approach [illegible] important topics in a uniform systematic fashion. The book has been [illegible] designed [illegible] [illegible] and related [illegible] [illegible] throughout the book.

[illegible]

[illegible]

The author [illegible] [illegible] in this book.

The author [illegible]

[illegible]

CONTENTS

1

ENVIRONMENTAL FACTORS

A seemingly infinite variety of environmental variables are associated with mating and the emergence of the next generation. Some of these variables are internal stimuli, which determine gonadal activity, and others are environmental factors or cues to which the reproductive cycle is adapted. This chapter deals with the role of these factors or cues.

NUTRITIONAL EFFECTS ON REPRODUCTION

Food has nutritional effects concerned with growth and energy metabolism and also specific effects on the reproductive system. Levels of quality as well as quantity affect reproductive ability in both sexes. In malnutrition or starvation the retarded output of pituitary hormones, including gonadotropins, causes inanition in both sexes. Sadleir (1969) emphasized the role of nutrition in reproductive seasonality of wild mammals, both in producing adequate growth in young individuals and in achieving proper support for gametogenesis in adults. Malnutrition delays sexual maturity and fecundity by depressing the activity of the hypothalamus. The study of nutritional effects on reproduction has generally emphasized the well-recognized needs for growth and maintenance and their relationship to the functions of the various parts of the hypophyseal-pituitary-gonadal axis.

There are many examples of proper nutrition promoting fecundity and of inadequate nutrition preventing reproduction. In most taxa food intake increase with need, but the sequential relationship of nourishment and reproduction, however, varies widely among groups of vertebrates. In midlatitudes reptiles, for example, nutrition in summer provides much of the energy for egg formation the following spring, and food intake may actually decline during the reproductive period. In most mammals both pregnancy and lactation place a high demand on the female, but the compensatory increase in nutrition must be synchronized not only with demand but also with hibernation, migration, and food availability. Some bears produce young during periods of fasting during periods of lethargy in the winter), and a number of pinniped species fast for several weeks while lactating. The actual details of nutritional requirements for reproduction are known for only a few species but the variation is substantial.

The nutritional quality of food has a paramount effect on reproductive performance. In the laboratory rat, inadequate protein content arrests both steroidogenesis and gametogenesis. The protein component, however, may not be absolutely essential in all kinds of mammals. In the domestic rabbit, for example, gonadotropins are synthesized even on a protein-fee diet, and such protein as is needed is apparently derived from nitrogenous products from the rabbit's own tissue. Vitamin E is essential for full development of reproductive structures in the laboratory rat. A deficiency causes sterility, and such a sterility is permanent. Vitamin E is not known to function as a cue for reproduction in nature. Experimentally, the collared lemming (*Dicrostonyx groenlandicus*) exhibits reduced fertility when maintained on a Vitamin D—depleted diet. During the winter, when there is no sunlight and when lemmings are subnivean, vitamin D must come from dietary sources.

Nutrition is a direct reflection of the individual's environment. The protein content of foods of the red-backed vole (*Clethrionomys rufocanus bedfordi*) appears in their serum protein and is related to the productivity of their habitat. The fecundity of this vole, moreover, varies with the protein content of its natural foods: protein content in vole foods is markedly lower on mature forest land than on plantations of young seedlings. In forests, where the protein of vole foods falls below maintenance levels from October to April, the voles cease to breed in late spring. In plantations, where protein in vole foods is optimal for six months of the year, populations are higher and reproduction extends into summer.

Specific differences in sensitivity to malnutrition are seen among species of comparable size and diet. A 30% reduction of an ad libitum diet was followed by reduced testicular activity in deer mice (*Peromyscus maniculatus*), amounting to azoospermia in some individuals, but with no measurable effect in house mice (*Mus musculus*). Hamsters exhibit an effect similar to that seen in deer mice: a reduction of food resulted in a reduction in testis size, independent of an intact pineal gland.

The domestic horse, a seasonally reproductive and photosensitive species, responds also to change in nutrition, but this may be masked by changes in ambient temperature, reflecting fluctuations in energy demands. An increase in energy content in the diet of mares in winter was followed by an earlier return to estrus than in controls. On the other hand, mares kept in winter at 10 to 15°C ovulated earlier than a control group kept at -15 to 9°C. One must conclude that horses are sensitive to nutritional levels and that nutritional needs are modified by ambient temperature, as in some wild birds.

In a broad spectrum of both freshwater and marine teleost fish, there is a general positive correlation of quality and quantity of food with the number of ova. The correlation is apparent in captive fish maintained on controlled diets as well as those in the wild. Food supply is usually density dependent, and in wild, free-ranging populations food abundance may be deceptive, for greater densities of fish mean less food for each individual. Wild populations of fish regularly face periods of plenty alternating with periods of poverty. The Atlantic herring (*Clupea harengus*), like many temperate ectotherms, has a marked annual cycle of fat accumulation during a nonreproductive period and fat depletion at spawning. Herring captured after spawning and kept without food for 78 days not only survived but also suffered no irreversible damage to reproductive capacity, although there was no gametogenesis as in well-fed fish. In *Gillichthys mirabilis*, a goby of brackish water, starvation results in prompt gonadal regression but does not prevent seasonal gonadal recrudescence if there has already been an accumulation of energy. The three-spined stickleback (*Gasterosteus aculeatus*), when kept on diets that varied only in amount and frequency, grew more rapidly with increased nutrition, and the weight of mature fish was correlated with increased frequency of spawning as well as number and mass of eggs produced. The convict (*Cichlasoma nigrofasciatum*), a Central American cichlid fish, becomes increasingly fecund with more frequent feeding: spawning is positively correlated with ratios of 1g of food given daily, three times a week or once a week. In the African catfish (*Clarias lazera*) there is a sexually dissimilar gonadal response to food and temperature. Males are more sensitive to temperature than to food, but females are more responsive to varying levels of nutrition. Captive hoddock (*Melanogrammus aeglefinus*) produced a mean of 16.6 (10-25) batches of eggs over a mean period of 33.2 (19-59) days while maintained under controlled feeding of diets that contained from less than 5 kcal/day to more than 13 kcal/day. Egg production and food intake were positively correlated.

Egg number does not always relate directly to nutritional levels. When kept on diets of various protein levels (15%, 31%, and 47%), the guppy (*Poecilia reticulata*) showed greater somatic development in response to greater protein intake. Although gonadal weights were positively correlated with body weight, an increase in ovarian size was not accompanied by an increase in number of ova in females fed higher protein diets.

Nutritional Variations in Nature

There are many examples of wild vertebrates failing to reproduce in years of general food scarcity. A few examples will illustrate adjustments in fecundity following natural variations in food supply. Many of these species live in high latitudes, where there may be few different kinds of food and the decline or failure of one food source or prey species may be catastrophic for the predator. Some raptors lay large clutches in years of prey abundance, and some passerine birds raise large broods when their favorite food abounds.

The lapwing (*Vanellus vanellus*) arrives on its nesting ground in southern Sweden at the same time every year, but the length of the prelaying period is negatively correlated with the abundance of lumbricid worms, the main fare of this plover.

Secondary Plant Compounds

The effect of food is complex and may reflect not only its general nutritional value but also, for herbivores, specific influences of secondary plant compounds on the reproductive system. Many plants contain materials that structurally and functionally resemble gonadal steroids. Some of these compounds are chemically unlike estrogens of animal origin, and many are not steroids, but they do stimulate or depress the reproductive system. Some have been compared with gonadotropins, but it is not known how their effects are produced. Although some of these secondary plant compounds are known not to function as estrogens, the term *phytoestrogen* is well established in the literature and has long been used as an inclusive name for sexual stimulants, as well as suppressants, that occur in plants.

Studies in the late 1930s revealed that extracts from fresh alfalfa leaves stimulate ovulation in the domestic rabbit, and it was noted that commercially prepared alfalfa feed is much less effective.

Early Chinese books included many botanical drugs that were administered to treat reproductive disorder. Although much of this information lies in the realm of folklore, some of these substances are known to be toxins, such as alkaloids, and sexual stimulants, such as isoflavones. The most famous Oriental herb is ginseng (*Panax ginseng*), which appears to be a general tonic, without discrete effects on the reproductive system.

Several of the precursors of gonadal steroids of vertebrates occur regularly in plants. Cholesterol is known to be widely distributed among plant taxa. In plants, as in animals, progesterone is derived from cholesterol, with pregnenolone as an intermediate product. Moreover, sitosterol, a common sterol in many plants, can also be converted to progesterone. Progesterone, a gonadal steroid as well as a precursor of other gonadal steroids in vertebrates, is found in plants and presumably gives rise to estrogens and androgens that have been reported to be in plants. Thus there exists in some plants the chemical sequence for deriving the steroid hormones that are produced by vertebrate gonads. It will be seen, however, that not all sexual stimulants in plants are steroids.

The existence of estrogenlike and androgenlike substances in plants has been known since about 1926, and their chemical similarity to animal steroid hormones was demonstrated in the early 1930s. A most surprising similarity was shown, experimentally, when animal hormones sprayed on a normally bisexual species, the caryophyllaceous herb *Silene=Malandrium dioecum*, determined the sex of the flowers that subsequently developed.

Phytoestrogens and Mammalian Reproductive Systems

There has been a sporadic history of plant-caused reproductive disorders in humans and domestic animals, but most effects are rather mild and therefore go unnoticed. For example, tulip bulbs were sometimes eaten in the Netherlands during food shortages in World War II, all this tissues is rich in phytoestrogens that cause menstrual irregularities. Also, certain strains of subterranean clover (*Trifolium subterraneum*) contain substances that alter reproductive performance in sheep. A wide variety of succulent and rapidly growing plant materials stimulate or depress various aspects of ovulation, pregnancy, and lactation when taken orally. Many plant taxa have in their leaves, stems, roots, flowers, and fruits substances that stimulate the gonads, endometrium, and mammary tissue. Some of these secondary compounds in legumes and grasses are known to be especially powerful.

Phytoestrogens from some plants may cause irregularity in implantation, distochia, and failure in sperm transport in sheep. They are deleterious because sheep are almost constantly feeding on these plants and because the effects are cumulative and irreversible. These substances occur in leaves as glycosides and, when crushed, are released as free isoflavones. They are not always potent in laboratory mice but may have a strong stimulatory effect in sheep. Coumestrol, a phytoestrogen in some clovers and alfalfa, is about 30 times as powerful as genistein when taken orally. Coumestans are of greater estrogenic potency than isolfavones are and appear to depress levels of ovarian estrogens in sheep, suggesting a negative feedback on the hypothalamus and/or the anterior pituitary. In most examples, the active compounds have a rather prompt effect and in this way differ from changes resulting simply from improved body condition and increased weight. They cause hypertrophy of the endometrium, presumably due to chronic stimulation caused by continuous ingestion of clover. After grazing on "estrogenic" clover for three years, ewes experienced a significantly higher ovulation rate than was observed in a control group, although there was no difference in body mass. Some of these materials in clovers induce lactation in nonpregnant ewes and even in wethers.

A difference in the estrogenic activity of some phytoestrogens results from the degrees of digestion. In the ruminant stomach of the sheep, both genistein and biochanin A are changed to estrogenically inactive phenols, but formononetin is degraded to equol, which is mildly estrogenic. Relatively weak phytoestrogens such as equol become stimulatory when concentrated in the plasma as a result of frequent ingestion. Thus, despite their rather low affinities at estrogen binding sites, their estrogenic effects are strong because of their continuous ingestion and high serum levels, and they may inhibit binding of estrodiol to uterine binding sties in vitro. Phytoestrogens are rather similar in structure to estrogens of vertebrate origin and seen to function in a similar manner when ingested. Both genistein and coumestrol, for example, compete with estrodiol at estrogen binding sites on the rabbit uterus.

In Australia the inclusion of lupin grain in the food of sheep significantly increases their fecundity. This effect is independent of the caloric content or nitrogen content of the feed. Lupin grain added to supplement normal feed of Merino ewes increases both ovulation rate and percentage of ewes twinning. Grain of uniwhite lupin (*Lupinus angustifolius*) increases the ovulation rate by 8 to 25 ovulations per

100 ewes, without an accompanying increase in the body mass of the ewes. If 50% more nitrogen and 25% more digestible energy is in the feed, there is no increase in ovulation. The effect of lupin on ovulation and overall fecundity seems to be directly on the reproductive system and is not an indirect effect of nutrition.

Isoflavones have been found to affect reproduction in wild populations of the California quail (*Lophortyx californicus*). In 1972 in San Luis Obispo County, California, when there was light rainfall and a meager growth of spring forbs, which contain high levels of genistein and formononetin, the reproduction level was low: there were 25 offspring per 100 adults, Isoflavones were nearly absent in the lush spring growth a year later, and there were 325 offspring per 100 adults. Isoflavones reduced fecundity in quail as in sheep, and the pathway in quail may be the same.

Occurrence of Phytoestrogens

The phytoestrogen content of plants varies with both the season and the growth stage. In very early growth of alfalfa there is a slight increase in levels of coumestrol, which soon decline to undetectable amounts until a very marked buildup from the dough stage of the seed through to the development of the mature seed.

There appear to be local and yearly variations in the presence of steroid hormones in plants, for efforts to locate them are not uniformly successful. Estrone, for example, has been reported to occur in high concentrations in seeds of the pomegranate (*Punica granatum* var. *nana*) by some researchers. The occurrence of isoflavones in plants is known to increase with a deficiency of phosphorus. Fungal infections also account for marked increases in the coumestan content of white clovers or lucernes (*Medicago* spp). Methylated coumestans and coumestrols occur in both leaves and stems, and levels are higher in those plants growing on sands than on gravelly soils. In *Medicago polymorpha* var. *denticulata* infected with a rust (*Uromyces striatus* var. *medicaginis*), the coumestan content of leaves of equal age increases over that found in uninfected leaves. Those methylated coumestans are probably not demethylated in the sheep's gut as isoflavones are known to be.

Several native African grasses have been found to contain reproductively stimulating compounds. Laboratory mice experienced a marked increase in uterine weight with only a trivial increase in body mass when fed extracts of veld that thay included *Ergrostis curvula*. Three common veld grasses (*Hyparrhenia filipendula*, *Setaria ciliolata*, and *Cynodon dactylon*) grew rapidly and developed a high level of estrogenic material, and their estrogenic activity was positively correlated with their growth rates. The active materials were found to be genistein and diadzein; biochanin A, formononetin, and coumestrol were absent. Among several British pasture grasses, such as ryegrass (*Lolium perenne*) and cocksfoot (*Dactylis glomerata*), estrogenic activity is highest in the spring, during early growth of the plant. In red clover (*Trifolium prantense*), however, which grows in the summer, there are repeated high concentrations of estrogens. In beans (*Phaseolus vulgaris*) phytoestrogens develop at the time of flower bud formation and are at maximal levels during bud growth and pod formation, at which time the greatest concentrations are in the leaves.

Some of the stimulating plant compounds are not estrogenic in their activity. There is in some grasses and germinating grass seeds a material, 6-methoxybenzoxazolinone, that is a powerful stimulant to reproduction in voles. This material is highest in the growing tissue of corn seedlings and other grasses. This and similar compounds in corn protect the growing plant from attacks of the corn borer (*Ostrinia nubilalis*). The presence of this reproductive stimulus in the early stages of grasses increases the likelihood of these compounds being proximate cues, for they would announce the abundance of food for young voles a few weeks later. When 6-MBOA was injected subcutaneously in laboratory rats or administered in Silastic capsules, there was a significant increase in ovary weight due to an increase in number of enlarged follicles. This increase occurred within 24 hours following subcutaneous treatment and three days after treatment with Silastic capsules. There was also a 100% increase in uterine weight in animals 30 days of age. Following 6-MBOA treatment, the number of corpora lutea increased from 13.4 ± 1.3 in controls to 19.2 ± 0.9 in those females treated with 6-MBOA.

Effects of Secondary Plant Compounds in the Field

One of the earliest reports of environment variation in litter size in wild mammals was the observation that the Levant vole (*Microtus guentheri*) of the Near East has a mean litter of 8.5 on irrigated land but only 5.2 on nonirrigated fields. This effect was attributed to "gonadotropins" in growing leaves of legumes, and similar effects are now known to occur in other voles.

Populations of a California vole (*Microtus californicus*) on perennial grasslands have smaller litters than those on annual grasses. When the two populations are maintained under the same conditions in the laboratory, however, there is no significant difference in mean litter size, indicating that differences observed in the field do not have a genetic basis. When the same voles are kept in outdoor enclosures of either annual or perennial plant cover, the mean litter size observed in the field are replicated: on perennial grasses the mean litter size was 3.45 in contrast to a populations of *Microtus montanus*, yearly variations in reproductive seasonality are correlated with the appearance of green shoots and freshly growing vegetation. In addition, winter breeding, although the exception, occurs with the unusual presence of green sprouts under snow cover.

Experimental Studies on Fecundity in Rodents

The experimental observations on *Microtus californicus* clearly indicate that at least some variation in mean litter size in voles is not determined genetically. Laboratory studies on *Microtus montanus* suggests pathways through which 6-MBOA may alter fecundity. This vole lives in the Great Basin of North America, a semiarid region characterized by year-to-year variation in rainfall and in the growth of forbs and grasses on which voles feed.

A diet including extracts from sprouting wheat produced a 41% increase in delta cells of the anterior pituitary of *Microtus montanus*. Delta cells are a site of gonadotropin production, and their increase may enhance the pathway by which certain foods stimulate reproduction. Moreover, when sprouted wheat was given to a nonbreeding wild (field) population of this vole, all of 11 females captured two weeks later were gravid, but there were no pregnancies in 11 females in a control area 200 yards distant. In captive voles of this species, sprouted wheat stimulated immediate onset of estrus in immature females, and a diet of spinach extract fed to 4-week-old females was followed by an increase in uterine weight and number of developing follicles.

An extract of growing wheat (20 cm) fed to *Microtus montanus* stimulated the reproductive tract of both sexes, and the active material was identified as 6-MBOA. Also, 6-MBOA was effective in the field when applied to rolled oats. When 6-MBOA was given to free-ranging populations of voles in the field in November and December or in January, males showed testicular hypertrophy and females became pregnant; a nearby control group given untreated oats showed no stimulation.

When intact and ovariectomized voles were fed rabbit chow supplemented with extracts from fresh lettuce, there followed a highly significant increase in uterine weights in intact and sham-operated voles but not in ovariectomized female, suggesting that the active substance in lettuce is not estrogenic. Lettuce is also stimulatory to testes in pocket mice (*Perognathus formosus*), without a significant increase in body mass. In captive *Microtus californicus* fresh green spinach enhances the stimulatory effect of long days and overcomes the suppressive effect of short days (10L:14D).

The effects of phytoestrogens and such materials as 6-MBOA are most apparent in deserts where rainfall is irregular and on agricultural land with a mosaic of irrigated and unirrigated cropland. Stimulatory qualities, however, are found in leaves, fruits, flowers, and underground parts of many sorts of plants and have long been known to occur in such diverse materials as palm nut oil and willow catkins. Phytoestrogens are in tissues of such common foods as spinach, apples, parsley, garlic, and cherries.

Negative Cues that may Terminate Reproduction

In many vertebrates the reproductive season is assumed to cease as a consequence of a photorefractory period, but there is evidence that some environmental factors have powerful suppressive effects on the reproductive system. Extracts from several species of *Lithospermum* (a genus of Boraginaceae) are antigonadotropic to equine chorionic gonadotropin. The effect occurs when extracts are administered subcutaneously to laboratory rats and is greatest when injected in moderate amounts. The antigonadotropic material is most concentrated in stems in autumn. Injections of water-soluble extracts of *L. ruderale* roots, consisting of polyphenolic acid salts, inhibit both gonadotropin and thyrotropin secretion in day-old cockerels. The mechanism of this inhibition is not known.

Toward the end of the growing season, some grasses accumulate compounds that retard or suppress activity of the reproductive tract. Inhibitors such as phenols and cinnamic acids increase as grass mature, develop fruiting heads, and start to brown. The dried and brown culms of salt grass (*Distichlis stricta*), an important food of the vole *Microtus montanus*, build up high levels of 4-vinylguaiacol, which experimentally suppresses both follicular and uterine growth but without any deleterious or apparently toxic effects.

Powerful ergot alkaloids, such as ergocryptine, produced by the fungus *Claviceps purpurea*, cause a marked suppression of prolactin secretion, causing hypoglactia (reduced milk flow). This fungus grows on cereal grains, especially rye, and also on wild grasses. It is sporadically encountered in cultivated fields. Ergocryptine is specific in its suppression of prolactin secretion and is about 100 times as potent as dopamine. As little as 1% barely ergot can inhibit lactations in sows. Bromocryptine, a synthetic ergot alkaloid, is sometimes used as a suppressant of prolactin, but the side effects are unpredictable, for this material resembles the hallucinogenic drug known as LSD. Ergot alkaloids, especially ergocornine, bromocryptine, and ergocryptine, inhibit implantation and lactation in the laboratory rat, and both effects are caused by a suppression of prolactin secretion. Ergocryptine is about twice as potent in this effects as bromocryptine is.

Although the ergot alkaloids are dramatic and extremely powerful, other fungi also have toxic effects. Toxic fungi (*Epicoccum nigrum*, *Cladosporium* sp., and Fusarium tricinctum) that occur on the tall fescus (*Festuca arundinacea*) are known to be deleterious to cattle on rangelands and to mice experimentally. These fungal infections are common in some pastures in the Mississippi Valley, and infected grass is not visually different from uninfected plants.

The evidence for nonphotic negative (inhibitory) factors in strongest in rodents, but probably most or all herbivores are affected. Negative factors or cues were not considered by Baker (1938), and they appear to be ignored by modern ecologists.

The response to these secondary plant compounds, regardless of whether they are estrogenic or not in their mode of action, suggests that they might be important regulators of the breeding season in some vertebrates. The stimulatory materials appear in some grasses during early growth, and in legumes they seem to be produced over a greater part of the growing season. In herbivores the stimulation for reproduction occurs as succulent forage becomes available for both adults and young. In contrast, suppressive materials develop at the end of the summer period of plant growth, when there are few tender plants to serve as food for young voles and other small mammals. It is also of interest that the wild mammals known to respond to the effects of secondary plant compounds are also those that are only weakly responsive to the photocycle.

Temperature

Ambient temperatures are usually reflected in the body temperatures of ectotherms. Temperature not only affects the teleost hypothalamus but may directly influence testicular steroidogenesis. In most fish, ambient and body temperatures are the same or nearly so. Amphibians, through evaporative cooling, are likely to be cooler than the ambient when in air. Some reptiles have the capacity, mostly through behavioral thermoregulation, to raise their core temperature will above the ambient.

Despite some degree of control over their body temperature, metabolism in these three vertebrate classes is by no means independent of ambient temperatures. These animals are generally exposed to rather regular annual changes in environmental temperatures, and it is not surprising that their annual reproductive cycles are frequently adjusted to seasonal temperature cycles. They are also usually exposed to the annual photocycle, and in some species both the pineal gland and the parietal eye are known to be photosensitive. The role of day length canbe important in a nocturnal species as well as in a diurnal animal, for the length of both the light and the dark periods can affect gonadal activity. The annual photocycle of ectothermic vertebrates acts together with thermal influences, in addition to the role of nutrition discussed earlier in this chapter.

Interactions between Daylight and Temperature

In water, light intensity is reduced and red light penetrates least. By secreting themselves at will, a fish may restrict its exposure to light within the light periods of the natural photocycle. Nevertheless, many examples of fish and other ectotherms are sensitive to changes in day length. One obstacle to the study of light and temperature as reproductive factors appears tobe that responses vary in the captive bred forms that are frequently employed in experiments. Photoperiodic responses of ectotherms are very commonly related to ambient temperatures, or responses to temperature are related to photoperiod, and it is helpful to consider the two combined as phtothermal factors.

Fish

Numerous lines of evidence indicate that the pineal gland plays an important role in the annual cycles of fish. Pineal influence in fish is, however, strongly affected by water temperature. The pineal and the retina, together with temperature, may mediate photoperiodic responses and thus suppress or limit gonadal activity during seasons of short days. In cyprinodont fish, however, temperature is an important regulator of gonadal activity, and both length and temperature may be paramount in salmonid and gastereosteid fish. In vernal-spawning species gonadal growth is most rapid in cocl water and with a schedule of increasing day lengths, and gonadal recrudescence may be greatly retarded if day length is shortened or temperature raised.

When pinealectomy is performed at the onset of the breeding migration in the lamprey (*Lampetra flaviatilis*), there is a delay in gonadal development and appearance of secondary sexual characteristics in both sexes. As in some endotherms, levels of hydroxylindole-Q-methyltransferase (HIOMT) in larvae of the lamprey (*Geotria australis*) have a peak in the dark phase of the photocycle some four or five times greater than inlight. Nevertheless, in lampreys gonadal growth and reproduction seem closely cued by water teimperature, and light does not appear to affect gametogenesis or spawning.

Observations on elasmobranchs indicate effects of both light and temperature on their gonadal growth. The pineal of the dogfish (*Scyliorhinus caniculus*) contains receptor cells with the morphological features of retinal cones, and the organ appears to be photoreceptor. The dogfish pineal, moreover, appears to have an antigonadal role, which can be greatly reduced by pineal extract injection, resembling the complex role of the pineal in endotherms. The apparent role of the pineal in the dogfish, however, is secondary to that of temperature. In nature there is a broad spectrum of reproductive seasonality in elasmobranchs. Some breed continuously throughout the year, and others have a rather diescrete period of reproductive behaviour.

The anterior brain region of at least some teleosts is sensitive to light, so that pinealectomy and enucleation together do not destroy light-induced responses. Thus it is difficult to isolate the role of the pineal gland in photoperiodic phenomena. The rainbow trout, for example, can detect light even when deprived of its eyes and/or the pineal gland, and plasma melatonin increases significantly in the dark phase of a 12L:12D cycle, independent of temperature. The rainbow trout is a spring-spawning species and experience autumnal gonadal recrudescence followng exposure to gradually *shortening* periods of light at both 8^0C and 16^0C, but no change in gonadal growth occur when this fish is maintained at *constant* constrasting light schedules (8L:16D and 16L:8D) at these temperatures. The rainbow trout (*Salmo girdneri*) has cycles of ovulation and testosterone peak approximately every six months when kept on a 18L:6D schedule. Melatonin occurs in the pineal of the salmon (*Oncorhynchus tshawytscha*), and levels in sea-dwelling immature fish were found to be about six times as high as in adults that had begun a breeding migration in fresh water. After they enter fresh water on their spawning migration, salmon are exposed to much brighter light than they experience in the sea, regardless of the photocycle. The spawning of the shelly (*Coregonus lavaretus*), a salmonid of the English Lake district, is temperature induced, and its eggs require cold water. The eggs are demersal and develop at surface temperature of from 4 to 6^0C, but there is no survival at 10^0C.

Many salmonid fish have a breeding migration, and it is not surprising that both migration and gonadal growth are affected by the same exogenous factors. Among several species that spawn in the autumn, gonadal activity is promoted by both short days and low temperatures. The brook trout (*Salvelinus fontinalis*) is a well-known autumnal spawner, and the gonads seem to be stimulated by short day length. Gonadal recrudescence in nature is apparently initiated under long days and accelerated by shortened day length. This has been confirmed many times. Experimental evidence in salmonids indicates that increasing day lengths trigger spermatogenesis and that decreasing day lengths promote sperm maturation.

Among minnows there is a general pattern of spawning during warm seasons, and experimental evidence indicates that minnows seem to be stimulated by increase in both daylight and temperature. One of the earliest demonstrations of photosensitivity in minnows showed a clear positive response of follicular growth independent of temperature.

A Eurasian minnow, *Phoxinus phoxinus*, has a temperature-modified response to day length. When these minnows are maintained at low temperatures and long days, there is no gonadal growth, but when kept at higher temperatures, the same fish experience a significant increase in gonadosomatic index in response to long days. In nature there are year-to-year variations in the time of vernal recrudescence, such variations being correlated with annual variations in vernal water temperatures. In the golden shiner (*Notemigonus crysoleucas*), a small minnow of eastern North America, long days and warm temperatures stimulate gonadal development. Under these same conditions, gonads show arrested or reduced development after pinealectomy, suggesting a stimulatory role for the pineal gland. Also, under a schedule of long days and low temperatures, pinealectomy was followed by retarded ovarian development. Under short days and warm temperatures, however, pinealectomy seemed to account for an *increase* in gonadal development, suggesting that pineal activity during times of short days suppresses gonadal growth. The pathway of pineal influence in fish is unknown, but the pattern seen in the golden shiner resembles that seen in hamster—the pineal is active in darkness, and long days result in "physiological pinealectomy."

Among Japanese species of bitterlings (Cyprinidae), *Rholdeus ocellatus* spawns from spring through summer, *Acheilognathus tabira* spawns in the spring, and *Pseudoperilampus typus* spawns in the autumn. Experimentally, gonadal recrudescence in induced by rising temperatures, regardless of photoperiod, in *Rhodeus* and *Acheilognathus*, and regression follows a shortened day length without a temperature change. In contract, in the autumn-spawning *Pseudoperilampus*, gonadal maturation is in response to a decrease in day length, and subsequently regression is induced by a lowering of temperature.

The goldfish (*Carassius auratus*) is sensitive to day length and temperature. Goldfish ovulate spontaneously when maintained at 20⁰C on a 16L:8D cycle. Volumetric and morphologic changes in photoreceptor cells of the pineal gland of the goldfish when exposed to a long-term 16L:8D schedule indicate that secretory cell organelles are affected by light. When exposed to 20⁰C and long days in the spring, goldfish experience gonadal enlargement, but when pinealectomized at this season and maintained under short days, goldfish have smaller gonads than sham-operated controls do. Maintaining goldfish in total darkness is, however, more suppressive to gonadal development than pinealectomy is.

As in salmonids, gametogenesis may be initiated under one light-temperature regimen, and gamete maturation may be initiated six months later under a contrasting light-temperature environment.

The medaka (*Oryzias latipes*), a cyprinodont fish of Japan, responds to long days (16L:8D or 14L:10D) with increased gonadal development, and blinded individuals show a similar but reduced effect, including a distinct reduction in number of eggs laid. The pineal in this species is photosensitive. When kept under long days, blinded and pinealectomized individuals have reduced ovarian growth, suggesting a nonsuppressive or permissive role for the pineal gland under long days. Under a natural short-day photocycle, however, blinded and pinealectomized medaka had greater gonadal development than controls did, which indicates a reduction of the antigonadal effect of both eyes and the pineal under short days. Melatonin inhibits ovarian growth when injected into medaka kept on a 14L:10D schedule at 26⁰C, but melatonin had no effect at 17⁰C. Similarly, injections of melatonin for 15 days two hours after the onset of daylight retards gonadal growth in the killifish (*fundulus similis*) in contrast to controls.

A possible error in the technique of pinealectomy in fish may account for the occasionally reported "progonadal" effect. A part of the pineal gland may adhere to the parietal bone as the latter is raised to expose the pineal for removal, and with replacement of the parietal bone, this segment of the pineal remains. Thus, what appears to be a pinealectomy may leave some of the secretory tissue remaining.

The mummichog (*Fundulus heteroclitus*), an estuarine cyprinodont of the North American Atlantic coast, is well known as a tidal spawner and lays eggs with a semilunar frequency in synchrony with spring tides. It is also stimulated by a long-day (L15:D9)schedule and fails to develop ova when exposed to a short (L9:D15) photocycle.

Several poeciliid fish, introduced in northern Australia, have seasonal patterns of breeding. These patterns are controlled by day length and, to a lesser extent, temperature. *Gambusia affinis*, *Xiphophorus helleri*, and *X. maculatus* show an increase in the rate of pregnancy corresponding to increased day length, but reproductive activity declines with day length in midsummer while temperatures are still favourable. Fecundity of poeciliid fish in nature tends to be maximal in summer. Two species, *Poeciliopsis gracilis* and *Poecilia sphenops*, in El Salvador (in nature) are exposed to natural extremes in day length from "short" days of 11.3 hours to "long" days of 12.9 hours. When these species are kept in these two photoschedules and at similar temperatures, the longer day length increases vitellogenesis and number of embryos, independent of the size of the female, and there was also an increase in testicular size.

The testes of the marine live-bearing teleost *Cymatogaster aggregata* are affected by both temperature and light, and the gonadal response varies with the time of year. In early winter increased day length, from 8L:16D to 16L:8D, concurrent with a change in water temperature from 10 to 20^{0}C was follwed by an increase in testicular growth and spermatogenesis, the change being due to temperature increase independent of day length. In late winter and spring the stimulatory effect of higher temperature was enchanced by increased day length, and shorter day length resulted in gonadal regression.

The role of photoperiod in tropical ectotherms is difficult to evaluate, for if a long day (12 hours) is stimulatory, tropical species are in a continuously stimulatory photocycle. It is plausible that the tropical photocycle bring an individual ectotherm to a subfunctional reproductive state, and it is then prepared to respond to one or more ultimate factors.

The pineal organ of the angel fish (*Pterophyllum scalare*), a Neotropical cichlid fish, is photosensitive, and its electrical activity varies with exposure to light. The four-eyed fish (*Anableps dowi*) breeds throughout the year in a tidal estuary in El Salvador. There is a correlation, however, between day length and number of embryo and the gonadosomatic index of the male, suggesting a photoperiodic response in this fish. Several species of *Sarotherodon* (*Tilapia*) have distinct period of spawning in tropical regions of relatively constant schedules of light and temperature. In these fish gonadal activity can be experimentally induced by warm temperatures, and in nature spawning occurs at the beginning of the seasonal rains.

In tropical species that inhabit waters of constant flow and composition, spawning may occur more or less continuously throughout the year. Very commonly, tropical water are altered by periods of drought and rains, and resulting conditions control reproduction in many low-latitude teleosts. These seasonal changes not only affect the physical and chemical aspects of the waters but also may time seasonality of planktonic bloom. Rainfall is frequently followed by changes in pond levels of stream flow, which is associated with spawning of a number of African and Asiatic teleosts. In the Murray-Darling River system of New South Wales several species of native fish in earth-lined ponds were induced to spawn by replenishing water that had been lost to evaporation. Apparently there were cues associated with water entering the dry soil.

The Neotropical electric fish *Apteronotus albifrons* and *Eigenmannia virescens* live in freshwater watercourse where there are only slight seasonal changes in light and temperature. In these species spawning can be experimentally induced by (1) a decrease in conductivity of water (2) an increase in water level, and (3) an imitation of rain.

Amphibians

Inasmuch as amphibian behaviour is responsive to both light and temperature and inasmuch as temperatures tend to fluctuate with the daily and annual photocycle, separate roles of temperature and light are not always apparent. Experimentally, few investigations on the role of the photocycle are designed with "cold" light, so that a cycle of artificial light usually produces a cycle of temperature as well. In some salamanders the hypothalamus appears to be the thermosensitive center controlling gonadotropin and prolactin release. There is evidence, moreover, that daily activity may have a thermoregulatory significance, and the pineal may be important in entraining daily cycles of activity. The association of peaks of melatonin with periods of rest in some amphibians does not preclude a direct role of this hormone in depressing gonadal activity of amphibians.

In addition to the pineal, anurans possess a frontal organ, also known as the *stirnorgan*, which is a homologue of the parietal eye. The forntal organ is absent in salamanders and caecilians. In anurans both the frontal organ and the pineal gland are photosensitive and respond both to visual signals and to ultraviolet wavelengths. As in fish the pineal gland secretes melatonin. A well-known effect of this hormone in amphibians is the contraction of melanocytes in the skin, producing a light hue. This response accounts for lightening of skin colour of some frogs in the dark.

Melatonin also retards gonadal development in some amphibians. Injections of melatonin (4 μg) given to tree frogs (*Hyla cinerea*) daily for 28 days, 10 hours after the onset of daylight, resulted in significantly smaller gonads than in controls. In the leopard frog (*Rana pipiens*), HIOMT levels are greater in the retina than in the frontal organ, suggesting a regulatory role for the paired eyes. In the larval tiger salmander (*Ambystoma tigrinum*) the pineal responds to darkness in a manner observed in some teleost fish, and there is a nocturnal peak of circulating melatonin in animals kept on a 12L:12D schedule. The melatonin peak disappears after pinealectomy, but melatonin still remains in the light period, possible from the retina.

Gametogenesis in amphibians and reptiles depends on gonadotropin release, as in other vertebrates, and the hypothalamus in amphibians is responsive to both light and temperature. During early winter, (November and December) gonadotropin in *Rana temporaria* increases with rise in temperature, but there is no change in spermatogenesis at that season. From January to March, however, primary spermatogonia become increasingly sensitive to gonadotropin support, and in the springtime the spermatogenic cycle progresses with increase in temperature. In the bullfrog (*Rana catesbeiana*), spermiation in response to gonadotropin hormone is depressed at low temperatures (15-20^{0}C), and optimal spermiation occurs at about 25^{0}C, with sensitivity to temperature being unaffected by day length. Similarly, under natural conditions day length seems not to be critical in timing of the ovarian activity of the toad (*Bufo bufo*) in Denmark. Temperature is a major influence on oocytes: warm summer temperatures seem tobe essential to their growth, but low temperatures that obtain during hibernation are required for their final maturation.

Day length plays a role in gametogenesis in at least one salamander. In the red-backed salamander (*Plethodon cinereus*), spermatogenesis proceeds rapidly at 20^{0}C and is retarded at 10^{0}C. At 20^{0}C the process is accelerated in increasing day length, but temperature is clearly more important than light. As pointed out previously, the hypothalamus of salamanders is temperature sensitive and monitors the release of both gonadotropins and prolactin. In the Salamandridae, prolactin initiates the movement to water, and its release is triggered by a sudden change in temperature.

Reptiles

In many species of lizards and in the tuatara (*Sphenodon punctatus*), there is a light-sensitive parietal eye immediately beneath a translucent scale on the top of the head. The parietal eye is lined with a cup-shaped retina, and a nerve leading form the retina course along the pineal gland. The parietal eye is found in all reptilian order except the crocodilians. The parietal eye occurs irregularly in lizards but is absent in snakes. In *Lacerta viridis*, a lizard of the Old World, the pineal is both photosensitive and secretory. Lizards of several families can be entrained to light schedules when blinded, and testicular recrudenscence occurs as in sighted individuals; the extraretinal receptor is not known. The parietal eye and/or the pineal might both serve as this receptor.

Reptiles have a complex of gonadal responses resulting from manipulation of temperature and day length, and the roles of each are not always clearly separable. The parietal eye of the horned lizard (*Phrynosoma douglassi*) affects both thermal and photic responses, and because body temperature is altered by the lizard's photic responses, the two roles are difficult to distinguish. The entrainment of a daily rhythm of motor activity, including basking and rest, results in important seasonal trends in changes of body temperature. In at least one lizard the effect of light and heat are sequential: temperature is the initial factor involved in the gonadal activity of the anole (*Anolis carolinensis*), but later photoperiod becomes dominant.

Under natural conditions, however, a pond turtle, Blanding's turtle (*Emydoidea blandingi*), nests annually at some time within a 17-day period from late May to early June, and oviposition is independent of mean May conditions but correlated with mean April temperatures. Apparently April temperatures constitute a proximate cue in this species, but there remains a possible role of light in the field.

There is a rhythmic secretory activity of the pineal in some squamates and chelonians, with HIOMT and serotonin increasing in the dark phase of the cycle. It has been suggested that serotonin in the sauropsidian pineal is converted to melatonin as in the mammalian pineal. Little experimental work has been done on the pineal gland of most reptiles, but the majority of observations suggest an antigonadal role for the reptilian pineal. In the Asian agamid lizard, *Calotes versicolar*, there is an inverse cycle of size and activity of the pineal on the one hand and of gonads on the other hand, and pineal hypertrophy in nature occurs in seasons of short day length. In this species pinealectomy accelerated recrudescence of the testes and also delayed their autumnal regression. Melatonin appears to depress gonadal development in the zebra-tailed lizard (*Callisaurus draconoides*) when maintained on a natural vernal light schedule. On the other hand, the role of subcutaneous melatonin may be to alter daily rhythms and affect gonadal growth only indirectly. As in amphibians, however, reproduction in the reptilia reflects appropriate combinations of day length andtemperature, and most investigators indicate that temperature has the greater role.

Temperature seems to control the secretion of reptilian gonadotropins and may also differentially alter sensitivity of target organs. Experimentally, gonadotropin and testosterone reach higher levels at 20^{0}C than at 3^{0}C in *Anolis carolinensis*.

Endotherms

Although the body temperature of most endotherms are usually sufficiently constant that they do not fluctuate with the ambient temperature, changes in the latter can alter the rate of deplection of energy stores. Exogenous heat does not, of itself, alter gonadal activity, but by altering the amount of energy, environmental heat may alter fecundity.

In northwestern Europe, nesting of the great tit (*Parus major*) is closely correlated with temperature as well as with leading of birch (*Betula* spp.) and the fall of the caterpillars of the winter moth (*Operophthera brumata*), both of which are related to temperature. Thus temperature indirectly provides for a greater food supply for these tits, which allows females to accumulate adequate fat for egg formation early in the year. There is an additional effect of heat that strikes the nest boxes in which the tits sleep. Those boxes exposed to late-afternoon sun have increased nocturnal temperatures, and females roosting in "warm" boxes during their prelaying period lay earlier than do those roosting in "cool" nest boxes. Presumably the lower nighttime temperatures place greater demands on the stored energy of the sleeping bird and retard accumulation of fat egg formation.

There is apparently a relationship between ambient temperature and dates of oviposition in the rockhopper pengiun (*Eudyptes chrysocome*). This species breeds annually, from late September until early December, depending on the region, and laying dates are closely correlated with mean annual seawater temperatures. It is not known if the marine temperatures relate to the abundance of food of the penguin or to thermoregulation of the penguins themselves.

Mammals generally do not seem to be extremely sensitive to changes in ambient temperatures. Perhaps this apparent immunity from the effects of exogenous heat is due to the nocturnal activity patterns of most species of mammals. This topic has not been a major realm of study. Deer mice (*Peromyscus maniculatus*) seem to be unaffected, reproductively, by differences in temperature. Among populations of deer mice from contrasting latitudes, ambient temperature has no role in reproduction. In the European vole (*Microtus arvalis*), however, fertility is lowered by high temperatures.

HUMIDITY

Environmental humidity can have distinct effects as either rainfall per se or the accumulation of water on or in the ground. There is also evidence that atmospheric humidity can regulate gonadal development and behavioural activity.

Rainfall

Many anecdotal accounts of reproduction of both birds and mammals in the tropics indicate that rain itself may be a proximate cue for reproduction. The exact nature of the stimulus or the mechanism of its action is not known. Even at the sound of thunder and before there could be any effect of rain, some Australian paserine birds have been known to build nests and lay eggs. The zebra finch (*Poephila guttata castanotis*), which occurs widely in arid and semiarid regions of Australia, is not affected by variations in the photoperiodic cycle, but it nests after, and sometimes before, rains. Experimentally, after the birds have been deprived of free water, the addition of drinking water is followed by testicular growth. Testicular growth is also induced by the increase of atmospheric humidity or the addition of fresh green grass, which is an important food of these birds during the nesting season. Water in one form or another seems to induce testicular growth in this finch. The European rabbit (*Oryctolagus cuniculus*), feral in Australia, seems to be preadapted to irregularity in rainfall: when rain broke an 18-month drought, there were suddenly pregnancies in rabbits within two weeks after the rain and before the resurgence of plant growth.

Some ectotherms are known to be both sensitive to standing water and dependent on soil moisture or standing water. The spadefoot toads (*Scaphiopus* spp.), which live in arid and semiarid areas of North

America, spawn in temporary ponds and must oviposit promptly with the accumulation of standing water. During the nonreproductive seasons adults lie buried deep in the ground. They are stimulated to emerge by low-frequency vibrations (probably below 100 Hz) in the soil (which can be produced by an electric motor, simulating the sound of falling rain) and also by the sound of rain on a roofed enclosure, but not by soil moisture. The African viviparous toad (*Nectrophrynoides occidentalis*) estivates in dry ground in nature, and in the laboratory saturated atmospheric humidity induces spermatogenesis. Although the triggering mechanism by which rainfall initiates a gonadal response may well vary among different classes, it is clear that rainfall constitutes a proximate cue.

Rainfall and the consequent soil moisture can also be an ultimate cue. The annual cyprinodont fish of South American and East Africa die as their ponds dry up with the approach of annual droughts, and the species survive as partly developed eggs in the dry soil. The eggs are embryonated, with a half-developed embryo, but they enter a diapause and cease growth until triggered by the return of the rains and temporary ponds. This is one of the very few cases of diapause in verebrates. When rainfall restores these ponds, the eggs hatch, and the adults quickly reach adulthood and spawn. Not all eggs resume growth at first. While one segment of the population matures and spawns, others develop some time later, thus protecting the species against the danger of having the entire population maturing in a pond that is too short-lived to provide water for oviposition.

Endogenous Reproductive Cycles

The persistence of a reproductive cycle in the absence of exogenous cues has been taken as evidence for an endogenous reproductive cycle. It has been suggested that some vertebrates have a naturally recurring circannual cycle in which the photocycle is unnecessary. This circannual cycles does not preclude the annual photocycle as a *Zeitgeber* but simply means that the gonads will undergo more or less annual recrudescence without photic stimulation. Although an endogenous circannual rhythm of reproduction and gonadal regression is apparently uncommon, it has been reproted to occur in a number of diverse texa.

Although the Indian catfish (*Heteropneustes fossilis*) is sensitive to day length, there is a strong endogenous circannual elements to its ovarian cycle. Captives in continuous light or darkness at 26^{0}C have annual cycle of ovarian growth comparable to those in nature. This fish is not insensitive to light, for if kept under a schedule of 6L:18D with the dark phase interrupted with one hour of light, there is a significantly greater ovarian enlargement than occurs in a control group of 7L:17D.

The extent of endogenous circannual rhythmicity among reptiles is unknown. In the desert iguana (*Dipsosaurus dorsalis*) there is a circannual reproductive cycle in individuals maintained under constant stimulatory conditions of light and temperature. *Cnemidophorus uniparens*, a parthenogenetic teiid lizard, also has a circannual breeding cycle. In the field, however, any existing rhythms are cued by exogenous *Zeitgebers*.

The existence of endogenous circannual rhythms in birds has been questioned, but experimentation suggests that some species have reproductive cycles that recur at nearly annual intervals. An African weaves finch (*Quelea quelea*), for example, preserved its reproductive cycle for two years when kept on a 12L:12D schedule. There are also apparently endogenous rhythms of migration of Old World warblers (Sylviidae). Circannual rhythms of testicular size and molt in a warbler (*Sylvia borin*) and a starling (*Sturnis vulgaris*) persisted up to four years when kept under light schedules of 11L:11D, 12L:12D, and 13L:13D.

It has been suggested that birds maintained on a 12L:12D cycle for a prolonged period may be able to record the passage of days and thus keep track of the seasons. This ability is clearly not present in all birds, for the white-crowned sparrow (*Zonotrichia leucophrys gambelii*) kept under constant stimulatory photocycles does not spontaneously recover from photorefractoriness.

Endogenous circannual reproductive cycles obtain for several species of ground squirrels. The reproductive system of the antelope ground squirrel (*Ammospermophilus leucurus*) is unresponsive to short-day displacement of the photocycle. Annual changes are unaffected by experimental exposure to extremes of daily cycles at all times of the year. Animals held in constant laboratory conditions for about two years showed spontaneous cycles of gonadal function with an approximately annual periodicity. This species reproduces only once a year, regularly, in the spring, and advanced preparation for this reproduction is under endogenous controls.

The hibernation cycle of the golden-mantled ground squirrel (*Spermophilus lateralis*) is also endogenous and recurs at some what less than 365-day intervals under uniform conditions. The mechanism for an endogenous reproductive cycle of males may lie in a varying sensitivity of luteinizing hormone (LH) secretion to negative feedback. Male golden-mantled ground squirrels (*Spermophilus lateralis*) exhibit a circannual decline in negative feedback depression of LH secretion by gonadal steroids during the breeding period of the spring, which results in an elevation of serum LH in intact males. In castrated males threre is no apparent seasonal rhythm to serum levels of LH. Secretion of LH in males, then, is predicted on an annual cycle of sensitivity to the negative feedback of gonadal steroids. A different endogenous rhythm of LH release in seen in ovariectomized females of the same species, with vernal peaks of LH independent of ovarian steroid feedback. The endogenous nature of LH release in females of this species and the seasonal (vernal) decline in sensitivity to negative feedback in males together could determine the endogenous reproductive pattern seen in these squirrels. There is nevertheless a role of the pineal gland in this species, for it seems to affect the circannual rhythms of both weight changes and onset of estrus: in pinealectomized females these circannual rhythms are shortened.

Interrelationship of Exogenous Factors

In spite of the great amount of experimental evidence showing the effect of light on the reproduction of many diverse taxa, the role of the natural photocycle in wild populations may not be the sole major factor. The photocycle operate together with other environmental influences, including temperature, precipitation, plant growth, social factors, and both nutrition and ingested secondary plant materials. As pointed out in this chapter, "food" is a complex of materials that have a multiplicity of effects on the reproductive system. Food should not be considered as only having a nutritional role for maintenance and growth.

Refractoriness

Refractoriness probably occurs in a number of ectoderm groups and is presumably part of any endogenous circannual rhythm. In many ectotherms gonadal regression occurs when both light and temperature are stimulatory, and these pattern will become apparent in the next chapter. Refractoriness in fish occurs in a number of vernal spawning species. They may respond to changes in light, temperature, or other cues separately or in sequence, and there is frequently a postspawning gonadal regression. Refractoriness has been reported for several teleost fish and also for several squamates.

The experimental research on day length is designed to preserve uniformity in all factors but light, but even in the laboratory small events can alter reproduction. For example, irregular sound can alter levels of gonadotropins; minor stress can stimulate the release of prolactin; and social factors are known to alter sexual maturity, ovulation, and preimplantation mortality. It is important to remember that in nature the photocycle is experienced with a host of other cues and that experimental studies on captive animals are obviously not intended to replicate the conditions under which a free-living animal exists.

Although day length is easily and unequivocally controlled in a caged animal, many wild creatures, such as mice, many reptiles, most amphibians, and many small fish, very easily modify their exposure to light. This is, many species can modify their activity so that it does not coincide with the natural photocycle. Such species can decrease their exposure to the daily period of daylight or darkness. Thus although many nonavian vertebrates are well known to be photosensitive, one must not expect laboratory experiments to be exactly predictive to their behaviour in nature. It is apparent that light is not a *Zeitgeber* at low latitudes, but seasonal reproduction in the tropics is triggered by cues other than the photocycle.

Among the constellation of cues or factors, both proximate and ultimate, discussed in this chapter, some are stimulatory and others appear to be suppressive. These factors become increasingly clustered and more closely associated with the calendar at higher latitudes. Stimulatory cues occur in the spring, and suppressive factors prevail in later summer and autumn.

2

Light and Temperature

Photoperiodic Responses of Birds

Knowledge of the influence of day length on avian physiology is not new. Artificial extension of daylight was used to promote lying in domestic fowl long before there was electricity. The exact mechanism of this response, however, has not been understood until rather recently, and research continues to provide explanations for such events as seasonality, photorefractoriness, molt, and migration.

Extraretinal Photoreception in Birds

Extraretinal reception of light in birds appears to lie in the hypothalamus. The expected gonadal response of birds to long days is seen in bilaterally enucleated individuals, but only if the top of the head receives full light. When light is obscured or blocked from reaching the crown of the bird, the photoinduced response is lacking. Even in intact (sighted) sparrows (*Passer domesticus*) the eyes do not transmit the visual signals that result in gonadal development, and the same phenomenon is seen in the white-crowned sparrow (*Zonotrichia leucophrys gambelii*) and the Japanese quail (*Coturnix coturnix*).

When deprived of their vision, domestic ducks respond to long days by gonadal recrudescence and to short days by gonadal regression. This response does not necessarily preclude a role for the paired eyes and retinal stimulation, but it does indicate the presence of other photoreceptors. After the realization that the paired eyes are not the essential photoreceptors that affect gonadal activity in birds, there followed a search for other sensory tissues or structures. Light directed to the pineal gland of blinded ducks stimulates testicular growth, but light directed to the avian pineal may well strike adjacent structures. The mechanism for this response is not apparent. Luminous beads implanted in the brains of sexually mature Japanese quail (*Coturnix coturnix*) had an effect when the birds were reared and continued under a nonstimulatory light schedule, suggesting that the hypothalamus is photosensitive. Within three weeks after the beads were implanted, the testes had achieved a greatly enlarged state in contrast to the minute testes of controls without implanted beads. A more precise determination of the photosensitive area was made by the use of fiber optics. By directing light-conducting fibers to basal hypothalamic areas, it is possible to illuminate specific sites. In the male white-crowned sparrow (*Zonotrichia leucophrys gambelii*), photosensitive areas, as revealed by fiber optics, lie at or near the ventral hypothalamus; these photosensitive areas may also produce gonadotropin-releasing hormone (GnRH). There appears to be a rhodopsinlike visual pigment in the tissues of this region.

Effect of Light on the Avian Reproductive System

Studies on the response of birds to varying light schedules have been conducted mostly on small passerine species and some domestic birds, such as the fowl, the Pekin duck, and the Japanese quail. Most research concerns males, in which the results are more readily observed.

In male birds the first effect of long days in the rise of serum luteinizing hormone and follicle-stimulating hormone. Secretion of gonadotropins in the Japanese quail is proportional to day length and is not a simple "on-off" response. The development and activity of the Leydig cells are promoted by LH, and an increase in testicular size follows the increase in serum levels of FSH and testosterone. This, in turn, is paralleled by growth and spermatogenic activity of Sertoli cells. If the bird is soon returned to a short-day schedule, LH levels remain high for from 4 to 10 days, but the cause of the "carryover" effect is unknown. In the Japanese quail the increase in testosterone occurs about a week after the onset of long day lengths, suggesting that LH receptors develop in about one week. After day length has reached a critical minimal stimulatory period, the rate of gonadal recrudescence has a log-linear relationship to the duration of light in passerine birds and in the Japanese quail.

There is no question that long days are stimulatory to gonadal growth in males of many passerine birds. In females light has a stimulatory role on early follicular development. In females, long days are essential for development of follicles up to the onset of yolk deposition, and this is the state at which female white-crowned sparrows arrive on their breeding grounds. Thereafter supplementary information, such as singing males, courtship, and proper trophic conditions, is needed to bring the ovary into a functional state.

Photosensitive Phase

there is a period of sensitivity, the photosensitive phase, during which long days can induce the release of gonadotropins, and light is stimulatory only during this period. The photoperiodic control of the endocrine system results from the coincidence between a photoinducible phase of an entrained circadian oscillation (which dtermines the PSP) and the natural photocycle. Thus stimulation of gonadotropin release is predicated on the duration of the coincidence of light with the PSP.

Photoperiodic Responses of Birds at Low Latitudes

Miller (1958) and Lofts and Murton (1968) have suggested that the role of photoperiod as a cue is insignificant in low latitudes and that daylight itself does not set the nesting schedule. The change in day length is, indeed, rather slight near the equator, but this does not preclude an important role for the photocycle in tropical avian reproduction. The tropical weaver finch (*Ploceus philippinus*) can discriminate between small differences in day length and has the potential to detect minor photoperiodic changes in nature. Although tropical day lengths are presumably adequate to bring the avian reproductive system to a "subfunctional" state, at least for species native to the tropics, actual nesting may depend upon other exogenous factors.

Several tropical birds exhibit increased testicular growth when exposed to long days. This has been shown in the rufus-collared sparrow (*Zonotrichia capensis*) by Miller (1959), in the zebra finch (*Poephila guttata*) by Marshall and Serventy (1958), and in several other species. The zebra finch, in fact, rarely has conspicuously regressed gonads.

There is no reason to assume that low-latitude vertebrates have lost the ability to respond to changes in day length or that tropical species have been immobile in the low latitudes for endless millions of years. Pleistocene faunal movements have been abundantly documented for the temperate regions, and in the low latitudes of both Africa and South America there were concurrent faunal movements in both latitude and elevation. In evolutionary time these events occurred only yesterday, and it would be very surprising if sensitivity to day length had been lost in this short time unless it had been deleterious.

It will become clear, however, that a basic difference between the tropical environment, in which the annual photocycle maintains the near-tonic subfunctional state, and the temperate latitudes, in which day length promotes vernal recrudescence, is the influence of many nonphotoperiodic factors. Although it would be unrealistic to underestimate the role of the highly contrasting photocycle in the mid- and high latitudes, one should not be so blinded by the effect of day length that the non-photoperiodic factors are neglected. In the tropics a broad spectrum of nonphotoperiodic factors is scattered throughout the year, but in temperate latitudes these factors appear close to the maximal influence of increasing day length in a vernal crescendo. Because of the ease with which day length can be studied in the laboratory, our understanding of proximate factors is almost totally confined to the annual photocycle.

Passerine birds have a multifaceted response to day length. Not only are there several gonadal, migratory, and feeding responses to day length, but there is a complex of interrelated feedback mechanism that modify the effects of the exogenous factors. This arrangement provides for a considerable latitude of physiologic and behavioural pathway, so that the individual is constantly adapted to a broad spectrum of environmental conditions.

Weakly Photosensitive Birds

Among both migratory and sedentary birds are some whose gonadal cycles seem to be independent of photoperiod. The short-tailed shearwater (*Puffinus tenuirostris*) is unusual in having a transequatorial migration and breeding in the Southern Hemisphere: it nests in Tasmania and annually migrates to the Aleutians. When kept under unnatural photoperiodic schedules for up to six months, the birds showed no change in gonadal development. Similarly, the spotted munia or mannikin (*Lonchura punctulata*), a nonmigratory ploceid finch of the tropical Orient, does not respond to photoperiodic stimuli but appears to have a tendency to follow an endogenously controlled reproductive pattern.

Refractoriness

Many vertebrates have a refractory period, a time during which the reproductive system does not respond to exogenous cues. This period is best known for birds but apparently occurs in all classes. Refractoriness prevents the continuation of reproduction into a season that would be unfavorable for the survival and growth of the next generation, as well as the survival of the adults.

Although the avian reproductive system is very sensitive to the increase of day length in the spring, it becomes unresponsive to the same day lengths later, usually in the late spring or summer. This refractory condition constitutes the postreproductive failure to be stimulated by factors that were stimulatory period to breeding. The refractory period varies with respect to the time at which it ends reproduction and the duration of its effect.

Refractoriness is best known in birds. There is an increased sensitivity of the hypothalamus (and perhaps also the anterior pituitary) to the negative feedback of gonadal steroids; such increased sensitivity accounts for the decline in gonadotropins in late summer. In addition to a decline in gonadotropins, refractoriness can also bring about a decline in gonadal steroids. Thus the drop in plasma steroids does not automatically result in a rise of gonadotropins. Photorefractory red grouse (*Lagopus lagopus scoticus*) had mean plasma LH between 0.28 and 0.34 ng/ml under a 16L:8D schedule. After castration circulating LH increased markedly in a 20L:4D schedule and increased only slightly in a 6L:18D schedule. Photosensitive and photorefractory male white-crowned sparrows (*Zonotrichia leucophrys gambelii*) treated with synthetic GnRH responded positively and equally, suggesting that refractoriness occurs at a level above the anterior pituitary.

There is also negative feedback from other steroids, perhaps from the adrenal cortex. This feedback is suggested by castration of photosensitive and photorefractory willow grouse (*Lagopus lagopus lagopus*). Castration of photorefractory individuals was followed by an immediate rise in LH, but in the photorefractory birds the increase in serum LH did not occur until some five weeks later, at which time they may have no longer been refractory. The continuation of refractoriness in castrated grouse suggests that steroids other than testosterone continued to depress the activity of the hypothalamus and perhaps also the anterior pituitary.

An alternative explanation for the development of refractoriness lies in a proposed seasonal time shift in the PSP. When exposed to light early in the natural dark period in October, house finches (*Carpodacus mexicanus*) did not experience gonadal recrudescence, but exposure at the same time in January was followed by gonadal growth. It is likely that different physiological conditions account for refractoriness among the many families of birds in which it is known.

The development of photorefractoriness in the Japanese quail depends upon serum thyroxin during exposure to long days. Thyroidectomized males, moreover, respond only slowly to a reduction in day length. In the starling (*Sturnus vulgaris*), thyroidectomy prevents photorefractoriness and subsequent molt, but short days nevertheless induce testicular regression.

A conspicuous characteristic of the refractory period is the eventual increasing sensitivity to stimulatory day lengths. This is seen most clearly when refractory individuals are kept under a schedule of short days. Hermit thrushes (*Hylocichla guttata*), when maintained on a 9L:15D schedule for six weeks, became responsive to long days. Gonadal growth followed exposure to 15L:9D days, whereas another group remained photorefractive from October to February when kept on a 15L:9D schedule. Similarly, the slate-coloured junco (*Junco hyemalis*), when subjected to short days (9L:15D) for five to six week in July and August, became sensitive to long days. By breaking refractoriness with exposure to short days, it became possible to obtain five annual periods of gonadal growth and fat deposition and two molts.

The Role of the Pineal in Birds

Because of the importance of the pineal in mediating light in mammals, it is reasonable to explore its role in birds. There is little evidence, however, to suggest a major role for the pineal in fecundity of birds. As previously indicated, the hypothalamaus is the structure of primary importance in the response of birds to the photocycle.

The avian pineal gland is nevertheless photosensitive and is an important source of melatonin. Melatonin is synthesized from serotonin through the action of *N*-acetyltransferase. An intermediate product is *N*-acetylserotonin, which, in the presence of hydroxylindole-O-methyltransferase (HIOMT), is changed to melatonin. HIOMT is localized in the pineal gland of both birds and mammals. The light-dark rhythmicity of *N*-acetyltransferase known for the laboratory rat is also seen in the avian pineal. The depressing effect of light is seen in the diurnal elimination of *N*-acetyltransferase, which appears in a circadian pattern in constant darkness. There is also a circadian rhythm of melatonin in the pineal of the Japanese quail, with an increase in the dark phase. In the rat this rhythm originates in the suprachiasmatic nuclei and is transmitted via the superior cervical ganglia, but in the chicken the rhythm arises in the pineal. Unlike the mammalian pineal, which follows retinal signals, the pineal of the domestic fowl is cued by an extraretinal photoreceptor. Similarly, the domestic duck, which also has a light-sensitive pineal gland, remains responsive to light when enucleated and pinealectomized.

Although the avian pineal is sensitive to light and exhibits nocturnal peaks of *N*-acetyltransferase, it is not known to affect gonadal activity in birds maintained under natural photocycles. In the sparrow (*Passer domesticus*) kept under certain "skeletal" photoperiods and following pinealectomy, there is a marked reduction of testicular growth. This effect is not consistent and may depend upon a seasonal change in sensitivity. The pineal in the house sparrow preserves circadian rhythm of motor activity, a function seen in some reptiles. In many nonmammalian vertebrates the pineal is important in generating endogenous circadian rhythms and may function in controlling the cycle of photosensitivity of the hypothalamus.

Photoperiodic Response in Mammals

In many major taxa of each of the classes of nonmammalian vertebrates, the pineal gland is photosensitive, but in mammals the pineal receives light signals from the retinas of the paired eyes. In all classes, therefore, the pineal gland is potentially a structure that can transmit photic information to the endocrine system. The pineal may be extremely important to the seasonal adjustment of the annual photocycle and may also have roles in other seasonal cycles of nonreproductive activity. A major product of the pineal is melatonin, a hormone that depresses gonadal activity in mammals.

Historical Aspects

Early in the twentieth century there were suggestions that the pineal gland was somehow related to a suppressed gonadal development. In the first 30 years of the 1900s several researches proposed that pathological changes in pineal size and activity were inversely correlated with sexual development and precocity in humans and that pineal products were antigonadal.

Until the early 1960s the pineal "body," as it had come to be known, had no clearly established functions, and it was certainly not accepted as an endocrine gland. From the early 1970s, research on the mammalian pineal gland has demonstrated that it is the source of substances that are carried to other regions of the body, with a subsequent change in metabolic rates of specific tissues. The pineal gland of mammals in indirectly cued to light, and the daily photocycle is a timegiver or *Zeitgeber* that entrains rhythmic production of some pineal enzymes and hormones. This has been shown by extensive experimental work on the laboratory rat, the hamster, and to a lesser degree on the ferret and some wild mice.

The concept of an antigonadal role for the products of the pineal gland and its relationship to the mammalian pituitary had long been expressed if not always accepted; but the effect of light on the pineal and the relationships of pineal secretions, especially melatonin, to reproduction remained obscure for many years. The demonstration that subcutaneous injections of melatonin depressed ovarian metabolism in the rat suggested a regulatory role of reproduction for the mammalian pineal. By the late 1960s the role of the pineal as an endocrine gland was suggested by the substantial amount of work conducted on the laboratory rat. Also, it was found that the preovulatory surge of LH is depressed by melatonin in the laboratory rat, even though the reproductive system of that rodent is not especially sensitive to day length. It later became apparent that although the gonads of the laboratory rat are only slightly responsive to photoperiodic changes, some other kinds of rodents clearly reflect changes in day length by fluctuations in pineal activity. The enzyme HIOMT leads to the formation of melatonin from serotonin, and this process is known to occur in both the retina and pineal gland of fish as well as in the pineal of fish, amphibians, reptiles, and birds in response to changes in light. Although the mammalian pineal is not photosensitive, both HIOMT and melatonin occur in the mammalian pineal and respond to light signals from the retina. In wild population of the Uinta ground squirrel (*Spermophilus armatus*), HIOMT increased after reproduction to its highest levels just before hibernation. Testicular size was maximal at emergence the following spring, when HIOMT levels were lowest.

Gradually the relationship of the occurrence of the pineal enzymes and their products to the daily light cycle was elucidated in the laboratory rat. During the daytime, serotonin levels are high; but its conversion to melatonin, in the presence of the enzymes *N*-acetyltransferase and HIOMT, is a nocturnal reaction inhibited by light. The pineal of the laboratory rat has long been known to increase HIOMT formation and melatonin secretion in the dark. Also early studies on the white-footed mouse (*Pero myscus leucopus*) suggested a reduction in pineal activity as a consequence of exposure to long day lengths.

Position and Function of the Pineal Gland

The pineal gland is an outgrowth of the posterior section of the dorsal surface of the diencephalon. As pointed out previously, the pineal is photosensitive in nonmammalian vertebrates. Blood enters the pineal form branches of the posterior choroid arteries and leaves via venules to adjacent menigeal tissue. Studies of pineal ultrastructure reveal endocrine secretory tissue with sympathetic innervation. The development and relationships of the pineal gland are described by Oksche (1965) and Quay (1974).

The Effect of Light on Pineal Activity

In mammals the pineal receives light information via a complex neural route from the retinas of the paired eyes. Impulses travel from the retina to the suprachiasmatic nuclei (SCN) of the hypothalamus. The second part of the pathway involves neurons going to the medial basal hypothalamus. The message then travels down the upper thoracic spinal nerves and, via the superior cervical ganglion, returns to the skull and to the pineal. During darkness norepinephrine is released by the sympathetic neurons within the pineal, where it stimulates the conversion of secretion to melatonin, at least in the laboratory rat. In nonmammalian vertebrates the pineal is itself photosensitive and release melatonin in darkness.

Although the pineal is clearly involved in reproductive response to day length in some mammals, there are many unanswered questions regarding the pathway through which melatonin becomes effective. Experiments indicate that different species of mammals are not equally responsive, and the basis for some observed differences are gradually being revealed by surgical elimination or destruction of the various critical structures.

Experimental work on the mammalian pineal clearly established it as the organ that, controlled by signals from the eye, reflects day length for the reproductive system. Although research has been confined to only a few species of mammals, it is the only structure in mammals known to have such a role. Studies of ectotherms suggest that the pineal, which is photosensitive in those vertebrates, also affects their reproductive activity. The pineal produces indoleamine and polypeptide hormones, and they probably have a variety of effects that are not well understood. The indoleamines in the pineal are derived from the amino acid tryptophan. In mammals at least, melatonin is perhaps the most important pineal secretion affecting reproductive activity.

Pineal activity is monitored by observing changes in levels of the enzymes that lead to the production of melatonin. There is a moderate peak of HIOMT and a very conspicuous peak of *N*-acetyltransferase during dark periods. The enzyme *N*-acetyltransferase and melatonin both respond to the cessation of light: their production commences soon after the onset of darkness and ceases promptly with the return of light. The circadian rhythm of *N*-acetyltransferase in the pineal has long been known to have a nocturnal peak as much as 30 to 50 times the diurnal low. Therefore, pineal activity is usually indicated by changes in levels of *N*-acetyltransferase.

Responses of Light

The testicular cycle of the golden, or Syrian, hamster (*Mesocricetus auratus*) is much like that of many wild mice. Testicular growth follows periods of long days, in the spring, and regression follows a decrease in day length, in late summer. In a regimen of changing light, the golden hamster requires at least $12^1/_2$ hours of light daily to preserve testicular function. Concomitant with the short-day induction of testicular regression in hamsters, there is a gradual decline in copulatory effort. Apparently the melatonin increase under short days causes a lower threshold (increased sensitivity) of the hypothalamaus to gonadal steroids, thus effecting a decline in secretions of LH, FSH, and prolactin from the anterior pituitary.

Like the gonads of many wild cricetid mice, the testes of captive hamsters remain small for the duration of most of a natural winter. In the Djungarian hamster (*Phodopus sungorus*), day length affects gonadotropin secretion, independent of gonadal feedback. In that species the critical day length is about 13 hours, and gonadal recrudescence is not induced by shorter day lengths. In the vole *Microtus agrestis* the critical day length is approximately 13.5 hours, and after six weeks testicular growth is greater in males kept at 15 and 16 hours than at 14 hours. In the white-footed mouse (*Peromyscus leucopus*), which has an extensive latitudinal distribution, the critical day, length increases with latitude.

It is not always clear how the annual photocycle is used as information in control of the annual reproductive cycle of some species of wild mammals. Many mammals are more or less nocturnal and spend daylight hours in burrows or hollow trees, in which they are well shielded from daylight. Nevertheless, in a strictly nocturnal species, such as the southern flying squirrel (*Glaucomys volans*), there is a clear testicular response to changes in day length in the laboratory. In such species of mammals, the critical aspect of the photocycle is not the period of day length but the period of darkness.

The active pineal is antigonadal in female hamsters as it is in males, but although there is both uterine regression and a reduction in ovarian activity, ovarian interstitial tissues increases and the ovaries are actually heavier than in sexually active females. Also, melatonin has some effects in the human female. Just prior to the estimated day of ovulation, there is a drop in circulating melatonin, based on daily measurements taken at 8 A.M. Also, exposure to 24-hour light stimulation at the midpoint of the menstrual cycle reduced variance in the duration of the cycle and increased the occurrence of 29-day rhythms.

In the ferret (*Mustela putorius*), which is a strongly seasonal species in nature, pinealectomy alters the timing of estrus and reduces the animal's responsiveness to light. Pinealectomized female ferrets fail to respond to artificially early exposure to long days or short days, although estrus occurs at the normal time for those under a natural photocycle. The effect of pinealectomy on the ferret, however, varies with the time at which the operation is performed: pinealectomy early in one estrous cycle delays estrus the following year. The pineal is essential for the female's reproductive cycle to become entrained to a natural photocycle.

A photoperiodic effect has been suggested in female monkeys. In Rhesus monkeys sexual activity generally increases during the short-day part of the annual photocycle, but social influences are also very important, for mating efforts do not necessarily parallel hormonal cycles. Although most conceptions in macaques (*Macaca* spp.) occur under regimens of decreasing day lengths (October–December), experimentally they seem not to be cued by short days. The ring-tailed lemur (*Lemur catta*) breeds in the winter of the Southern Hemisphere in Madagascar. A light schedule of 14L:10D inhibits ovulatory cycles, and a 5.5-hour reduction in the light period was followed by a return to ovarian cyclicity.

Circadian Sensitivity to Melatonin

In addition to the rhythm of melatonin production, there is a circadian pulsation to the sensitivity to melatonin, and this sensitivity occurs in the dark period or in the hours immediately prior to expected darkness. For example, both male and female hamsters experience gonadal growth under long days, and injections of melatonin are effectively antigonadal only if given in the latter part of the light cycle. The photosensitive phase of the golden hamster begins about 30 minutes prior to the individual's daily activity period and extends to approximately 12 hours. The circadian period of sensitivity to melatonin in mammals seems to be comparable with the PSP of birds, and it provides for a restricted period of responsiveness corresponding to the time during which melatonin is secreted. These afternoon injections of melatonin do not depress gonadal development in pinealectomized animals. This suggests the possible secretion of additional melatonin at a later time in the daily photocycle. There is also the possibility that the pineal produces other antigonadal hormones.

Because the rhythm of greater sensitivity is circadian, any light schedule other than one totaling 24 hours, or in 24-hour increments, will introduce part of a stimulatory period in the unresponsive phase of the animal. That is to say, a light-dark schedule repeated every 36 hours might expose the individual to light in its responsive phase (or PSP) every other day. Thus, ahemeral regimens (other than a nycthemeron, or 24-hour, period) of 12 or 36 hours, for example, produce results inconsistent with a 24-hour schedule with the same light-dark ratio.

The explore the effects of melatonin on the golden hamster, exogenous melatonin has been administered both by injections and by subcutaneous implants. Results vary, and melatonin does not always cause the gonadal regression seen in hamsters kept under a regimen of short days. Hamsters under a 14L:10D schedule are unresponsive to melatonin when it is given by injection early on the light period, but injections are antigonadal when administered late in the light period. Pinealectomy or superior cervical ganglionectomy results in continued fecundity in male hamsters, and when males and females that have undergone pinealectomy or superior cervical ganglionectomy are housed together, fertile matings follow. Thus pinealectomized hamsters resemble those kept under a long-day regimen, and destruction of the superior cervical ganglia produces the same effect.

Similarly, subcutaneous implants of melatonin in beeswex in hamsters kept under short day lengths are counterantigonadal. Apparently chronic administration of as little as 3.6 mg/day of melatonin suppresses the antigonadal effect of pineal melatonin and is tantamount to pinealectomy. Chronic exposure of receptor organs may maintain refractoriness to the suppressive effect of pineal melatonin. In other words, subcutaneous implants of melatonin have a counterantigonadal (instead of progonadal) effect and therefore resemble pinealectomy.

Thus, although short days and melatonin administered late in the light period both depress gonadal activity, this effect disappears when the animal is continuously exposed to melatonin by subcutaneous implants. It seems plausible that above a certain exposure to melatonin, its receptors become insensitive to the melatonin's effect, accounting for the failure to respond early in the daily light period after a nighttime of exposure to melatonin. Hamsters kept under stimulatory schedules (14L:10D) and given 25 μg at 1700 hours underwent gonadal involution when saline had been administered at 1100 hours, but not when 1 mg of melatonin had been injected at 1100 hours. These results support the concept that exposure to melatonin renders receptors temporarily insensitive and also that injection of melatonin early in the light period prolong this insensitivity. The nature or location of these receptors is still undetermined. In the white-footed mouse (*Peromyscus leucopus*), receptors appear to be in the suprachiasmatic nuclei.

Suprachiasmatic Nuclei

The role of the hypothalamus in mammalian rhythmicity was suggested by the loss of discrete circadian rhythms in laboratory rats following lesions in the ventral median area of the hypothalamus. More recent studies of circadian control have centered on the specific region of the suprachiasmatic nuclei (SCN). The SCN of the rat lie in front of the anterior hypothalamus above the rear margin of the optic chiasma. The SCN receive a major afferent nerve tract from the retina through the optic chiasma. Other afferent nerve tracts are received from the midbrain raphe nuclei and form the ventral lateral geniculate nuclei. Thus there is a variety of nervous input to the SCN, and within the SCN there is a large complement of synaptic types. Efferent fibers from the SCN are not well known.

The SCN seem to preserve circadian rhythms in some mammals, for its destruction is generally followed by arrhythmicity, including the loss of pineal rhythms. There is a clear correlation between circadian rhythmicity of motor activity (such as wheel running) and the regularity of estrous cycles in the golden hamster, and they appear to be regulated by the SCN.

The SCN of the domestic rabbit contain neurosecretory cells that show increased activity following ovariectomy and also after copulation. These neurosecretory cells lie adjacent to capillaries and in some rodents may be the source of GnRH.

Autumnal Breeding Mammals

In contrast to many species of mammals of the temperate regions, several kinds of ruminants mate in the autumn, under a regimen of decreasing day lengths. Autumnal mating occurs in several kinds of deer (*Odocoileus hemionus*, *O. virginianus*, and *Cervus elaphus*) and domestic sheep. Perhaps because of their rather large size and the precocial condition of the young at birth, fall mating allows the development of young by the following spring. The roe deer (*Capreolus capreolus*) constitutes a conspicuous exception. It mates in the summer, but because of delayed implanation, the young are nevertheless dropped the following spring. In the roe deer, testicular growth and high levels of testosterone are concurrent with maximal day lengths.

Experimental work with domestic sheep has been designed to clarify the pathway of testicular recrudescence under short days. When Soay rams were moved abruptly from four months of long days (16L:8D) to short days (8L:16D), there was a resurgence of gonadotropins in less than two weeks and an increase in testicular size in from two to four weeks. LH and FSH peaked daily in the dark part of the 24-hour period, which may reflect a circadian rhythm of the hypothalamic-pituitary-gonadal axis. In another group of Soay rams kept under alternating periods of long (16L:8D) and short (8L:16D) days, gonadotropin release and gonadal activity were conspicuous under the short-day schedule. At the ninth week of short days there was a marked rise in both LH and FSH with a subsequent increase in both testicular growth and testosterone output; after week 9 gonadotropin levels declined, but testicular activity continued to increase. When the rams were returned to long days, gonadotropins quickly dropped to a minimum level, but testicular activity declined only slowly. In the red deer (*Cervus elaphus*) serum LH levels rise to a peak in August, and testosterone levels are maximal from September to November, the period of mating.

In sheep, a decrease in response of the pituitary and the hypothalamus to the negative feedback of estradiol is associated with decreasing day length. Once this response decrease results in a sustained rise of LH for two to three days in autumn, ovulation will ensue. Conversely, in the nonbreeding season of the domestic sheep, estradiol suppresses the release of LH through an increased response of the hypothalamus and anterior pituitary to the negative feedback of estradiol.

To examine the role of pineal melatonin in the photoperiodic control of seasonal breeding, pinealectomized and ovariectomized ewes were provided with physiological levels of estradiol from Silastic capsule implants, and serum LH was monitored. Pinealectomy eliminates melatonin secretion, and therefore melatonin was infused intravenously. A simulated long-day secretion of melatonin was followed by an LH decline typical of long days and similar to the decline seen in pineal-intact ovariectomized ewes. The pattern of melatonin secretion is an aid in measuring day length and transduces the photoperiodic signal into a change in the response of LH to estradiol negative feedback. It has also been suggested that gonadal activity depresses melatonin secretion, for in ovariectomized ewes melatonin is three times that in intact ewes.

Many vespertilionid bats have a modified autumnal mating, with spermatogenesis in the spring and summer and copulation in the fall, by which time the testes have regressed and spermatozoa are stored in the cauda epididymides. The male pallid bat (*Antrozous pallidus*) maintained from April or May until August on either short (10L:14D) or long (14L:10D) schedules revealed a sensitivity to day length. By August, males on short days had regressed testes, and all had epididymal spermatozoa. Males on long days showed some spermatogenesis, and only 55% had epididymal spermatozoa. By September all males had regressed testes and epididymal sperm. The effect of day length may be a result of melatonin secretions in this bat. When males were provided in July with melatonin subcutaneous Silastic capsules or with empty capsules and thereafter maintained on short days (10L) or long days, the long-day (14L), melatonin-treated individuals had smaller testes and were more likely to have epididymal spermatozoa than those long-day bats that had received empty capsules. Melatonin seems to mimic the short day effect on testicular activity.

Refractoriness

In mammals refractoriness is usually a failure to respond to the suppressive influence of darkness, which is called scotorefractoriness. This results in a spontaneous recrudescence of the reproductive system in individuals kept under a schedule of long nights (or short days).

When kept under short days (less than 12.5 hours of light), hamsters undergo testicular regression. Eventually, after about 30 weeks, there is spontaneous recrudescence, indicating that the reproductive system had become insensitive to the suppressive influence of short days. In the late winter and early spring the animals have already become scotorefractory.

Sensitivity to short days returns, however, after activity is restored to the reproductive system. Exposure to long days (14L:10D) for 20 weeks resulted in renewed sensitivity to the suppressive influence of short days in the golden hamster. Spontaneous testicular recrudescence has been found to occur also in the white-footed mouse (*Peromyscus leucopus*) when kept for 10 to 20 weeks under short days (10L:14D). The hamster and the grasshopper mouse (*Onychomys leucogaster*) showed similar responses.

After three months of exposure to long (16L:8D) or short (8L:16D) days, two groups of pine voles (*Microtus pinetorum*) did not differ in weights of body, testes, or seminal vesicles or in testosterone levels. Voles tend to be rather weakly responsive to seasonal changes in daily photocycle and are clearly capable of becoming scotorefractory.

The mechanism by which mammals become scotorefractory is not yet established. There may be an increasing threshold of sensitivity of the neuroendocrine axis to melatonin. When given melatonin injections (25 g) daily just before the end of the light period of a 14L:10D schedule, hamsters experienced a sudden decline in testicular size. However, in males that had either been (1) pinealectomized or (2) exposed to 30 weeks of a 10L:14D schedule and had spontaneously recrudesced testes, refractoriness to the antigonadal effects of melatonin was exhibited in both groups.

The intact individuals had regressed testes after of melatonin was exhibited in both groups. The intact individuals had regressed testes after seven weeks of melatonin treatment, but their testes recrudesced spontaneously after an additional fourteen weeks of melatonin treatment. This suggests that spontaneous testicular recrudescence in the hamster results from an eventual failure of melatonin to suppress gonadal activity.

Although most research on the relationship of light on gonadal recrudescence has emphasized specific photoperiods, in mammals as in birds, there is an apparent sensitivity to *changes* in duration of light. Mating in the Australian dasyurid *Antechinus stuartii* is correlated with the rate of change in significantly different in two populations from different latitudes.

A return to fecundity in the absence of long days suggests a mechanism for the appearance of an endogenous circannual rhythm. Fecundity returns not only after a prolonged exposure to short days but also, in hamster, when blinded shortly after the winter solstice. It appears that during the winter, hamster develop a gradual increase in scotorefractoriness and a concomitant increase in sensitivity to long days. Although photorefractoriness (to long days) and scotorefractoriness (to short days) might be viewed as opposite sides of the same coin, the spontaneous return of testicular activity under short days indicates an increase in scotorefractoriness in the absence of long days. Brief interruptions of short (winter) days by long (summer) days do not reverse the scotorefractory condition, but a sensitivity to short days returns after exposure of between 10 and 22 weeks of summer days.

As previously pointed out, sheep breed as the days become shorter—at the end of summer and early autumn. Sheep appear to become alternately refractory to both long and short days. Soay rams experience testicular regression under long days but become photorefractory after about 16 weeks, and thus testicular recrudescence begins somewhat prior to the onset of short days. Similarly, after some 16 weeks of exposure to short days, the system becomes scotorefractory and involution commence. The role of melatonin release in Soay rams is equivocal. It is clear whether the effect of melatonin depends upon the duration of the dark period or the time of the melatonin peak relative to the light-dark cycle. After 25 weeks of exposure to short days (8L:16D) or to long days (16L:8D), the peaks of melatonin in Soay rams become somewhat dissociated form the dark period, perhaps reflecting slow alterations in the timekeeping mechanism. This slow adjustment might underlie scotorefractoriness in mammals in both short-day and long-day breeders.

Weakly Photosensitive Mammals

Some mammalian reproductive seasons depart from oscillations of the natural photocycle. This does not necessarily mean that such species are not sensitive to the effects of light observed in other mammals, but simply that they may be cued by several exogenous factors. Voles and lemmings are notoriously variable in their fecundity, both in time and magnitude. Voles (*Microtus* spp.) do respond to day length, and gonadal systems do recrudesce under long days. When subjected to several light schedules in the laboratory, the field vole (*Microtus arvalis*) displayed an increase in testis mass, seminal vesicles, and epididymal sperm as well as increased ovulation rates with increased day length, but it appears that light alone did not control reproductive seasonality in the field.

The photoperiodic response of *Microtus agrestis* in British Isles resembles that of *M. arvalis*. When reared form birth under long days (16L:8D), *M. agrestis* becomes sexually mature at 40 days; when kept under short days (8L:16D), fertility was attained at six months of age. Although testicular regression occurs in voles kept under short day lengths, regressed testes of captive voles do not reach the extremely small size of those of wild mice. Factors other than day length promote seasonal changes in gonadal size and activity.

Lemmings live in the Arcitc and are subnivean for much of the year. Although they live where photoperiodic changes are extreme, lemmings are virtually unresponsive to changes in day length. The collared lemming, (*Dicrostonyx groenlandicus*) when reared under contrasting light regimens (6L:18D and 20L:4D) from weaning to day 70, exhibited no difference in gonadal development in either sex, and individuals under both set of light schedules became sexually mature.

The cotton rat (*Sigmodon hispidus*), a crecitid rodent of the New World, experiences little effect from various photoperiodic regimens. When kept on short days (19L:14D) form birth to day 50, there was slower gonadal growth than in animals on a long day schedule (14L:10D). Gametogenesis, however, was not statistically different in the two groups. Quite likely, reproductive seasonality in the cotton rat in nature is controlled by factors other than the photocycle. Similarly, seasonal reproduction is not modified by day length in the feral house mouse (*Mus musculus*), and the house mouse is nonseasonal in its reproduction in nature. Wild house mice respond negatively to strong light, whereas laboratory stocks are unaffected by light intensity. Captive-born young of wild stocks breed in constant darkness (with only a very dim red light), but when exposed to a 14L:10D schedule including light of 1000 lux, there is a marked reduction in litter size. There is no apparent mechanism for this response, and domestic mice are unaffected.

Although reproduction is associated with seasonal or sporadic rainfall in many Australian marsupials, the pouched "mouse" (*Sminthopsis crassicaudata*) experiences gonadal recrudescence under a long-day schedule (15L:9D). The hopping mouse (*Notomys alexis*), a murid rodent of Australian deserts, is only slightly affected by photoperiod: short days prolong estrus, but do not influence ovarian or uterine weights.The degree of photosensitivity is unknown in most desert rodents, desert-dwelling rodents at both low and midlatitudes appear to breed in response to factors other than day length.

It has sometimes been suggested or assumed that domestication somehow dilutes the sensitivity of the gonadal system to day length. As we have seen, however, some wild rodents are not cued by the photocycle, and some domestic animals (eg., sheep) are quite dependent upon day length for gonadal stimulation.

The mare experiences a winter anestrous phase, presumably reflecting the biology of wild or feral individuals. Exposure to a 16L:8D cycle in midwinter is followed by estrus almost two months prior to that of controls, with a simultaneous increase of LH and estradiol. Data on the hamster suggest recent changes in its photosensitivity, perhaps a result of its history of domestication.

Testicular activity of the laboratory rat is known to be minimally responsive to variations in photoperiod. The pineal gland in rats does, however, have a depressing effect on the gonads when rats are kept under short days, but there is no effect in intact animals. When male rats are deprived of their olfactory bulbs, however, the typical short-day response of testicular regression and decreased testosterone levels occurs; but this effect does not occur when bulbectomy and pinealectomy are performed concurrently. That is, olfaction in this rodent may limit the antigonadal activity of the pineal. This state of affairs raises more questions than it answers. Possibly there are olfactory signals that are of social significance and more powerful than day length in influencing gonadal activity.

3

BIOTIC FACTORS OF FRESH-WATER ENVIRONMENTS

FRESHWATER HABITAT

Freshwater environments are also divided into two major categories, lotic (lotus=washed, or running water), and lentic (lenis=calm, or standing water) habitats. Lotic habitats are those existing in relatively fast running streams, springs, rivers and brooks. Lentic habitats are represented by lakes, ponds, swamps etc. The above classification of the freshwater environments is based on tow conditions: currents are the radio of the depth to surface area. Since lakes and ponds often contain currents or at least wave action and since streams often harbour quiet pools or calm backwaters the difference between lotic and lentic waters is not very precise. However, temperature, light currents, concentrations of respiratory gases, and concentration of biogenic salts are important limiting factors influencing organisms of all fresh water habitats.

Lentic Community

Lentic waters are divided into three major zones or sub habitats littoral, and profundal.

Littoral zone

The littoral zone adjoins the shore (and is thus the home of rooted plants) and extends down to a point called the light compensation level, or the rate of respiration. Within the littoral zone producers are of two main types: rooted or benthic plants, and phytoplankton, or floating green plants, which are mostly algae.

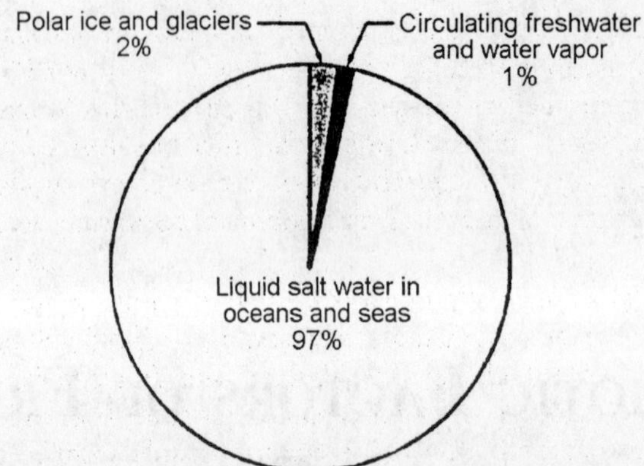

Fig. 3.1. World's water supply. Life processes on land depend on the 1 percent of freshwater and water vapour in circulation.

The littoral zone is the home of a greater variety of consumers than are other zones. The zooplankton of the littoral zone is rather characteristic, and differs from that of the limnetic zone in preponderance of heavier, less buoyant crustacean which often cling to plants or rest on the bottom when not actively moving their appendages. Importent groups of littoral zooplankton are large. Weak swimming species of Daphnia and Simocephalus, some species of copepods, many families of ostracods and some rotifers. The nekton of littoral zone is often ricin species and numbers. Adult and larval diving beetles and various adult Hemiptera are conspicuous.

Various Diptera larvae and pupae remain suspended in water, of ten near the surface. Pond fish, frogs, turtles and water snakes are almost exclusively members of the littoral zone community. Tadpoles of frogs and toads are important primary consumers, feeding on algae or other plant material. Periphyton of the littoral zone exhibit a zonation paralleling that of the rooted plants, but many species occur almost throughout the littoral zone. Among the periphyton forms, for example; ponds snails, damsel fly nymphs and climbing dragonfly nymphs rotifers flatworms, bryozoa, hydra, and midge larvae rest on, or are attached to stems and leaves of the plants. Another group containing both primary and secondary consumers may be found resting or moving on the bottom or beneath silt or plant debris—for example sprawling odonata nymphs (which have flattened rather than cylindrical bodies) crayfish, isopods and certain may fly nymphs. Descending more deeply into the bottom mud are burrowing odonata, and ephemeroptera, clams, true worms, snails, chironomids (midges), and other diptera larvae.

Limnetic zone

The limnetic zone includes all the waters beyond the littoral zone and down to the light compensation level. The community of the limnetic zone is composed only of plankton nekton, and sometimes neuston (organisms resting or swimming on the surface). Phytoplankton producers consists of diatoms, green algae, blue green algae, and the algae like green flagellates, chiefly the dinoflagellates.

The limnetic zooplankton consists of few species but the numbers of individuals may be large. Copepods cladocerans, and rotifers are generally of first importance. But their species are largely different from those found in the littoral zone. The limnetic nekton consists almost entirely of fish. In ponds the fish of the limnetic zone are the same as those of the littoral zone but in larger bodies of water a few species may be restricted to the limnetic zone.

Profundal zone

The bottom and deep water area which is beyond the depth of effective light penetration is called profundal zone. This zone is often absent in ponds. Since there is no light, the animals of the profundal zone depend on the upper zones for basic food materials. In return the profundal zone provides rejuvenated nutrients, which are carried by currents and swimming animals to other zones. The major community constituents are bacteria and fungi and three groups of animals consumers:

(a) blood worms, or haemoglobin containing chironomid larvae and annelids,

(b) small clams,

(c) and phantom larvae or chaoborus (corethra).

The first two groups are benthic forms, the last are plankton that regularly move up into the limnetic zone at night and down to the bottom during the day. All animals of the profundal zone are adapted to withstand periods of low oxygen concentration, whereas many bacteria are available.

Lotic Community

The biotic community in streams and rivers is quite different from that in lakes and ponds. The difference in the community are largely due to the differences in physical and chemical conditions of their environments. The presence of definite and continuous current is the main characteristic of lotic habitats. Streams are usually relatively shallow and, therefore have a large surface compared to their depth.

Land water interchange is relatively mor extensive in streams, resulting in a more open ecosystem. This means that streams resulting in a more open ecosystem. This means that the streams are more intimately connected with the surrounding land than are most standing bodies of water. For this reason a significant portion of their (streams's) nutrients falls into them from their banks.

Another large nutrient source for many streams is the detritus washed in from connected lakes and ponds. Third important difference between streams and ponds is that streams are ordinarily well supplied with oxygen. For this reason stream animal generally have a narrow tolerance and are very sensitive to decreases in the oxygen content of the water. Current is a major limiting factor in streams. It might be assumed that plankton would be absent from streams, since such organism are largely at the mercy of the current. Only in slow moving parts of streams and in large rivers plankton is able to multiply and thus become and integral part of the community.

Stream have their own producers, such as fixed filamentous green algae, (Cladophora) encrusted diatoms, and aquatic mosses (Fontinalis). But these are usually insufficient to support the numerous consumers found in stream. Many of the primary consumers in streams are detritus feeders that depend in large part on organic materials which are swept or fall in from terrestrial vegetation. Generally benthic invertebrates are more numerous in rapids communities, whereas stream nekton and borrowing forms, such as clams, burrowing odonata and ephemeroptera, are more abundant in pools. Stream fishes usually live in pools and feed in or at the base of rapids communities. Most stream animals are easily recognizable from their adaptations for living in swift current.

Characteristics of Freshwater Habitat

Following are the characteristics of freshwater habitat or the limiting factors of freshwater habitat.

Temperature

In freshwater, the range of temperature variation are smaller and temperature changes occur more slowly as compared to those in air. This is due to the following unique thermal properties of water.

(i) *High specific heat,* that is a relatively large amount of heat is involved in changing the temperature of water. One gram calories (g cal) of heat is required to raise 1 ml (or 1 g) of water by 1°C (between 15° and 16°C). Only ammonia and few other substances have values higher than 1.

(ii) *High latent heat of fusion.* Eighty calories are required to change 1 g of ice into water and *vice versa.*

(iii) *Highest known latent heat of evaporation.* Five hundred and thirty six calories per gram are absorbed during evaporation of water. The process of evaporation occurs more or less continuously from vegetation, water and ice surfaces. A major portion of the incoming solar radiation is dissipated in evaporation of water form the ecosystems of the world, it is this that moderates climates and makes possible the development of life.

(iv) *Water is densest at 4°C,* expands and hence becomes lighter both above and below this temperature. This unique property of water prevents deep bodies of freshwaters to freeze solids.

Temperature is one of the major limiting factors for aquatic organisms as they are stenothermal (having narrow range of temperature tolerance). Temperature changes produce characteristic patterns of circulation of water and its stratification which greatly influence aquatic life. These will be described later in this chapter. Large bodies of water greatly modify the climate adjacent land area. The freshwater habitats occupy a relatively small portion of the earth's surface as compared to marine and terrestrial habitats but their importance to man is far greater due to the following reasons:

1. They are the most convenient and cheapest source of water for domestic and industrial needs (we go more water from the sea but at great cost in terms of energy required and salt pollution created).
2. The freshwater are important in the hydrological cycle.]
3. Freshwater ecosystems provide more convenient and cheapest waste disposal system.

Physical factors play an important role which for the most part are responsible for the distribution of animal life in fresh water habitat.

Depth

Obviously, depth has an important bearing on the physico-chemical prospects of water. The penetration of solar rays varies according to the degree of turbidity as well as depth. The photosynthetic regulation and water circulation patterns are very much dependent upon sunlight. On one hand, productivity in shallow ponds is facilitated as light reaches the bottom while on the other, in very shallow one the same factors inhibit the survival of bio-organic materials.

Water Movements

Water movement plays a vital role in shaping the biocoenosis, because of its effect on the substratum. When the movement is vigorous it will carry away the finer particles leaving only stones. When the movement is weak sand will settle and as it slackens more particles will drop to the bottom. Moss grows on the stones that are not turned over frequently and the higher plants take the root in sand and mud. In the lake, the effect of waves diminishes as the depth increases. The sand effect of waves diminishes as the depth increases. The sand region is often sparsely inhabited since the animals are confronted with a difficult substratum with little to eat. Beyond this fragments of dead leaves and other fine particles of organic matter fall to the bottom and provide a supply of food.

Animals that live in running water show adaptations of structure and behaviour and inhabit under stones and plants where there is decrease in the water current. They are most numerous faster water. Animals that depend on the current to bring food, usually select an optimum speed. Some animals desire a current of certain seed to supply them the oxygen they require. Selection of a substratum of particular type is common in both still and running water. It may be related to the feeding and reproduction of the animals.

Molar Agents

Molar agents may be in solid or fluid state. Wind currents, waves, and tides are important molar agents which make the habitats.

Wind

By inducing large scale water movements wind facilitates dispersal of aquatic organisms. Many examples could be ciled. Indeed, the distribution of immature larval forms is largely due to the churning action of wind on the water bodies. Moreover, the direction of wind is such as to enable the aquatic organisms to obtain their natural food.

Currents

Currents are caused due to winds and rotation of the earth . They help in distribution selection and dispersal of vast number of smaller animals. Currents may be seasonal, permanent of adventitious. In freshwater, currents distribute the food and oxygen supply to the organisms. Dissolved oxygen content of rivers, streams and lakes is correlated with the rate of flow of water currents. Stream and pondwater forms react to currents according to the content of oxygen in the water The rivers having sluggish current show large abundant bottom flora and fauna. Currents above moderate speed limit the plankton population. The fish species in streams are roughly arranged in a graded series according to the swiftness of the current. The caddis fly larvae are also distributed according to speed of the running water. Sponges develop their shapes according to speed of the currents. In hill districts water current may increase suddenly as a result of heavy rain and decrease again quickly when the rain stops.

Effects on plankton

Majority of animals that dwell in running water also show adaptive behaviour to life in sheltered places between stones or in vegetation, where there is little current. In rapidly flowing water, where bottom is stony, larvae of simuliidae are morphologically adapted for life in that they have means to cling to a smooth surface and for filter feeding from water to obtain food. Net spinning tricopterans also make use of the current to obtain their food. *Ancylus*, *Rhithrogena* and *Gammarus* are three commonest species which are found fairly fast streams. Many fishes are well adapted to life in torrential streams.

Seiches

In standing water wave movements play an important role. They cause larger displacements than the travelling waves. In travelling waves, the water particles move through the same distance but differ in phase. However, in standing waves, phase of all particles is the same but extent of the movements is different. Phenomenon of standing waves occurs in all lakes. This is called seiches. However, only in large water bodies, a periodic rise and fall of the water surface is observed. Such changes are extensive enough of measuring instrument. Standing waves are not only limited to interface of water and air but also occur within water body mass of different strata having different densities get developed. These are called internal seiches.

Internal seiches occur in lakes at interface of epillimnion with lighter water and hypolimnion with heavier water. Phenomenon of seiches is conveniently studied by temperature measurements taken at short intervals of time and hence standing waves are also called as temperature seiches. Importance of internal seiches to limnology lies both in vertical horizontal movement of water. Seiches depend upon morphology, density, temperature, fluctuation of lake water, etc. Oscillations are more rapid in summer stagnation and slacken in autumn circulation and disappear entirely when the density differences get diminished.

Tides

Lunar tides, also called true tides are found in large lakes. Amplitude of lake tides is naturally smaller. Due to the presence of complicated seiches difficulties occur in observing and measuring lake tides. Seiche synchronised with lunar period would enhance the tide range, and conversely when it is out of phase, would inhibit tidal effects. Tides exert influence on river flow which serves to increase turbulence in the lower stream course and to develop periodic flooding of the valley.

Waves

Currents, wind or combined effect of both cause development of water waves. A series of trichoidal circles produced by surface waves in water become smaller with increase in depth.

Evaporation

Cold air has very little capacity to hold moisture. However, when air becomes warmer, it attains high evaporating power. Winds accelerate the evaporation and drying of water. The rate of evaporation of water depends upon surface of water. Deficiency of moisture in the upper atmosphere, temperature, winds, current, etc. The evaporation also depends upon the difference in concentration also depends upon the difference in concentration of water molecules in the interfacial layers of both air and water.

Temperature

Water is a poor conductor of hear. Because of its specific heat, it takes longer time to warm up than the air and slower in cooling so that the temperature fluctuations are neither so great nor so radical in water as in the air. This is of importance to organisms that entirely depend on their surroundings for their body hear. Large variation in temperature is particularly well evident in ponds and brooks, which freeze quickly in severe winter and dry out as fast in the summer heat.

In running water habitat there is smaller annual and diurnal temperature variation than is usually encountered in stagnant water maintains it live organisms at even 4°C. This ice cap cover protects the lakes from adverse effects of wind Stratification of temperatures can be seen only in those expansive lakes which suffer large diurnal environmental variations. A factor of the greatest biological importance, in large deep lakes is the existence in summer, of layers of water at different temperatures.

The absorbing power of water prevents the sunlight and hear penetrating into the lower depths, so that two main layers are formed viz., the upper warmer and lighter waters and colder heavier layer at the bottom. Winds circulate the water to a certain extent but their effect is limited to the upper stratum of the lake and an intermediate layer outside the influence of the wind comes into existence wherein the temperature change is very rapid. This layer is called 'Thermocline', water below is called the 'Hypolimnion'. Thermocline acts as an effective barrier, preventing the free mixing of the waters, dissolved chemicals and gases between upper and lower layers thus rendering the two layers quite different in their ability to sustain life. The hypolimnion is cold and dark with little plant life.

On the other hand, the epilimnion is well lighted, warm and well oxygenated in which multitude of microscopic animals and plants can produce food and multiply. On the death of animals and plants, they fall into the hypolimnion where they decompose by the action of bacteria and use up the oxygen present. As there is no circulation between the two layers, the products of decomposition do not become available to enrich epilimnion. This may sometimes limit the animals the lake can support a lake where such conditions exist is called as Eutrophic lake. On the other hand, where there is scarcity of substances which are needed for the plant growth, a sparse plankton is produced in the upper layers and the decomposition of its remains at the bottom does not use up all the oxygen during the season. Such lakes are called as Oligotrophic lakes. A third type of lake is called Dystrophic lake which is generally a lake with bottom composed of humus ooze and soft acid water, which is brown in colour.

Lakes are often seen changing their characteristic due to seasonal fluctuations. The large quantities of dissolved suspended inorganic and organic matter are being washed in the autumn and the winter flood together with the products of decomposition brought into circulation at the overturn and provide conditions favourable for the growth of animals and plants. Such are the main exciting characteristics of lake biology and the study is not only a absorbing interest but also better understanding contributes to the benefit of mankind.

Turnover

Above 4°C the highest temperature occurs at surface of water while coldest remains at the bottom. This produces a thermal gradient. However, below 4°C the colder water occupies the surface and the thermal gradient in which water gradually becomes warmer increase with depth. Thermal gradients of such type occur not only in winter but also are reversed during summer. A transition between these two gradients occurs in autumn and spring and is known as a turnover or overturn. Strong spring winds stir the water from top to bottom. This phenomenon is called spring overturn.

Summer and water stagnation

In sheltered situations in a lake when during long summer periods the upper strata gets warmed rapidly preventing mixing by wind with lower levels and thus stability is quickly established. This is well known as summer stagnation. During low temperatures the surface layer of ice cover protects the lake from the wind action. Such a permanent stratification in which there is a sharp rise in the temperature of lower strata for long periods is called as winter stagnation. Mixing of water is prevented when the difference in temperature exceeds 10°C.

Lights

In natural waters, blue light of spectrum is main illumination which is transmitted in water at the depth of 100 meters. Red and orange colours are transmitted in coloured waters more deeply but their intensity reduces quickly.

Transparency of Water and Light Penetration

Standing freshwater is often not as transparent as it contains lot of suspended material which obstructs the passage of light through it. This slows down the photosynthetic activity of green plants which the habitat harbours. The fauna which depends on the flora is also affected. Turbidity, especially when caused by clay and slit particles is often important as a limiting facto. Conversely when turbidity is caused by living organisms measurements of transparency become indices of productivity. Shallow standing freshwater habitat e.g., small ponds receive light to the very bottom this results in an abundant growth of phytoplankton and other rooted plants. The fauna which feed on these plants are also abundant e.g., growth of zooplankton is controlled by the growth of phytoplankton upon which they feed.

Light is the principal factor which controls the diurnal (day and night) movement of plankton although other factors like dissolved O_2 free CO_2 temperature food and gravity also have a slight influence. The diurnal movent may be of various kinds. For example, the total plankton population may be in the surface waters in the afternoon and migrate deep downwards at night or it may be movements only upto

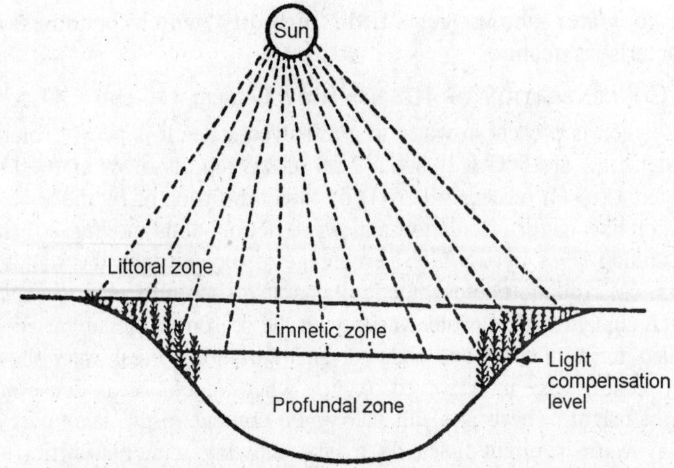

Fig. 3.2. Diagrammatic sketch showing the three major zones of a freshwater body as a lake.

thermocline or only a few members of the whole population may move, others may not move at all.

Colour

Pure water bodies appear nearly black as they absorb all light components of spectrum,. The lake water containing suspended material is seen blue in colour due to the scattering of light by water molecules. However, color of water ranges from green blue through blue green with yellow, yellow-brown to brown. Colour of water which appears due to material in colloidal state is known as specific colour. Apparent colour is that which is result of interplay of light on suspended particulate matter together with other factors such as reflection of water base and sky. Further, planktons are responsible for colour of water. For example, presence of abundant blue green algae gives dark greenish colour to water. Water containing diatoms shows yellowish colour. Zooplankton like micro-crustaceans gives tint of red colour to water. Undecayed organic matter gives brown colour to the water. Colour within any lake may not be uniform from surface to bottom.

Turbidity

Degree of opaqueness developed in water by means of suspended particulate matter is known as turbidity. Substances like humus, silt, organic detritus, colloidal matter, plants and animals brought into lake from outside are called as allochthonous. Substances which are produced within lake and produce turbidity are called autochthonous. However, running water shows very little turbidity which becomes its characteristics feature.

CONCENTRATION OF RESPIRATORY GASES: O_2 AND CO_2

Oxygen is present in water in a dissolved state. It is poorly soluble in water e.g., at 15°C only about 7 ml of oxygen dissolves in 1000 ml of water. Oxygen reaches water (1) by direct diffusion or by movements of water like water circulation and wave action, thus moving streams and running river have a high percentage of oxygen than the standing waters. (2) Aquatic photosynthetic plants also supply oxygen to water. Oxygen content of freshwater varies with the season (seasonal variation) and also during the day and night (diurnal variation). Freshwater shows a conspicuous rise in dissolved oxygen when vegetation is dense and sunlight bright. There is a fall in oxygen content in the later part of the day when sunlight fades. In a lake showing winter stratification oxygen may be completely absent.

The reduction of oxygen is caused in a number of ways the major being respiration and decomposition of dead organisms. In standing pools, there is a lot of decaying vegetation, hence such pools may become completely devoid of oxygen at times. In such cases oxygen forms a limiting factor for the organisms.

Carbon dioxide being soluble in water is present in a dissolved state as carbonic acid and as carbonated and bicarbonates of various elements like magnesium and calcium. Carbon-dioxide is formed during respiration of plants and animals and as a result of decomposition of organic matter. Some CO_2 is fatal to organisms. Molluscs, some insects and some bacteria can precipitate carbonates which accumulate gradually to from marl deposits on the bottom of certain lakes. When these deposits become huge there is decrease in animals life present in the lake particularly the bottom dwellers.

Buoyancy

Buoyancy depends upon the density and temperature of the water. Water becomes denser and viscous at low temperature and gives greater buoyancy. Aquatic animals are slightly heavier than water and these animals tend to sink in less dense water. Buoyancy of water varies with seasonal changes. Many aquatic animals change their forms according to buoyancy When water has greater buoyancy, animals have compact bodies, but when the buoyancy is less, animals develop organs which help them in flotation or impede sinking. Protozoans rotifers and many crustaceans show compact "winter" forms and attenuated or spinose "summer" forms a phenomenon known as cyclomorphosis which has been dealt with late. Buoyancy plays a vital role in minimising the locomotory adaptations of the aquatic animals.

The density of water relieves the aquatic organisms of the necessity of supporting their own weight and enables them to move easily through the water even with such feeble means of locomotion as cilia of protozoans and lashing hairs of flatworms. Many minute forms of both plants and animals spend much of their life floating in the water. The aquatic media are often used as excellent transmitters conduction sounds and vibrations. Some insects such as Corixid bugs and Hygrobia have stridulating organs by means of which they can produce sounds which are believed to be the means of attracting the opposite sex for mating.

Density

The density of distilled water to 0°C, and 760 mm Hg is 775 times greater than air. The density of large water bodies, lakes and rivers varies at different places and also at different times. These differences in the water density are due to variation in temperature and salt contents of the water. There is a liner increase in water density with increase in dissolved substances of salts. The changes in water density due to temperatures fluctuation are more important. The density changes more quickly at higher temperatures than at lower ones.

Viscosity

Viscosity of water is one hundred times greater than that of air. Viscosity is cause of the frictional resistance which water offers to the moving animals. The magnitude of viscosity function is proportional to the extent of surface of animals body which is in contact with water, the speed of movement of animals and to temperature.

Surface Tension

The surface tension acts at the water air interface and develops a biotope. Many plants and animals and their parts are affected by surface tension in many ways, depending upon whether parts of animals and plants are wettable or not. For example, Potamogeton young leaves and Cladocera shell are water repellant, hence water molecules always shy away from shell or leaves, etc., and the surface tension not only prevents wetting but many aquatic animals and plants are able to float and avoid submergence precisely because of this properly of their integumental epidermis.

Other Dissolved Cases

Other dissolved gases are methane (marsh gas), hydrogen sulphide, ammonia and nitrogen.

Hydrogen sulphide is formed due to decomposition in the absence of oxygen, i.e., anaerobic decomposition of organic matter containing sulphur. It characterizes stagnant fresh water pools where oxygen is depleted or in streams and lakes contaminated with sewage. This gas is highly toxic to organisms.

Methane is formed due to decomposition of organic matter at the bottom and is seen coming out of the surface as bubbles. It is not toxic to the organisms in small amounts.

Nitrogen gets diffused into water from the atmosphere. It is not toxic to organisms in general.

Ammonia is formed due to the decomposition of organic matter or due to industrial waste being thrown into water. It is toxic in high concentrations.

Concentration of Biogenic Salts

Salts are present in water in dissolved state. Except for certain mineral springs. Even the hardest freshwater have a salt content of less than 0.5 part per thousand compared with 30 to 37 parts per thousand for sea water. The important salts are compounds of nitrogen, phosphorus, calcium and other minerals present in traces.

Salts reach water by erosion, inflow and by decay of aquatic forms. From this, it follows that salinity will vary in different fresh water depending upon their place of occurrence.

Calcium occurs as carbonates and bicarbonates. In soft water calcium content is about 9 mg/ litre, in hard water it may reach 26 mg/liter. Molluscs need a lot of calcium for shell formation, thus they are abundant in hard water than in soft water. Calcium plays an important role as it is essential for metabolic activity of organisms activity of organisms and controls the pH of aquatic environment.

Phosphorous forms a limiting factor for organisms. For example if phytoplankton become abundant they utilizes most of the phosphorus and this results in a total elimination of the plants, which require it.

Nitrogen salts occur as nitrates and nitrites. The growth of phytoplanktonic organisms particularly that of blue green algae depends on nitrogen content of water. Excess of ammonium salts in water is fatal to organism. Thus nitrogen may be a limiting factor.

Other minerals present in water which are required by organisms in traces are magnesium, manganese, iron, sodium, potassium sulphur and zinc. These minerals are important for the physiological activities of the organisms.

Salinity and Osmoregulation

The concentration of salts in freshwater is lower than that of body fluids of animals i.e., fresh water is a hypotonic medium. On account of this difference in salt content water tends to enter the body of organisms by osmosis. Fresh water organisms have therefore developed special mechanisms to eliminate excess of water. The marine animals have not been able to invade freshwater due to difficulties in osmoregulation. A few exceptions are the bony fishes, some birds and mammals who have reinvaded the sea. Their body fluids have a slat content lower than that of sea water (=hypotonic) but they have evolved a metabolic osmoregulation mechanism which involves excretion of salt and retention of water.

pH

The pH of freshwater is influenced by soluble organic and inorganic substances and forms a limiting factor for organisms.

Wind and Currents

The wind influences the standing freshwater habitats as it causes water currents. This effect of wind depends upon the extent of exposed water surface presence or absence of an upland protecting the standing waters and its configuration. Water current determines the distribution of phytoplankton and zooplankton salts and gases throughout the body of water.

Classification of Lakes

(a) Classification based on water circulation patterns.

(b) Classification based on primary productivity.

(c) Special types of lakes

Based on Water Circulation Patterns

The lakes of the world can be divided into the following categories:

1. *Dimictic lakes* (mictic=mixed): Two seasonal periods of general circulation (=overturns).
2. *Cold monomictic lakes:* water never above 4°C as in polar regions. Seasonal overturn in summer.
3. *Warm monomictic:* Water never below 4°C as in warm temperate or subtropical lakes. One period of circulation in winter only.
4. *Polymictic lakes:* More or less continually circulating with only short (if any) stagnation periods as in high altitude equatorial lakes.
5. *Oligomictic lakes:* Barely or very slowly mixed *i.e.*, thermally stable e.g., many tropical lakes.

6. *Meromictic lakes:* Permanently stratified, mostly due to chemical difference in hypolimnion and epilimnion waters. The chemical difference is due to salts liberated from sediments which reach water and create a permanent density difference between surface and bottom waters. In this case boundary between circulating and noncirculating layers is the chemocline instead of thermocline. Examples are big Soda Lake in Nevada and Hemmelsdofersee lake in Germany.

Based on Primary Productivity

It depends on regional drainage, geologic age and depth, lakes may be 'Oligotrophic' (=few foods) and 'Eutrophic' (=good food) lakes.

Special Types of Lakes

The special types of lakes may be classified into the following seven types:

1. *Desert salt lakes:* In these lakes due to lot of evaporation, salt concentration is high. Examples are Great Slat lake in Utah. It has Brine shrimp (Artemia) which tolerates high salinity.
2. *Desert alkali lakes:* These occur in igneous drainages in dry climates. They have high pH and Carbonates e.g., pyramid lake in Nevada.
3. *Dystrophic lakes:* The include brown water, humic and bog lakes. They have a high concentration of humic acid in water. Bog lakes have pear filled margins (where pH is usually low) and develop into peat.
4. *Polar lakes* have surface temperature below 4°C or temperature rises above it only during brief summer when ice melts and water circulation occurs. Plankton population grows during this, they store fat for long winters.
5. *Volcanic lakes* are acid or alkaline lakes associated wit active volcanic regions. They have extreme chemical conditions and restricted biota e.g., some Japanese and Philippines lakes.
6. *Chemically stratified lakes* (=meromictic or partly mixed lakes) have been already described.
8. *Deep Ancient lakes:* Lake Baikal in Russia is the deepest lake in the world, it was formed by earth movement during mesozoic era. This lake has many endemic (found nowhere else) species.

Classification of Lakes Based on Thermal Cycles

The lakes are classified into following types depending upon thermal cycles.

Amictic lakes—are covered over by ice and hence get protected from weather and other external physical factors. Usually such lakes are found at high altitudes and in temperate zones.

Dimictic lakes—occur in temperate zone and in higher altitudes in subtropical regions. In these lakes spring and autumn circulation occurs every year. Thermal stratification is inverse in winter and direct in summer. The temperate oscillates on either side of 4°C during the year.

Oligomictic lakes—are common in tropical zones. Temperature remains always higher than 4°C. Circulation occurs rather irregularly.

Polymictic lakes—are located in high mountains of equatorial regions. The temperature remains a little over the 4°C and mixing goes on continuously. Uniform stratification does not occur in such type of lake.

Meromictic lakes—Some lakes have sharper concentration between epic and hypolimnetic zones and in sufficiently deep basins set up stabilities which prevent fall turnover. Circulation occurs only to a limited depth. Such lakes are called meromictic lakes.

Cold monomictic lakes—are found in polar regions and have temperature 4°C any depth. They exhibit inverse temperature stratification in winter. Even when stratification occurs, temperature never goes beyond 4°C in summer also.

Warm monomictic lakes—are formed in warmer latitudes water temperature, at any depth never falls below 4°C only one circulation per year takes place.

Classification of Animals

Neustonic Animals

Surface of the water bodies, as an interface between air and water is responsible for harbouring an interesting and diverse community of plants and animals. The organisms which occur on this surface film of water are called neustonic organisms and the ecological area as neuston. The animals and plants which live under surface of water film are called as infraneustonic. Infraneuston often includes such organisms as snails and flatworms. Seasonally, it may also include larvae and pupae of Dipterous insects, species of hydra, Entomostraca and variety of organisms. There is free floating plant community including plants like Lemna and Azolla. Number of insects coexist with these plants.

There also occur swimming whirligig beetle. Such organisms occurring on the upper surface of water film are called Supraneustonic.

The large animals in supraneuston are water skaters, water striders, Hybrids, Springtails, certain spiders, amphibian tadpoles and Aplocheilus types of fishes Certain of these animals have power to break the surface of film water and swim below for a short period. The minute supraneuston include bacteria, fungi and certain algae.

Nektonic Animals

Organisms of large size which are capable of swimming freely independent of water currents and determine their own distribution regardless of water movements are called as nektonic organisms or nekton. Most of nekton of inland water is composed of fishes, amphibian larvae, reptiles, etc. In limnetic nekton, animals occupy almost whole of the open water. Few reptiles like snakes and turtles are also observed, swimming in water but they generally confine to the shores. In southern regions alligators and crocodiles occupy freshwater bodies. These nekton is more widespread and in larger numbers in littoral zones of water. Greatest diversity of species in nekton is observed here in littoral zone.

In addition to fishes number of free invertebrates including insects are present in this zone. In littoral zone, where "pond weed" vegetation is abundant and contains largest nekton population, certain birds like gulls, ducks and other swim temporarily in water. If the term nekton includes temporarily swimming animals, these birds should be included in nekton. Distribution season physiological state of animals and stage in their life history. Certain fishes occupy surface region others intermediate regions and still others are found in deeper and coldest water. Fishes such as lake trout and catfishes occur in the deeper water of great lakes.

The Aufwuchs (=Periphyton)

The organisms which are attached to or creep upon a submerged substrate are called as aufwuchs community. Ecologically base of aufwuchs is made up of unicellular and filamentous algae. On this base in between plants various protozoans, bryozoans and rotifers thrive. Free living organisms like roundworms, crustaceans, annelids are also found in the aufwuchs. The number and kinds of animals vary with substrate, water movements, depth and also chemical properties of waters. For example, bryozoans generally inhabit the substrate formed of a sulphur containing algae like Chara.

The Pelmatohydra are found in aufwuchs having substrate of dense vegetation including Vallisneria and algae. Thus, kinds of animals depend upon the nature of substrates. Aufwuchs on stone in brooks of moderate current and aufwuchs on leaf as substrate show corresponding difference in animal composition. On stones aufwuchs show dense growth of algae and diatoms. Animals which inhabit such surface are mayflies, nymphs and blackfly larvae, immature odonata as also snails and flatworms. Aufwuchs on leaves show dense association of filamentous organisms such as stalked protozoan and hydra. In lakes and ponds aufwuchs found on submerged plants or their parts are very rich and variable. Water which contains high concentration of bicarbonates shows absence of certain sessile rotifers. Dissolved oxygen, carbon dioxide and other chemicals are limiting factors for distribution of periphyton animals. This observation provides and illustration of how chemical constituents affect the animals life.

Ecological Classification of Freshwater Organisms

Based on their Position in the Food Chain

The freshwater organisms are classified as follows:

1. *Autotrophs (Producer):* Green, plants and chemosynthetic microorganisms.
2. *Phagotrophs (Macroconsumers):* Primary secondary etc. These include herbivore, predators, parasites etc.
3. *Saprotrophs (Microconsumers)* or *(Decomposers):* These are subclassified according to the nature of the organic substrate decomposed.

Based on the Mode of Life

The freshwater organisms may be classified as follows:

1. *Benthos:* It includes organisms which are attached to the bottom or rest on the bottom or live in the bottom sediments. The benthic animals may be subdivided according to the mode of feeding, not filter feeders (e.g. clam) and deposit feeders (e.g., snail).
2. *Periphyton:* Organisms (both plants and animals which are attached to or cling to stems and leaves of rooted plants or other surfaces projecting above the bottom.
3. *Plankton:* Floating organisms whose movements are more or less dependent on currents. Depending upon the size of organisms, plankton may be of the following 2 types
   (a) Net plankton (Macroplankton and microplankton) which is caught in a plankton net (fine meshed net) when it si towed slowly through water
   (b) *Nannoplankton* which is too small to be caught in plankton net, it is therefore collected in a bottle or by means of a pump.

4. *Nekton:* Swimming organisms able to swim at will, hence capable of avoiding plankton nets, water bottles etc. Examples, are fish, amphibians, large swimming insects and so forth
5. *Neuston:* Organisms resting or swimming on the surface.

Organisms of Lentic Habitat

Biota of Littoral Zone

(i) *Producers:* The producers of littoral zone are of two main types: (a) Rooted benthic plants which are mostly spermatophytes (Dn. Spermatophyta). (b) Non-rooted plants which comprise the phytoplankton (floating green plants) and filamentous green algae (pond scum).

(ii) *Consumers:* (a) Periphyton, (b) Zooplankton, (c) Nekton, (d) Neuston.

Producers

(a) Rooted benthic plants typically form concentric zones within the littoral zone. A representative arrangement proceeding from a shallow to deeper water is given below but if should not be assumed that all three zones will be present or they will be arranged in the order given here.

Zone of Emergent Vegetation: Consist of rooted plants with principal photosynthesis surfaces projecting above water. They obtain CO^2 for photosynthesis from air and raw materials from beneath the water surface e.g., cat tails (genus Typha), bulrush (scirpus) arrowheads (Sagittaria), and burreeds (spargamum), spike rushes (Eleocharis) and pickerel weeds (Pontederia). The emergent plants together with those on the moist shore form an important link between water and land environments. They are used for food and shelter by amphibious animals and provides a convenient means of entry into and exit from water for aquatic insects which spends part of their lives in water and part on land.

Zone of Rooted Plants with Floating Leaves. Ecologically this zone is similar to the zone of emergent vegetation but here the leaves float horizontally on the surface and this reduces light penetration into water e.g., water lily Nymphea which has four species and water shield Brasenia. The under surfaces of lily pads provide convenient resting places for egg deposition by animals.

Zone of Submergent Vegetation comprise rooted or fixed plants completely or largely submerged. The leaves of vegetation in this zone tend to be thin and finely divided or adapted for exchange of nutrients with water. In this zone, pond weeds (Potamogetonaceae) are prominent. Potamogeton is the largest genus having sixty five species which occur in the temperate zones of the world. In the two species of potamogeton (*P. diversifolia* and *P. pectinatus*) have been shown. Other common weeds are Ceratophyllum, Najas and Vallisneria Chara (muskgrass) and the related genera Nitella and Tolypella are generally classes as algae yet they are attached to the bottom and their life resembles that of higher plants. In view of this chara may be classed with submergent vegetation. Chara often marks the inner boundary of littoral zone.

(b) *The Non-rooted Plants: Phytoplankton,* comprise many species of algae found floating throughout the littoral zone. Some species of algae are found attached to or associated with rooted plants, these are especially characteristic of littoral zone. The chief types of algae are the following:

Diatoms (Bacillariaceae) with yellow or brown pigment in the chromatophores which are so abundant that they mask the green colour of chlorophyll. These are good indicators of water quality e.g., Navicula, Fragilaria, Asterionella (which floats in water like a parachute) and Nitzschia.

Green algae (Chlorophyta) having both floating and attached forms. In these, the chlorophyll is not masked by other pigments therefore populations of green algae have a bright green appearance e.g., Scenedesmus, Coelastrum, Richteriella and Closteria.

Blue green algae (Cyanophyta) having diffused chlorophyll in the algal cells (and not concentrated into chromatoplasts) which is masked by blue green pigment. In polluted ponds and lakes this may develop into blooms. These algae are able to fix up gaseous nitrogen into nitrate (on land the bacteria of soil fix up atmospheric nitrogen into nitrates).

Many of the blue green algae are not palatable, hence not eaten by consumers. This is one of the reasons for the formation of large blooms. The cells of these living algae excrete metabolities which are toxic and five and bad taste and dour to drinking water. After death, during decay the breakdown products of these algae are also toxic and five a had taste to water, the common genera are Anabaena, microcysts, Gloeotricha, Oscillatoria and Rivularia may cover the bottom or may be found attached to stems and leaves of submerged spermatophytes (seed plants). A sample of free floating phytoplankton of littoral zone reveals numerous diatoms, desmids and other green algae, chlorophyll containing protozoa like Euglena and its many relatives.

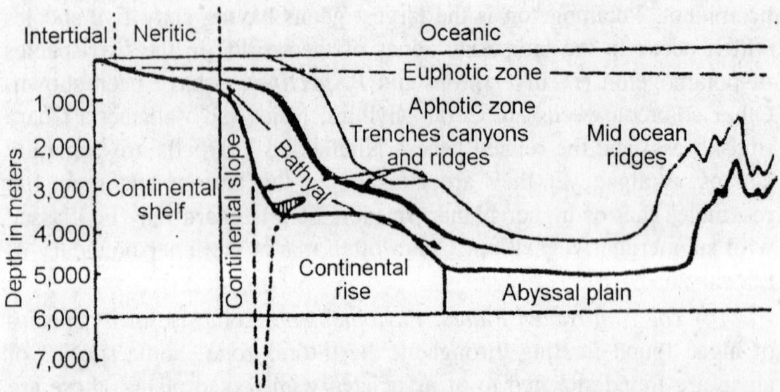

Fig. 3.3. Diagrammatic sketch showing horizontal and vertical zonation in the sea.

Filaments Green Algae or Pond Scum comprise the common genera namely spirogyra zygonaema and Oedogonium. Other filamentous algae are found attached to rooted plants like Chara, this association is a commensal or mutualistic relationship. Such algae assume a periphyton life. When a pond or lake is polluted with excess of nutrients, the filamentous green algae of ten develop huge blooms, that rise to the surface as they are buoyed up by the entrapped oxygen. They are now said to form pond scum, Now the oxygen produced by photosynthesis largely escapes to the atmosphere. When the bloom dies oxygen in water is used up, this results in mass killing of fish, blooms are thus very harmful to pond life.

Consumers

Periphyton exhibits a vertical zonation in contrast to the horizontal zonation of rooted plants although many species occur throughout the littoral zone. One group of periphyten comprises both primary and secondary consumers like pond snails, damselfly nymphs, climbing dragonfly nymphs, rotifers, flat worms bryozoa, hydra, midge larva which rest on or attached to leaves of large plants. The second group of periphyten also comprises both primary and secondary consumers which rest or move on the bottom or below silt or plant debris e.g., odonata nymphs, crayfish, isopods and certain May fly nymphs. The third group of periphyten is met with on descending down more deeply into the bottom mud. These include burrowing odonata and ephemeroptera, clams, annelids snails and especially chironomids (midges) and other Diptera larvae which live in minute burrows.

Zooplankton of littoral zone is characteristic and differs from that of limnetic zone in that the invertebrate groups of zooplankton are large, weak swimming species of water fleas (cladocera) like Daphnia and Simocephalus, copepods, ostracods and some rotifiers.

Nekton of littoral zone is rich in species numbers it comprises of insects and vertebrates. The conspicuous insects are adult and larval diving beetles and various adult Hemiptera, some of these are carnivores and other are partly herbivorous or scavengers. Larvae and pupae of various diptera remain suspended in water of ten near the surface. The vertebrates are mostly amphibious like frogs, salamanders, turtles and water snakes. These cold blooded vertebrates become more numerous as one moves towards south. Tadpoles of frogs and toads are important primary consumers. The pond fish move freely between littoral and limnetic zones but they spend most of their time in the littoral zone. Common fishes are sunfish, minnows like gambusia, bass, gars etc.

Neusten comprises three surface insects (1) the whirling beetles (fam Girinidae) whose eyes are divided into two parts one part for vision above water and one under water. (2) Large water striders (fam Gerridae). (3) Smaller water striders (fam veliidae).

Organisms of Limnetic Zone

Producers

The producers of Limnetic zone are only phytoplankton which consist of algae of three groups listed in the littoral zone and algae like green flagellates chiefly dinoflagellates, Euglenidae and volvocidae. The limnetic zone is the one of variation where water level, temperature, oxygen content etc. vary from time to time. Depending upon the available conditions, the growth of phytoplankton and hence the consumer differs. For example in the northern temperate lakes and ponds like those in northern U.S.A., during early spring there is an abundant growth of phytoplankton which persists only for a short time. This is know as 'bloom' or 'phytoplankton pulse' or 'spring flowering'. These is another usually smaller pulse in the autumn.

The diatoms. are responsible for the spring bloom while Anabaena: the blue green algae are responsible for autumn pulse in temperate regions, the seasonal variation in population density of planktonic organisms is due to the following: During winter, the rate of photosynthesis is low due to low water temperature and reduced light but nutrients like nitrates, phosphorus and silicates keep on forming continuously due to the action of microorganisms thus they increase in concentration.

In spring when the temperature and light conditions are favorable plankton show a rapid sudden increase since there is an abundance of nutrients very, soon, the nutrients are exhausted so the bloom disappears. Again nutrients accumulate and another bloom is seen in the autumn when conditions are favorable. Turbulence or upward current movements of water caused by temperature differences in limnetic zone and in keeping phytoplankton near the surface where maximum photosynthesis can occur.

Consumers

Zooplankton consists of only few species but the number of individuals may be large. Copepods, cladocerans and rotifers rank first in order of abundance, then come the crustacea but the species are different from those found in the littoral zone. For example, the copepod Diaptomus and Cyclops are common genera, the cladoceran Diaphanosoma, Sida and Boomina are common. Mott of the crustacea are strainers, they filter bacteria netritus and phytoplankton by means of combs of setae on their thoracic appendages.

The zooplankton are predators, they 'graze' the phytoplankton in the same manner as cattle do on land. Zooplankton also exhibit bloom along with phytoplankton or immediately after phytoplankton blooms as they largely depend on them. Vertical diel migration is a characteristic feature of limnetic zooplankton.

Nekton consists almost entirely of fish. In ponds, the limnetic fishes are the same as littoral fishes but in larger bodies of fresh water a few species may be characteristic of limnetic zone. Most adult freshwater fishes feed not on plankton but on animals of good size, only a few like Dorosoma and Signalosa are plankton feeders.

Organisms of Profundal Zone

All the animals of Profundal zone are adapted to withstand periods of low oxygen concentration and many bacteria are anaerobes i.e., able to carry on without oxygen.

Producers

The producers are absent in the profundal zone as there is no light.

Consumers

Consumers of profundal zone depend on littoral and limnetic zones for food. In return, the profundal zone provides rejuvenated nutrients which are carried by current and swimming animals to other zones. The consumers belong to three groups:

(a) Blood worms or haemoglobin containing chironomid larvae and annelids

(b) Small clams (fam, sphaeriidae)

(c) Phantom larvae (chaoborus): Plankton forms which have a float of four air sacs. The air in sacs also provides a reserve supply of oxygen. These larvae move up into the limnetic zone during night and down to the bottom during day. The adult of these larvae are land dwelling midges (Diptera).

Decomposers

Decomposers comprise bacteria and fungi. These are especially abundant in the water mud interphase where organic matter of littoral and limnetic zones accumulate. The break down the organic remains into simple inorganic substances which are made available to the producers again.

Organisms of Swamps

The edge of lake or pond consists of swampy (water logged) ground outer to which is moist soil. Not much of literature is available on life in swamps. The following is a brief account.

Producers of the swampy ground are plants like arrowheads, cat tails (Typha) and spikerush (Eleocha) and tall grasses known as reeds. Rice plants and plants like polygonum and scirups grow very well in marshy areas. This plant is in fact marshy grass and rice fields are merely artificial ecosystems. The moist soil bears water grasses and plants like Ranunculus, polygonum and scirpus.

Consumers of swamps consist of protozoan, nematodes, crustaceans, insects, snails, live fishes (=air breathing fishes which have assessor respiratory organs for breathing and storing air) and turtles. Swamps are often visited by amphibians and ducks. Animals living in swamps have adaptations to tide over periods of draught. For this, some have resistant eggs and cysts, some like the African lung fishes survive draught by burrowing in mud and getting surrounded by a slimy muddy cocoon. Water pollution is another hazard in swamps, under such conditions only autotrophic ciliates and holophytic flagellates survive. Marshy areas seen in village near human habitations, contain drainage water, containing disease germs and provide breeding places for mosquitoes.

The Pond

Types of Ponds

Depending upon their origin, ponds are of the following three types:

1. *Artificial ponds* are those which are created by man by damming of stream or basin or by animals like beaver. Artificial ponds are

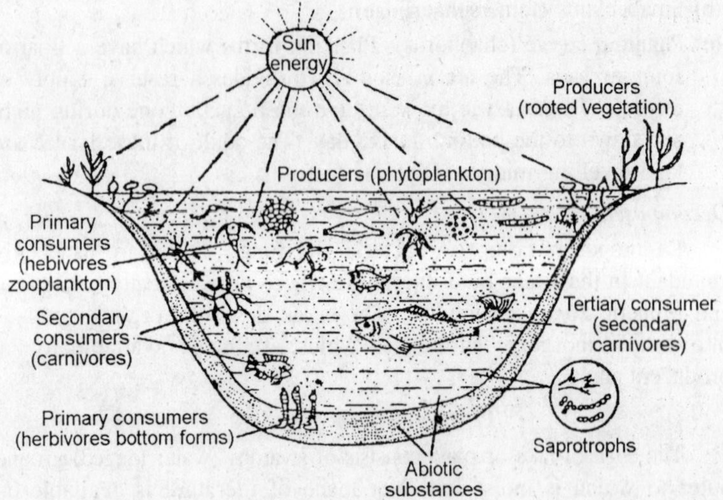

Fig. 3.4. The pond ecosystem.

rarely deep enough for stratification. These ponds are of no interest to ecologists as the variety of organisms is reduced to favour large number of desired species particularly fish. The beaver ponds were abundant in America when the European man arrived.

2. *Small ponds* which are not connected to any body of water right from the time they originated e.g., natural ponds which develop in limestone regions where depressions develop in the rocks.
3. *Ponds Derived from Large Lakes or Streams.* These ponds are continually being formed e.g., when a stream shifts its position it leaves the former bed isolated. These ponds are quite productive and attract a number of fishermen.

Depending upon their duration, ponds are of two types:

1. *Temporary ponds* are those ponds which remain dry for a part of the year. These ponds support a unique community. The organisms of such ponds reproduce in a short time when water is available and tide over the dry period either by moving out and reaching another pond having water or by lying in a dormant state in the dry soil and becoming active when water is available e.g. Fair shrimp (Eubranchiopoda). When a temporary pond exists only during spring, it is called a vernal pond, when it contains little water throughout the year but freezes during winter, it is called aestivial pond.
2. *Permanent ponds* are those which contain water throughout the year.

The Lake

Lakes may be natural or artificial (=man made). The artificial lakes have turbid water and fluctuating water level. Benthos is lesser than those of natural lakes. Primary productivity and fish yield differ in different artificial lakes. Following is an account of 'stratification' of natural lakes (a feature which is so characteristic particularly of temperate lakes) and classification of lakes.

Stratification in Temperate and Tropical Lakes

Stratification in Temperature Lakes

In temperature lakes, temperature variation are pronounced, hence they show summer and winter stratification as follows:

During Summer, the surface waters become warmer than bottom waters hence only warm top water (epilimnion or surface lake) circulates, This top circulating water does not mix with the more dense colder bottom water (=hypolimnion or under lake). This state of affairs results in the creation of a zone with a deep temperature gradient (=Thermocline) between upper warm and lower cold water.

During hot months of summer, there is a strong drop in temperature at the thermocline) at least 1° per meter depth of water). As the temperature of epilimnion kises further, the different of temperature between epilimnion and hypolimnion also increases. Since thermocline acts as a barrier for the exchange organisms migrate into the hypolimnion during the summer months. As the thermocline is usually below the compensation level (=range of effective light penetration), the green plants and so the oxygen supply are cur off in the region of hypolimnion this is referred to as the period of summer stagnation in the hypolimnion. If the water of lake is very transparent and the thermocline is above the compensation level, light reaches the upper part of hypolimnion, the phytoplankton now grows here and oxygen may be present in greater abundance than on the surface (cold water hold more oxygen). It is to be noted that the euphotic zone does not necessarily coincide wit the epilimnion.

Euphotic zone i.e., producer zone is based on light penetration and epilimnion on temperature. These zones however coincide roughly during summer stratification of lake. The extent of depletion of oxygen in the hypolimnion depend son the amount of decaying matter and the depth of thermocline. The lakes which are highly productive (e.g. Great lakes of U.S.A.) are subject to greater oxygen depletion than the poor lakes as in the former the rain of organic matter from the limnetic

Table 3.1. Difference between Oligotrophic and Eutrophic Lakes

Oligotrophic lakes	*Eutrophic lakes*
1. These lakes have changed very little since the time of their formation i.e., they are 'geologically young'.	They are geologically old as they have changed a lot since their formation.
2. Deeper	Shallower
3. Hypolimnion is larger than epilimnion	Hypolimnion is shallower than epilimnion.
4. Low primary productivity.	High primary productivity.
5. Littoral vegetation scarce.	Littoral vegetation is more abundant.
6. Plankton population density is low although number of species may be large.	Plankton population is denser.
7. Plankton blooms are rare as nutrients rarely accumulate.	Plankton blooms are characteristic as nutrients accumulate in large quantities.
8. Hypolimnion is not subject to severe oxygen depletion hence stenothermal-cold-water bottom fishes like trout are restricted to the hypolimnion of these lakes.	Hypolimnion is subject to severe oxygen depletion so much so that summer stagnation may exclude cold water-fishes.
Examples: Great lakes and Finger lakes of New York.	Examples: Lake Mendota and Linsley pond.

and littoral zones into the profundal zone is greater. In productive lakes, cultural eutrophication, hastens oxygen depletion in profundal zones.

In poor lakes the stenothermal fishes (low temperature tolerant) can survive in the hypolimnion due to the presence of oxygen. In productive lakes, these are the first animals to go on eutrophication. This is referred to as the 'Summer kill of the fish'. In contrast to fish, the lower animals of profunda zone are adapted to withstand oxygen deficiency for long periods. With the onset of winter, temperature, of epilimnion drops until it is the same as that of hypolimnion. Now water of the entire lake begins to circulate and oxygen supply is returned to the hypolimnion. This is the 'fall overturn' as now the animals feed and reproduce. During colder months of winter, surface water cold below 4°C, it expands, becomes lighter so remain on the surface and finally freezes (water is densest at 4°C). The bottom water is still at 4°C, thus providing a suitable habitat, for stenothermic animals. This is winter stratification.

Table 3.2. Differences between Ponds and Lakes

	Pond	*Lake*
1.	Pond is a reservoir of rain water found in many regions with adequate rainfall.	Lake is a natural source of water among hills.
2.	Pond is much smaller than a lake.	Lake is large in expanse.
3.	It is shallow.	It is deeper than a pond.
4.	Littoral zone is large while limnetic and profundal zones are either small or absent.	Limnetic and profundal zones are large in comparison to littoral zone.
5.	Thermal stratification is either very little as in deep ponds or absent.	Lakes show thermal stratification during summer and winter.
6.	Exposed water surface is much lesser as compared to that of a lake hence wind does not cause water current i.e., pond waters are quiet.	Exposed water surface is large hence wind causes water currents.
7.	Ponds may be temporary or permanent.	Lake are always permanent.

During winter stratification, photosynthetic rate is slowed down as the producer do not receive enough sunlight due to a layer of ice at the top. Oxygen supply is however not greatly reduced as water holds more oxygen and respiration of organisms and bacterial decomposition are slow at a low temperature. This is referred to as the period of winter stagnation. This is referred to as the period of winter stagnation in the hypolimnion. Winter stagnation is generally not so severe as summer stagnation. Sometimes, when snow covers the ice at the top, light is cut off and photosynthesis stops completely resulting in oxygen depletion and 'winter kill of the fish'.

With the onset of spring ice melts, water become warm, when temperature reaches 4°C it being heavies sinks to the bottom and water of the entire lake begins to circulate once again. This is the spring overturn a now the animals feed and reproduce. In America and Eurasia two overturns in a year are seen as described above but not in all temperate lakes.

Stratification in Tropical Lakes

In sub tropical lakes, the surface temperature never falls below 4°C. These lakes show a distinct thermal gradient from top to bottom.

There is only one general circulation period. In deep tropical lakes the surface temperature is high (20 to 30°C). General circulation is irregular, it occurs mostly in cooler seasons. The very deep tropical lakes tend to remain only partly mixed i.e., permanently stratified.

Lentic Habitat Versus Lotic Habitat

The characteristic feature of running water is the movement of water in only one direction. The velocity of water current varies with depth. Normally the maximum velocity is observed in the first 1/3rd depth of river. The velocity becomes gradually reduced as depth increases and after a certain depth the velocity is almost zero i.e., standing water conditions results. For example a river may have a lazy flow at a place and huge water fall at the other where the slope of the ground is very steep.

The current velocity may differ at two different faces of a stone which is submerged in water of a river or even in the interspaces of stones. Current is the chief factor which accounts for the big difference between lentic and lotic habitats, it is also responsible for the differences in the difference parts of a lotic habitat. At rocky or sandy shores of lakes especially when rooted plants are absent, velocity of water may be even double that of stream. In such cases therefore the organisms which are considered as pond organisms might live in stream and *vice versa.* The thermal an chemical stratification is very little or absent in lotic habitats.

Although green plants are not abundant oxygenation is rich and almost uniform throughout due to the large surface exposed to air and constant motion of water. This aids in easy respiration, it si due to this that the animals of running water are susceptible to reduced oxygen content and cannot live in standing water or organically polluted part of stream where oxygen supply is reduced.

The depth of water and cross-section area of streams is much less than that of lakes, the land water junction is relatively larger in proportion to the size of a stream. This means that a stream is more intimately associated with the surrounding land than are most standing water bodies. Many of the streams depend on land areas and their ponds, back-waters and lakes for the supply of phytoplankton. From streams energy goes to land and also to lentic habitats in the form of emerging insects and stream dwelling animals removed by the predators. Thus the streams are intimately associated with terrestrial and lentic habitats. A running water exhibits two major zones: the rapids and the pool.

Organisms of Lotic Habitat

As mentioned earlier, the lotic habitat has two zones: the rapids and the pool. The species composition of communities in the rapids differs totally from those of ponds, lakes and also of pools. Since pool communities contain some organisms which also occur in the ponds, the organisms of rapids are said to be typical stream organisms. The organisms (and their density) of rapids and pools are different due to the different types of bottom available whether sand, silt, pebbles, clay, rubble rocks (loose stones etc. as covering of some rocks) etc. For example if the bottom is of sand and silt (least favourable bottom), the number of species and that of individuals (plants and animals) is very small. If the bottom is of rubble rock the number of species and density of bottom organisms is the highest. If the bottom is soft as in pool areas, only benthic burrowing form will occur. In rapids the benthic invertebrates have a higher density than that in pools. In pools, the nekton and burrowing forms are denser in comparison to those in the rapids. The stream fish links the rapids and pools as it goes to pools for refuge and feeds in or at the base of rapids.

Organisms of the Rapids

The plants and animals inhabiting the rapids have unique morphological, physiological and behavioral adaptations to suit the currents which form the major limiting factor of the rapids.

Producers

Phytoplankton is absent but it may be swept from the neighbouring bodies of standing waters, if it happens to pass through the rapids it si soon destroyed as its is at the marcy of currents. In pool (slow moving parts of stream) however, it multiplies and becomes an important source of food. The producers of the rapids are the following:

(a) Filamentous green algae such as cladophora which grow attached to rocks.

(b) Encrusting diatoms which cover various surfaces

(c) Aquatic mosses such as Fontinalis which cover stones.

Consumers

Zooplankton is absent. Some of the consumer animals are plant inhabiting forms which live attache to roots, leaves or stems of mosses and other producers. Some are rock inhabiting forms which inhabit the underside or exposed parts of submerged rocks. Certain animals have taken to burrowing in order to withstand current while the fishes are capable of moving against the water current. In general animals of the rapids show the following adaptations.

Streamline body

All animals from insect larvae to fish have streamlined body which offers minimum resistance to water currents.

Flat body

Many animals of rapids have extremely flattened bodies to enable them to find refuge by clinging to under surfaces of stones and in crevice. Examples are stonefly nymphs and mayfly nymphs.

Small size

Many fishes have a small size which enables them to move fast against water currents.

Sticky undersurface

Many animals like snails and flatworms have sticky undersurfaces by which they adhere to surfaces.

Hooks and sucker

These are fixation devices by means of which the animals of rapids grip the smooth surfaces like those of leaves and steams: For example black fly larva: simulium has a sucker at the posterior and end and a headnet at the anterior end which is used for straining food from water while the Caddis larva: Hydropsyche cements a animals and plants suspended in water. Some fishes like loaches have a sucker which represents a pair of modified fins.

Reduced gills of many fishes make them unfit for life in ponds where oxygen content is low. This adaptation is in response to high oxygen of water in the rapids.

Positive rheotaxis (rheo current, taxis = arrangement)

This is an inherent behavior pattern found in the animals inhabiting the rapids. These animals almost always orient themselves upstream and if they are capable of swimming movements. They continuously move against the current. It is to be noted that animals of standing waters merely drift with the current and do not move against it.

Positive thigmotaxis (Thigma=touch or contact)

This is also an inherent behaviour pattern found in the animals inhabiting the rapids. These animals have a tendency to clip close to a surface or to keep the body closely in touch with the surface. Thus if some stone fly nymphs are kept in a petridish they either lie in contact with the dish or cling to each other.

Organisms of the Pool

In the pools the temperature of water is higher than that of rapids, the current is slow and the nature of bottom is different due to the accumulation of silt and decaying organic matter at the bottom. The pools harbour a variety of animals due to the rich nourishment they have.

Producers

The producers are the phytoplankton transported from the neighbouring bodies of standing bodies of standing waters and backwaters, hence they show a mixture of planktonic forms.

Consumers

Zooplankton abounds in protozoa's. The invertebrate animals consist of insects, fresh water muscles snails etc. The vertebrate animals consist of fishes such as carps and car fishes reptiles, like crocodiles and turtles and amphibians.

SPRINGS

Springs differ from other habitats in the relatively constant chemical composition, velocity and temperature of water. They are said to be ecologist's natural constant temperature laboratory. This is because as the spring water is altered by photosynthesis and respiration, it passes downstream and is replaced by water having the original composition from underground. In the spring therefore, the communities can be studied under known conditions. Knowledge is available about the following springs:

1. Hot springs with high salinity found in volcanic areas such as in Iceland, New Zealand, U.S.A. and Africa.
2. Hard water springs (having average temperature) found in the limestone districts of Florida, Denmark and Germany.
3. Soft water springs which emerge through sandstones, crystalline rocks and shales.

In the hot springs with high salinity, as one goes downstreams, the temperature and salinity decrease steadily. The community composition changes accordingly, hence study of communities of a spring run shows a natural ecological gradient. The study of hot springs have established the upper limit of temperature tolerance for the following organisms.

Blue green algae	88°C
Protozoan	54°C

Insects	50°C
Fishes	50°C
Bacteria	88°C

The study of spring shows that many species occur in a particular region just because the springs provide them with a favourable environment. For example arctic insects occur in German springs due to their low summer temperature, similarly the hot springs of iceland harbour many organisms which normally occur only in the regions of warm climates. It seems that springs have offered refuge to aquatic organisms during geological periods when climate changes were taking place.

Chemical Conditions of the Bottom Soil in the Aquatic Media

A well aerated pond is relatively self supporting ecosystem for production of nutrients. Total and available quantity of raw materials, suitability of environmental conditions and supply of nutrients are other vital factors influencing productivity of that ecosystem. Some major chemical factors influencing productivity are:

Hydrogen ion concentration

pH of the substrate soil governs the adsorption and further release of ions. The adsorption and release both take place at the soil water interface. The soil become acidic in a stagnant oxygen deficient pond. This increased acidity inhibits the bacterial action thus suppressing the continuous decomposition of organic material. Seldom, the soil is naturally buffered.

Chemical systems

It appears there are certain established chemical links and systems which are delicately balanced to promote natural productivity. Any tilt in these complicated long chains will affect the productivity schedule. Some of the well identified systems are dealt with to indicate involved factors. Some delicately balanced chemical systems exist in the soil. The types of system depends upon availability of minerals and pH of waters. Only the common well identified systems are treated here.

Iron-sulphur-phosphorus system

Iron-sulphur-phosphorus system is one in which H_2S is formed at the bottom by the anaerobic reduction of sulphates by bacteria. The products liberated during these decomposition are rapidly converted by bacteria and iron (Fe) is deposited as $Fe(OH)_3$ (Ferric hydroxide).

Calcium carbonate-phosphorus system

This system exists in ponds having alkaline water and higher calcium content. In such cases most the added phosphates is quickly are precipitated with calcium. In these ponds phosphorus is quickly adsorbed by the bottom mud.

Iron humus-phosphorus system

Bottom humus contains Humic acids. But the exact composition of Humic acids is to so well understood. Humic acids form complexes with heavy metals by chelation. In an aerated pond, iron is in ferric form and forms large particles. The iron linkage of humic acids is counteracted by calcium concentration. It is believed that such large quantities of iron are responsible for highly coloured (blue) waters. Chelation plays a significant role in lending about the deeper blue green colour.

Iron or aluminium silicon humus system

The linkage of phosphorus and silicon to iron and aluminium humic acid complexes is known to occur in some ponds.

Phosphorus

Phosphorus is necessarily taken up by bacteria and algae. The uptake and liberation follow a definite well organised cycle. In the presence of bacteria amount of phosphorus remaining in water at equilibrium is more than if bactericides like antibiotics are used. The total amount of dissolved phosphorus present in the pond is not so important to aquatic productivity as the amount of phosphorus freely available. Phosphate ions form insoluble compounds with iron and aluminium in an acidic environment and form compounds with Ca in an alkaline environment.

Phosphorus occurs in both organic and inorganic state. Organic compounds containing phosphorus are, phytin and its derivatives, phospholipids and nucleic acids. Remaining phosphates are adsorbed by the mud, resulting in precipitation. Phosphorus is liberated to epilimnion. During decomposition of littoral vegetation this free phosphorus is taken up by phytoplankton and littoral vegetation. Phytoplankton steadily looses phosphorus in the form of water soluble compounds. These compounds ionise and exist as free ionic phosphates. Bacterial putrefaction of the phytoplankton results in sedimentation. Phosphorus is liberated from the sediments to hypolimnion and settles in the sludge or soil-water interface. Remaining phosphorus from the sediments slowly diffuses into water.

Nitrogen

Organic matter settled on pond bottom is broken down by a number of micro organisms. End products of this decomposition are carbon dioxide, water and ammonia, Ammonium salts are formed from ammonia, which by bacterial action are converted into nitrates. Denitrifying bacteria reduce the ammonium salts and nitrates to free nitrogen. Bottom soil absorbs remaining NH_2 ions. Extent of absorption depends on the concentration of ammonium salts already in the soil. Some ammonium salts and nitrates being highly soluble are leached out with the percolating water. Ammonia release during decomposition evaporates in high temperature. Main source of nitrogen to the pond ins bottom soil. But its efficiency is often in doubt as some recommend to add artificial fertilizers to the pond.

Organic carbon and C/N ratio

Percentage of organic carbon is an important facto in fisheries development. Though it can not be staged definitely, in practice it was observed that 0.5-1.5 percent organic carbon yields average fish production while 1-5 2.5 percent is optimal. A ration less than five indicates poor production more than 15 also is less favourable while 10-15 ratio is optimum for good production.

Potassium

Potassium is always present in water and being necessary for submerged weeds is quickly taken up by them. But the optimal concentration is not so well ascertained.

Calcium

Calcium is important for its role in translocation of carbohydrates. It is also vital integral component of plant tissue system. Calcium reduces toxic effects of single salt solutions of other elements and increases availability of other ions. There are certain micronutrients which even in minute quantities promote metabolism of plants and animals. Such naturally occurring substances are called "trace elements" which impart significant differences in the productive capacity of fresh waters. Indeed, they play a vital role in the bionics i.e., the economics of any bodies community. It appears that manganese (Mn) promotes photosynthesis heterotrophic growth of phytoplanktonolgal in particular and increase in crustacean fauna.

4

INFLUENCE OF RADIATION

The mammalian testis has been subjected to numerous investigations concerning the effects of ionizing radiations on this organ. Rontgen's initial discovery of x-rays in 1895, Albers-Schonberg's observation in 1903 that radiation causes a rapid regression of the testis, Muller's classic work in the late 1920's on mutation rates caused by radiation, the detonation of the atomic bomb in 1945, and the possible encounter of solar or nuclear radiation in space travel, has all provided a further impetus for these studies. A number of excellent reviews have been written in an attempt to correlate the histological, morphological, and cytological changes that occur in the testis after irradiation. On the other hand, it is surprising to note the paucity of information that exists on the biochemical or endocrine changes that occur in the testis after this treatment as compared to the histological and morphological data. Moreover, it is also surprising to note that there are very few reviews attempting to correlate the biochemical or endocrine changes with the histological, morphological, and cytological changes after irradiation. The author is aware of one other recent review attempting to correlate these changes.

Since the testis is an endocrine organ producing hormones that have pronounced effects on nitrogen deposition in tissues, fat metabolism, the erythropoietic system, the reticuloendothelial system, the kidney, etc., this organ through its secretory products could affect radiation sensitivity and the general radiation syndrome. It is also possible that other somatic changes could influence the effect of radiation on the testis.

As a result of the numerous histological and morphological studies, the testis is now recognized as one of the most radiosensitive organs of the body. Due to its extreme radiosensitivity, the testis has been proposed as a tissue dosimeter in the event of accidental exposure to radiation. After exposure of the organism as a whole, or the testis as an organ, to irradiation, a series of physiological events occur. Most easily recognized is a marked change in testicular size due to a maturation depletion of various cell types within the testis (i.e., decreased fecundity). Sterility is evident with moderate doses, but it is considered to be somewhat more prolonged with high doses. The inability of the organism to produce viable offspring is brought about by the formation of dominant lethal chromosomal aberrations.

Libido does not appear to be diminished after radiation, and gross changes in androgen production are not necessarily evident unless young animals or special techniques are used in the investigations. Several decades ago, this observation promoted the treatment of human males to render them temporarily sterile for the early period of marriage, with no apparent effect on libido. Unfortunately, there was no concern by the physician as to the genetic damage that may result from this treatment.

Germinal Elements

The mammalian testis contains many cell types. The cells of the capsule, the interstitial cells, the two-cell layer that comprises the limiting or basal membrane of these miniferous tubules, and the Sertoli cells are now thought to be radioresistant, at least from a histological standpoint. From a biochemical standpoint, however, recent data suggest that some of these cells may be biochemically and physiologically radiosensitive as well be discussed later. It is now generally accepted that the numerous cell types within the seminiferous tubules have very different sensitivities to specific types and dosages of ionizing radiation, as far as cellular death is concerned. The timing of the stages of development that take a primordial stem cell to a mature spermatozoa are now known with considerable certainty for some species.

A differential is sensitivity exists within the type A spermatogonia group. The Ao or dusty spermatogonia appear to be less sensitive to radiation damage than the type A cells which have darker and more defined nuclei. Irradiation just prior to the last division has somewhat less effect on those that differentiate into intermediate and then to type B spermatogonia. Similarly, a mixed sensitivity has been observed for stages where all type A spermatogonia are in interphase. Thus, intermediate and type B spermatogonia are extremely sensitive to irradiation, while type A are a heterogenous group of cells with their sensitivity dependent on mitotic activity and stage of development. Damaged spermatogonia usually degenerate as they reach late interphase or early prophase of their first postradiation division. Degeneration may also occur, however, after more than one cell division. In this respect, Monesi (1962) concluded that not all irradiated spermatogonia die at the time of the insult, but do so as they reach a definite critical stage of the cycle. For type A and intermediate spermatogonia, the stage of cell death is late interphase or very early prophase. For type B spermatogonia it is anaphase or early telephase. Spermatids and spermatozoa show no visible or histological changes with doses up to 1500 R in mice; but this does not mean that no genetic damage has occurred, nor that such damage will not be expressed after fertilization.

Irradiation of testes with excessively high doses of x-rays results in meiotic chromosomes that lose their shape and identity by becoming fluid, sticky, and less viscous. The chromosomes tend to flow along the more rigid spindle fibers to such an extent that they may never regain their original configuration. The pyknotic nucleus is seen most often as the end point of radiation damage to the germ cells. Cells exposed to radiation at stages when they have no formed chromosomes can develop the abnormal configurations at the time of mitosis or meiosis.

Rugh (1960) has offered the following values in suggesting a relative order of resistance of the several spermatogenetic stages purely with respect to destruction and resultant sterility, but genetic effects are not taken into account here: type B spermatogonia, 1 (highly susceptible); type A spermatogonia, 70; spermatids, 300; spermatocytes, 400; spermatozoa, 10,000. It has been shown that intermediate and type B spermatogonia are about 70 times more radiosensitive than the type A spermatogonia.

After destruction of a specific cell in the developmental sequence the remaining cells disappear through a maturation depletion. Thus, the wet weight reduction of the testis after radiation is due to an actual cell loss, with the extent of loss being dependent on the dose. To further complicate the issue, the weight loss of the testis after radiation is confounded somewhat by the accumulation of fluid in some of the spaces once occupied by germinal cells.

Apparently as the dosage is increased, more and more of the type A spermatogonia are killed, leaving a very sluggish cell that is comparatively radioresistant and very slow to divide. This is the type Ao or dusty spermatogonia. These sluggish, but radioresistant cells, in turn are responsible for repopulating the testis. The duration of sterility is now thought to be determined by how many of these cells survive and their ability to divide and repopulate the testis. What it is that makes some of the type A cells more sluggish and slower to divide, needs to be explored in more detail. If a means could be found to activate these cells, the period of sterility could be greatly diminished. Certain types of infertility might also be effectively countered if the rate of cell division of these sluggish cells could be increased. It is now thought that when regeneration occurs, at least after low levels of radiation, it starts before the peak of hypoplasia occurs.

Fertility Changes

Although the spermatids and mature sperm appear to be very radioinsensitive morphologically, genetic damage occurs. In this respect, attempts to produce parthenogenetic mammalian embryos resembling those reported in amphibians after radiation have failed. It was observed that when the testes of black male mice were exposed to 41,000 R and the mice were mated with normal white females, eight viable black offspring were born. The F_1 generation exhibited no gross anomalies, but exencephalia and other anomalies were evident in the F_2 generation.

Spermatids and spermatozoa can effect fertilization for a period of days or weeks after a radiation insult. Freund and Borrelli (1964) completed a dose-time response study on semen production by the guinea pig. They observed that radiation markedly affected sperm concentration, but not ejaculated volume after the seventh week. Sperm output was severely depressed at all dose levels from 8 to 11 weeks, and the extent of depression was directly related to the dose. The 7-week lag between time of treatment and change in ejaculate corresponds to the 7 weeks required for spermatogonia to develop and pass through the epididymis and vas deferens. This 7-week period is in agreement with data collected for the rat (48-50 days); the bull (45-50 days); and the rabbit (56 days). The period of depressed sperm output was directly proportional to the dose—7weeks for 75 R, 21 weeks for 150 R, and 42 weeks for 300 R. These data are in close agreement with data collected for the bull and boar.

Effects of ionizing radiations on male fertility have been studied with a variety of techniques. The bulk of the investigations have centered around histological studies of the testis of animals sacrificed at various time intervals after radiation. This technique has demonstrated a large variability among tubules, with a perfectly normal tubule often seen next to a degenerated tubule. This is particularly so in the rat and mouse. Considerable variation characterizes serial sections from the same testes as well as sections from the two testes from the same animal. To complicate matters, differences are also reported between sections taken from two control animals. The data now show that some cellular death occurs in the normal testis under natural conditions with, of course, some species variation. Current findings in our laboratory suggest that radiation, like some other, pathological conditions, induces cellular death through accelerated, but natural mechanisms. We also have seen indications of mechanisms associated with testicular function that can be activated to protect against and some to facilitate cellular death caused by radiation. Observations of abnormal tubules next to normal tubules in a given tissue section suggest a differential radiosensitivity between cell types in given tubules at the time of treatment. These marked differences in cell types and their corresponding sensitivities could account for the differential in tubular survival in the testis depending on the cellular stages of development at the time of radiation. For a further discussion of factors that determine radiosensitivity as they relate to cell division and DNA synthesis, see Ellis and Berliner (1969). Spermatogenesis does not proceed in well-defined and organized waves in man. This would explain why there is not always a clear demarcation of damaged areas adjacent to normal areas in man as there is in rodents.

Attempts have been made to study male infertility through the use of mating testes. Under biological conditions, of course, the radiation of the female would be expected to be additive to that of the male. Mating tests do not account for individual differences in sexual activity and behaviour, or for the effects that repeated breedings might have no mortality or subsequent fertility after irradiation. In this respect, Rugh and Gruup (1960) have shown that sexual activity can increase mortality in the rat after radiation. This limited information seems to contraindicate what some current groups advocate—i.e., in the event of nuclear was "make love instead of war." To further complicate matters, Willham and Cox (1961) and Cox and Willham (1961) observed that boars irradiated with 300 R had the same conception rate as control animals 200 days after treatment, even though sperm production was only 50% of that of the controls. Because of the complex relations with irradiation effects, it is difficult at this time to distinguish or even speculate concerning the interactions of the following; sperm production, mating activity, general activity, conception rate, fetal survival, and long-term survival of the animal.

G.W. Casarett and Hursh (1956a, b) and G.W. Casarett (1950, and 1953) have reported detailed experiments on the effects of radiation on semen production in dogs. In addition, fertility studies have now been reported for men after accidental exposure to radiation, after removal of one testis and irradiation to the lymphatic drainage after exposure of the groin area, and after radiation of testes of volunteer prison inmates. Fertility studies have also been reported for bulls and or boars.

The author is aware of a previously nonreported case history where one testis from an man was removed because of a seminoma and the area of lymphatic drainage from the testis was irradiated with 3000 R. The dose to the remaining testis was calculated, and on this basis the man was told that he would never be able to father any more children. Because of this, his wife divorced him. Three years later he approached his physician in bewilderment wondering why he was about to be the father of another child. As pointed out by Sandeman (1966) and others, sperm counts do not appear to correlate at all with proven fertility after radiation for man. He reports that one man was fertile between the time he had a sperm count of 50,000 and a report of only a few visible spermatozoa. This leaves the entire question open as to what constitutes sterlity and how long it may last after radiation. At least with human patients, conclusion as to what constitutes sterility and how long it may last must be drawn with extreme caution. The data do suggest, however, that the degree of impairment of spermatogenesis in all animals is related to dose, while the probability that regeneration will occur after treatment increases with time after exposure.

Litter Size

Litters resulting from mating immediately after exposure of spermatozoa from murine species to high-level irradiation are not altered in size as litters of eight or more have been obtained following a dose of 41,000 R (Rugh, 1960). Moreover, the litters produced by males with recovered testes also approach the normal size. A reduction of litter size at other time intervals after irradiation is usually attributed to severe chromosomal aberrations that in turn cause developmental abnormalities and intrauterine death or abortion. Impaired translation of the genetic code could also reduce litter size. Such impairment could explain how development can be initiated and carried to a critical phase such as gastrulation, and then stop with intrauterine death. It is difficult to destroy motility or fertilizing capacity of spermatozoa unless the damage is incurred prior to their full maturation process. Motility and fertilizing capacity, however, are not adequate measures of genetic damage.

Sertoli Cells

Sertoli cells do not appear to be affected histologically be even high doses of radiation. Irradiation of mice *in utero* has shown that even the embryonic precursors of the Sertoli cells and interstitial cells are radioresistant. This is in sharp contrast to the extreme radiosensitivity of the germinal elements at this time of development. This apparent lack of histological damage, however, does not preclude the possibility of ultrastructural damage, biochemical damage, or a loss of physiological function. Nevertheless, a dose of either 10,000 or 50,000 rads locally to the testes of rats has failed to produce any actual damage that could be seen even at the highest of resolution with the electron microscope. Similar results were observed with the cell of the boundary tissue comprising the basal membrane. Nobel (1959) studied irradiated mouse testes and observed that degenerating spermatogonia were engulfed by the Sertoli cells. Similar observations also have been recorded by Lacy (1964), while Clegg and MacMillan (1965) have observed that normal Sertoli cells can take up vital dyes and particulate matter. In this respect Bateman (1958b) suggests that there may be an altered function of the Sertoli cells after irradiation. He proposed that radiation of the testis appears to induce an "aging" effect on the Sertoli cells, and that Sertoli cells accumulate lipid material after radiation and various other types of injury. The Leydig cells, on the other hand, have very little lipid in them under these same conditions. Furthermore, some of these lipids react positively for progesterone or its metabolites as judged by the Hooker-Forbes test. Since progesterone and 17 α-hydroxyprogesterone accumulate at certain time intervals after radiation in incubations of irradiated tissue, it has been postulated that the characteristic lipid droplets that accumulate in the Sertoli cells after radiation may promote an accumulation of progesterone and/or its derivatives. Similarly, Christensen and Mason (1965) and Ellis and Berliner (1969), have demonstrated that tubular tissue can synthesize androgens and as can be seen in the tubular synthesis of androgens is altered by irradiation. It has been postulated that the accumulated 17 α-hydroxyprogesterone and progesterone serve as antiandrogens to inhibit normal functioning of the seminiferous tubules after radiation. Thus, it would appear that, as postulated by Lacy (1967), abnormalities in Sertoli cell function may partially explain some of the altered testicular function after radiation. Some early workers had postulated that "indifferent cells" come from Sertoli cells for the repopulation of the testes. More recent studies, however, indicate that the spermatogonia do not arise from the indifferent cells, but rather they arise from gonocytes that are very radioinsensitive and sluggish to divide.

Leydig Cells

The Leydig cells are characteristically stable in the testes after a number of conditions in which the germinal epithelium is damaged. In this respect, they are similar to the Sertoli cells and the boundary cells of the seminiferous tubules. The Leydig cells are generally considered to be the main hormone-producing cells of the testes, especially for androgens, but the observation by Christensen and Mason (1965) and the upward revision of the androgen contribution by the seminiferous tubules, would lead us to question how important the Leydig cells are for spermatogenesis.

One aspect of testicular physiology often overlooked is that after irratiation, the testis shrinks and becomes smaller due to maturation depletion. In this respect, calculations have been made on the length and diameters of seminiferous tubules after radiation. It was concluded that both the diameter and the tubular length shrink to give a concentration of the Sertoli cells and other remaining germinal cells. As this shrinkage occurs, the endocrine tissue becomes concentrated in a smaller mass. Quite often, the androgen-biosynthesizing capacity per unit mass of the irradiated testis is the same as that observed for the normal testis. As a result, the total decrease in androgen synthesis *in vitro* brought about be radiation may be offset by the shrinkage of the tissue so that the net synthesis of steroids per unit of tissue remains unchanged after radiation. Nevertheless, in studies where gonadotropins were injected after radiation or testosterone was injected after radiation, these substances exerted a protective action on the testis. These data suggest that there is a general lack of testosterone for optimal spermatogenesis to occur.

Endocrine Changes

In Vivo Changes

Abbott (1959), in reviewing the literature on androgen production by the testis after radiation, concluded that, except for the immature animal, androgen production is not altered by this treatment. In 1932, however, Witschi, *et.al.*, presented evidence for an increased level of circulating gonadotropin in x-irradiated animals. Fogg and Cowing (1951) also cite evidence presented by other workers for an altered pituitary-testis axis. As pointed out above, it is now known that androgens or pituitary gonadotropins have a radioprotective effect on the testis. Moreover, the testis can respond to exogenous gonadotropin as indicated by gravimetric assays on accessory sex organs, but in most cases the response is diminished over that of the control animals.

When 16-week-old were insulted with 470 R whole-body x irradiation, there was an increase in prostatic weight and ^{65}Zn uptake by the dorsolateral prostate 5 days after treatment. A marked decrease in both entities was observed at 13 days with a return to normal by 20 days. Localized radiation of the testes with microwaves produced a similar response. At the lower dose, however, there was no significant decrease in the weights of the dorsolateral prostate glands, but ^{65}Zn uptake by this tissue was significantly altered. Previous work by these workers on the action of the male hormone on ^{65}Zn uptake by the prostate gland has shown that it is a more sensitive assay than the gravimetric test. It is highly possible that early workers failed to demonstrate a diminished androgen synthesis after irradiation of the testis because of the low precision of their assay, coupled with a high degree of variability among animals.

Irradiation also has been shown to inhibit steroid biotransformations by the adrenal gland *in vitro*. This phenomenon provides a sensitive assay for determining the effects of radiation on the enzymic pathways for the synthesis of androgens. As a result it has been shown that whole-body radiation alters the androgen biosynthetic pathways in the testis, as does also direct radiation to the testes. Localized radiation to the head can also diminish androgen synthesis at a specific interval after treatment.

In this respect, steroid biotransformations in testicular tissue from 12-week-old mice treated with 540 R of whole-body radiation showed a similar response to that observed by Gunn *et. al* (1960) for rats. The data show that 17 α-hydroxylase and the corresponding lyase activity were increased day 1 through 10, depressed on day 15, and activated again on day 30. Some time differences are evident between those two studies, but a general similarity prevails in the overall response. One must, however, use extreme caution in accepting this disphasic response as the only response of the testis to radiation. It has generally been assumed that testicular androgen synthesis is a constant phenomenon in mammals, but this is now known to be false. Seasonal variations are known to occur in androgen production by rat testes. Results of our present investigation correlate well with the observations of Gunn and Gould (1958) regarding the seasonal response. Evidence obtained in our laboratory also indicates that stress associated with radiation may affect the response obtained after radiation. These data emphasize the necessity for using proper control animals in a study involving androgen biosynthesis. The data show an increase in androgen synthesis over the controls from day 4 through day 34. Both group showed a similar cycle, but the controls differed from the treated animals during this period. Furthermore, a biochemical lesion is evident in the synthesis of androgens by testicular tissue after radiation. The data are consistent with the concept that the endocrine cells are inefficient in the biotransformation of steroids after this treatment. The increase in gonadotropins in irradiation animals suggests that the organism tries to overcome this defective condition by increasing gonadotropin secretion in an attempt to stimulate the enzymic mechanism for androgen synthesis to bring androgen synthesis to the desired level. This theory is supported by the observation that the radiated testis fails to respond quantitatively to gonadotropins in a manner similar to the control animals. Moreover, when androgen synthesis is stimulated endogenously the biochemical lesion becomes more evident. This would explain why some workers, as quoted by Abbott (1959), fail to observe a difference after radiation while others such as Witschi *et. al.* (1932) have observed a difference after the parabiotic union of animals, where there is an increased output of gonadotropins in the normal, but castrate parabiont, due to the nature of the animal preparation. Inefficiency in androgen synthesis because of a biochemical lesion would also explain why the administration of testosterone and gonadotropins after radiation is beneficial.

One can only speculate on reasons for the early increase in androgen synthesis that occurs 1 day after treatment. Conceivably, this could be an abortive attempt by the organism to protect the testis from radiation damage , or it could be associated with the destructive action of ionizing radiations. In this respect, massive doses of HCG and FSH can produce almost immediate histological damage to the testes of adult animals. Lastly, C.G. Heller *et.al* (1967) observed that plasma testosterone levels were unaltered in man following direct radiation, but urinary gonadotropin excretion increased 30-125 days following treatment. This increase in gonadotropin was also proportional to the degree of denuding of the germinal epithelium. C.G. Heller *et. al.* (1967) concluded that the normal output of gonadotropin is not used because of the decreased germinal cell content of the testes and hence the hormone spills over into the urine. This increase in gonadotropin excretion is also consistent with the concept that the disrupted androgen biosynthesis requires more gonadotropin to bring androgen levels in the blood to a level that would inhibit gonadotropin synthesis and release by the hypothalamic centers and the pituitary. Under these conditions, it would be logical for the increased levels of gonadotropins in the blood to spill over into the urine. If the endocrine cells of the testis are capable of responding to the increased gonadotropins in the blood, plasma levels could be near normal. When androgen synthesis is increased, due to seasonal variations, etc., a greater demand is placed on the pituitary to produce gonadotropins and on the endocrine cells to produce androgens. Under such circumstances any lesion in androgen synthesis would become more obvious as seen on days 19-25. This is why very uniform animals or subjects must be used with sensitive assays containing a high precision to be able to measure the subtle differences that exist in androgen production after irradiation.

When steroid biotransformations are plotted on a per mg of tissue basis, sharp inflections in androstenedione and testosterone production are observed on day 1 and day 30. These two increases apparently are due to a true stimulation of steroid biosynthesis, which would be consistent with the increase in pituitary gonadotropin secretion postulated by Foog and Cowing (1951), Witschi *et. al.* (1932), and Wall (1961). TSH activities of the pituitary undergo changes similar to the gonadotropins after irradiation. As judged by ascorbic acid depletion, cholesterol levels, and phospholipid content of the adrenal, ACTH responds to as little as 30 R. With *in vitro* and *in vivo* techniques, Echaute *et.al.* (1961) have shown that adrenal secretion is increased from 2 1/2 hr to 3 days after treatment. Of particular importance is the fact that the increase in adrenal secretion rate appears to be mediated through the adenohypophysis. A number of workers have now shown that the adrenal response to radiation is diphasic. Even when the adrenals were shielded, the diphasic response was observed, suggesting that the effect on the adrenal is mediated through humoral mechanisms. Of further interest is the fact that a radiomimetic drug, tris- (S-chloroethylamine-HCl), produces a diphasic response of the adrenal closely resembling that of radiation.

A similar diphasic response has been described for the thyroid after irradiation. Thyroid function was increased even when the gland was shielded. An initial increase in thyroid function immediately after irradiation also has been reported by Evans *et. al.* (1947), Monroe *et.*

al. (1954), and Botkin *et. al*. (1952). An increase in estrogen secretion has also been reported within several hours after irradiation of the transplanted sheep ovary.

Available data indicate that an immediate increase in hormone production by testis, adrenal, thyroid, and ovary occurs during the first day following whole-body x irradiation. The response after the initial increase may vary somewhat depending on the season of the year the animal is irradiated and the treatment the animal received just prior to radiation. Such things as taking the animals directly form their cages into the radiation room and immediately returning them to their cages, as contrasted to placing all of the animals into several large holding cages, taking them some distance to the radiation facility, and then returning them as a group after sham irradiating the controls are relevant. These factors could be extremely important in determining specific responses and may explain why some workers report different findings under seemingly similar conditions.

The response of the thyroid, adrenals, and testes after radiation, particularly after the first day, resembles very closely the situation observed under stress. In fact, many workers now consider radiation to be a stressor. If stress is a factor in determining the generalized response of the animal to radiation, then a humoral mechanism must be involved.

Teased-Tubular Synthesis of Androgens

When seminiferous tubules are teased from testes of control and irradiated rats and then incubated with radioactivity labeled pregnenolone and progesterone, androgens are formed. We conclude that teased-tubular preparations can synthesize androgens, and the response of tubular androgen synthesis to irradiation is very similar to that observed for minced tissue. Our data corroborate the observation of Christensen and Mason (1965), but show that much more pregnenolone than progesterone is converted to testosterone by this tissue. When androgen production is diminished relative to the controls, after radiation, progesterone and 17α-hydroxyprogesterone accumulate. This observation substantiates the concept of Lacy (1967) as to the function of Sertoli cells after radiation. These data suggest that since androgen synthesis does not diminish in a direct relationship to germinal cell loss, androgen production must occur in a stable cell in the tubules such as the Sertoli cell.

From incubations with teased-tubular tissue we have found a 5α-reductase enzyme for progesterone, and a 20α-reductase enzyme for

17α-progesterone. Since both these enzyme systems are increased after radiation when progesterone and 17α-hydroxyprogesterone do accumulate, the enzymes seem to be present in the tubules to inactivate 17α-hydroxyprogesterone and progesterone so they will not function as antiandrogens and thereby interfere with spermatogenesis.

Factors Influencing Extent of Radiation Damage

Age Effect on Testis

Russel (1954), after reviewing the pertinent literature concerning prenatal radiation, presented evidence for morphological alterations or functional changes in the fetus that may not be discernable during pregnancy or lactation. Among the delayed effects that may become evident later in life were impaired fertility and testicular atrophy. Ershoff (1959) observed that x irradiation during certain stages of prenatal development resulted in testicular injury and infertility in the male rat. Damage was greater on the 18th day of pregnancy than before or after that date. Some damage resulted if the dam was irradiated on the 14th day of pregnancy, but none occurred if she was treated on the 10th day. No reduction in testicular injury was noted in mice when the dose was fractionated during fetal life. A dose of 200 R did have more effect on the 16th day of development than it did on the 17th day. A curvilinear relationship existed between the portions of seminiferous tubules classified as either normal or sterile, the dose, and the fetal age at the time of exposure when pregnant rats were exposed to doses of 25R to 2 times 1004 R between the 13th and 20th days of gestation. Of particular significance is the rat testis at approximately 15 to 16 days of prenatal life as shown by histochemical techniques and by *in vitro* steroid biotransformations. An increased sensitivity of the testis to radiation *in utero* is apparently associated with the onset of androgen secretion.

Reduction in testicular weights and atrophy of the accessory sex organs later in life has been reported after neonatal irradiation. The reaction of the testes to radiation was not altered by the administration of gonadotropin to these animals. Similarly, a reduction in numbers of primitive germ cells before and after puberty was observed in swine after they were treated with 200 R of gamma rays. Crenated nuclei and a delay in morphogenesis were observed among the Leydig cells before puberty, but the urinary steroid output equalled or exceeded that of nonirradiated boars later. As indicated by accessory sex organ weights, exposure of rats of 400, 300, or 1200 R at 3 to 36 days after birth resulted in a markedly impaired secretory capacity of the testes.

The dose and the age when exposed seemed to determine how much the secretory function of the testis was affected. The reduced secretory ability of the testis of these animals was affected. The reduced secretory ability of the testis of these animals was associated with a decreased volume of intertubular tissue. As the animal neared the age of puberty, less effect was noted on the secretory function of the testis. These limited data suggest that the Leydig cells from immature animals may be more sensitive to radiation that the corresponding cells from mature animals. There is no known mechanism to suggest why this might be the case, but it is consistent with the age effect for the germinal epithelium, only at a different sensitivity.

After parturition the radiosensitivity of germ cells appears to decrease sharply. In this respect, Shaver (1953b) observed that animals irradiated 1 and 2 days after birth were completely sterilizes. Similar animals irradiated at 3 to 5 days of age showed some signs of regeneration. Changes in germ cells of rat testes were observed by Harding (1961) when doses of x-rays to 430 R were given 4 to 8 days after birth. An intrinsic delay in radiosensitivity of the testis was noted between the 21st day of pregnancy and 4 days postpartum, with a prolonged inhibition of mitosis after irradiation postulated to explain the findings. A decrease in radiosensitivity was noticed between 1 and 4 days after birth when rat testes were exposed to 150 and 460 R of x-rays. A change in endocrine activity of the rat was observed by Binhammer (1967) after the animals were exposed to 1000R at 30, 60, 90, or 150 days of age using the parabiosis technique for measuring androgen secretion of the testis. The effects of radiation on the testes and ventral prostates were most marked if the rats were irradiated when 30 days old.

The effects of radiation *in utero* will apparently be most pronounced when given on the 16-18 days for rats. A reduced sensitivity appears to be evident immediately after birth with another period of sensitivity occurring at 30 days of age, when there is a marked increase in the ability of the testes of rats to biosynthesize androgens. Thus, this change in radiosensitivity is also associated with an increase in androgen biosynthesis. Of particular interest is the observation that testosterone porpionate has been reported to have a radiosensitizing effect in the male and a radioprotective effect in the female. Moreover, female mice have been reported as being less radiosensitive to radiation than male mice. These data suggest that androgen secretion by the testis has a very dynamic relationship to the radiosensitivity of animals at various stages of development.

It is very likely that changes in hormone secretion can modify radiosensitivity in animals. In support of this concept lethality after radiation was observed to increase from 33 to 73% as a result of an increase in sexual activity following treatment. The change in radiosensitivity was attributed in part to the increase in testosterone secretion which occurs in the rat after breeding. The data strongly suggest that androgen secretion after radiation is much more important in terms of radiosensitivity, radiation sickness, and repair of radiation damage than heretofore recognized.

Species Differences

In a recent review of species comparisons in the response of the gonads to radiation, Oakberg and Clark (1964) point out one pattern they consider to be consistant for all species. After exposure to a sublethal dose of radiation, males demonstrate a fertile period until the spermatozoa and spermatids complete their development and the epididymis becomes emptied. This is followed by a temporary sterile period due to maturation depletion at the spermatogonial level, with an evantual return of fertility due to repletion from a few surviving spermatogonia. They also point out that the presterile period can be reduced to less than the normal transport time (from the early spermatid stage to the ejaculate) by increasing the dose. This decrease in time is apparently due to loss of the early spermatids. A species with a longer duration of spermatogenesis and a longer transport time should have a longer presterile period as evidenced by the data of Craig *et.al.* (1961) for the rat and Strandskov (1932) for the guinea pig. On the other hand, a relatively long presterile period for a given set of conditions is usually attributed to a greater survival of primary spermatocytes. Lower doses that still cause extensive killing of spermatogonia may give no sterile period since the regeneration of the germinal epithelium will occur before the maturation depletion occurs.

A dose of 300 R or more has been estimated as necessary before a sterile period will develop in the male, although there is some suggestion that strain difference may modify the dose required for this response. An inverse correlation exists with the surviving type A spermatogonia and the sterile period within species. The dynamics of spermatogenesis for a specific species will largely determine the rate of spermatogonial repopulation. One interesting assumption is that the normal sequence for the development of the spermatozoa is altered after radiation injury so that more type A spermatogonia or stem cells are formed before maturation and differentiation occurs. The time required for the passage of spermatozoa through the reproductive tract will also affect the duration of the sterile period. For comparative purposes, difference in the length of the sterile period and of the presterile period would not be good indices of radiation sensitivity. Comparisons based on the rate of spermatogonial reproduction are also questionable as this appears to be dependent on the spermatogenic dynamics for the species.

Physiological and Environmental Factors

Tables on the relative amounts of radiation that are required for germ cell hypoplasia (450-500R), arrest of spermatogenesis (700-800R) and permanent sterility for various species such as have been reported by Rugh (1960), emphasize that a number of factors may be responsible for differences in sensitivity. One particular factor that could effectively alter the sensitivity of the testis to radiation is the archaic endocrine cycle that has been explored in detail by Gunn and Gould (1958, 1959a, b). It is generally concluded that cellular damage after irradiation is very closely related to DNA synthesis and sages in the cell cycle. If the seasonal cycle does depress cellular division, as indicated by the lessened sensitivity of the animal to exogenous hormones, this could account for some radioprotection to the gonads. Data from this laboratory now suggest that the archaic endocrine cycle in the rat may be mediated at least in part by the pineal gland. Certainly, melatonin can inhibit androgen biosynthesis, and we have observed an inverse relationship between melatonin synthesis by the pineal gland and androgen synthesis *in vitro*. It is now known that melatonin is taken up by the testis as well as by other endocrine organs. Pineal synthesis of melatonin is diminished 1 day after radiation, and this diminished synthesis of melatonin occurs when there is an increased secretion of thyroxine, adrenal corticoids, estrogens, and androgens. When rats were irradiated with 350 R of x-rays in this laboratory, and some of them placed on a 24 hour constant darkness (DD) regime after treatment, there was a decrease in the testis weights of the irradiated and light-treated group of animals. There was, however, a radioprotective effect when the animals were placed on a 24-hr (DD) regime prior to and after radiation. Of particular interest is fact that melatonin has been reported to protect the testis against radiation. After exposure of mice to 1300-1600 rads of protons, no signs of somatic injuries were observed in the F_1 generation of irradiated males and females. Thus, it would appear that alteration of pineal function prior to radiation can have a protective effect on the testis presumably through the reaction of melatonin with essential sulfhydryl groups so that they are not destroyed by the free radicals and peroxides formed by energy deposition in the tissue.

In this respect, both melatonin and serotonin are now recognized as being among the most effective naturally occurring radioprotective agents. Although serotonin is an effective vasoconstrictor, and its radioprotective effect has been attributed to this action by some workers, it is radioprotective in organisms that do not contain a circulatory system. It also functions *in vitro*. Lastly, Dukor (1962) has shown that radioprotection by serotonin cannot be due to hypoxia or free radical scavengering. He proposed that complex formation between the radiosensitive target and the protective molecule might be responsible for the observed protection. If a radioprotective agent does react with a functional group on the molecule, it would in turn interfere with the kinetics of the biochemical reaction. This is exactly the situation that we find with melatonin and serotonin *in vitro*; for steroid biotransformations by adrenal and testicular tissue. We find that the effect of serotonin and melatonin on steroid biotransformations is a direct effect on the enzymes and not necessarily on cofactor availability that would reflect an effect on glucose metabolism through 3', 5' cyclic AMP. Our data suggest that these compounds might exert their radioprotective effect on the testis by complexing with essential sulfhydryl groups thereby preventing their destruction or conversion to disulfide bridges. It is now known that steroid biosynthetic enzymes contain essential sulfhydryl groups. Recently investigations show that the non-heme iron protein associated with steroid hydroxylations contains four sulfhydryl groups per atom of iron as determined by *p*-mercuribenzoate. Lohamann *et. al.* (1966) have reported that serotonin may react with iron in catalase and thereby demonstrate a radioprotective effect *in vitro*. The non-heme iron protein, with iron complexed to sulfhydryl groups, or both could be affected by serotonin and melatonin. On the other hand, Winstead (1967) showed that serotonin can be radioprotective to enzymes that have neither a sulfhydryl group nor a disulfide bridge. This would suggest that serotonin may function as a radioprotective agent through more than one mechanism. Other reports indicate that histamine may be a radioprotective agent, but less effectively than serotonin. Histamine does, however, rank as one of the best biological amines for radioprotection. Since serotonin and histamine are both associated with stress and inflammation, and are radioprotective, this would explain why almost any stress presented to the animal prior to radiation would afford some protection, as the availability of these compounds would increase after exposure to a stressor. After radiation, one would expect persistent stress and irradiation damage to be additive. This might explain why subtle variations in radiation procedures would give differences in LD 50/30 values and in responses of the testes to radiation damage.

Radiation is usually listed as a stressor, but, as pointed out by Mole (1956), this stressor must function by some unconscious or para-Freudian sense since there is no apparent overt injury or stimulation of sense organs. Nevertheless, irradiation does cause cellular damage that normally results in the release of compounds such as serotonin, histamine, epinephrine, and various kinins. These compounds are considered to be mediators of the inflammatory response. The effectiveness of cold as a radioprotective treatment could be due partially to a release of compounds such as serotonin, epinephrine, and histamine. Although serotonin, epinephrine, and histamine can increase adrenocorticosteroid secretion, and adrenalectomy increases the radiosensitivity of the organism, the radioprotective effect of serotonin appears to be independent of the adrenal gland. Moreover, the toxicity of serotonin is increased in the adrenalectomized animal. It is conceivable that the synthesis and release of melatonin by the pineal gland and the uptake of melatonin by tissues, including the testis, could result in seasonal differences in radiosensitivity of the organism or the testis. We now feel that when results are reported relative to LD 50/30 responses and testicular damage after irradiating, the season of the year and details of the radiation procedure must be included.

The role of stress in determining testicular function is additionally apparent when one realizes that sertonin and epinephrine have deleterious effects on the testis similar to those of radiation. Even the polynuclear spermatids observed after irradiation damage are seen after serotonin administration. Moreover, epinephrine can alter androgen synthesis by the testis *in vivo*. Recent investigations in this laboratory show that serotonin, melatonin, and histamine directly affect androgen synthesis *in vitro*. The involvement of serotonin is steroid biotransformations is not limited to the testis as the adrenal is also affected. At low concentrations, serotonin increase steroid biotransformations in both the testis and the adrenal, but inhibits them at higher concentrations. The pineal gland has been reported to influence adrenal secretion rates *in vivo,* while melatonin has been reported as having a deleterious or inhibiting action on the testis, the thyroid, the ovary and LH secretion by the pituitary.

The normal adult rat testis contains 64 ng of serotonin per gram of wet tissue. We also find the radioactive serotonin, when injected into rats, is taken up by the adrenals, thyroid, and testes. Similarly, Axelrod and Wurtman (1966) found that the ovary, pituitary, testis, thyroid, and adrenal glands all took up radioactive melatonin, and this ability was considerably greater than for other tissues. It now seems reasonable to assume that melatonin synthesis is decreased immediately after radiation. This reaction of radiation would remove a braking action normally exerted on the endocrine organs so that both the output of gonadotropins and the secretion of hormones from the adrenal, thyroid, ovary, and testis all increase, as has been observed after radiation. Since serotonin in small amounts can stimulate androgen synthesis and adrenocorticosteroid synthesis, radiation a a stressor may conceivably release this compound, which then could act synergistically with the removal of inhibition by melatonin to effect an increase in steroid synthesis. With higher radiation dose rates, the damage could be so great that serotonin would be released in copious amounts and could thus inhibit steroid synthesis. The intrinsic cellular damage from radiation plus the action of large amounts of serotonin would be most effective in decreasing hormone synthesis, as has been reported after a large dose of radiation for the thyroid.

Increased adrenocorticosteroid secretion might also be due to an action of serotonin, epinephrine, and histamine on the hypothalamic-pituitary complex by increasing ACTH secretion. One logically assumes that the direct action of serotonin on the adrenal would complement the action of ACTH. If it did not a higher concentrations, then ACTH would have to override any inhibiting action that serotonin might have on the adrenal gland— a very important consideration in evaluating adrenal reserve function. The ability of serotonin, epinephrine, and histamine to stimulate adrenocortical activity through the hypothalmic centers, while simultaneously inhibiting testicular function, could partially explain the typical response of animals to stress. It would account for the enlarged adrenals and regressed testis during exposure to prolonged stress. In this respect, Dina *et. al.* (1957) have shown that postural- and formalin-provoked stress both result in testicular injury and reduction of the germinal epithelium. In light of these observations, it is not surprising to find testicular damage in bulls after they are transported some distance by truck. In addition, warmth-induced aspermia has been reported for the bull, indicating that adrenal steroid secretion increased as testicular regression occurred. Of importance is the observation that traumatization of the testis will increase libido. Conceivably, this increase in libido could be due to an increased synthesis of androgens brought about by the release of serotonin and products of inflammation associated with the trauma.

An increase in adrenal function per se is most unlikely to be responsible for the destructive changes in the testis following stress, since adrenocorticosteroids have been reported as having an augmenting action, a depressive action, no action, or a protective action on the prostatic responsive to androgens in the rat. The above observations tend to corroborate the hypothesis of Sayers (1950) that the adrenal exerts a "normalizing function" on the testis rather than having a pure destructive action. On the other hand, epinephrine, histamine, serotonin, and melatonin as melatonin as regulators of steroid biotransformations either at the level of the hypothalamus or on the endocrine organ itself could mediate the response to stress.

Dose Fractionation and Sose Rate Effects

Conflicting data have been presented suggesting that fractioning the dose potentiates or has no effect on the production of sterility. Relatively large doses of radiation (400-800R) can be delivered to animals without affecting fertility if the dose rate is low enough. The intervals between fractions are very critical in determining the response. In this respect, Oakberg and Clark (1964) have postulated optimum doses and fractionation schedules for obtaining maximum effects.

It is generally concluded that dose rate is more important than total dose in determining the depressing action of radiation and the regeneration capacity of the testis—especially when low dose rates are used. In this respect, Eschenbrenner et. al. (1948) report that chronic irradiation damage was diminished when the dose rates were greatly reduced. They concluded that, at lower dose rates, equilibira could be established between the destructive actions of the radiation and the regenerative capacity of the germinal epithelium. It has now been demonstrated that a reduced effect on the seminiferous epithelium is possible when the doses are divided into small fractions and after continuous irradiation at low doses. What is not clear is whether the reduced effect is due to improved survival of individual cells, possible because of repair, or if regeneration capacity increases to compensate for the accelerated cell death with the radiation.

Mode of Radiation

Several workers have concluded that doses of 500 R given directly to the testes produced no apparent histological damage. Shaver (1953a), however, reports that whole-body radiation with 500 R produces a unique type of injury due apparently to combined effects of localized effects and secondary or mediated effects. On the other hand, Ferroux *et. al.* (1938) shielded the bodies of rats while irradiating only the testes. Observations were made from 7 to 300 days after single doses of 50-3300 R. With doses below 130 R they observed changes in the germinal epithelium that were due to radiation. An increased number of spermatogonia were killed with doses above 140R. Only dusty spermatogonia remained with doses above 400 R. Von Wattenwyl and Joel (1941a, b; 1942) irradiated the testes of shielded rats with doses of 60-2400 R. either or both testes were treated, and the animals were sacrificed at 1 to 300 days after treatment. With 60 R, histological changes were observed after 4 days, with regeneration evident at 25 days, and complete recovery occurring by 50 days. Maximal injury resulted at 30 days with partial recovery evident at 50 days with 150 R. Above 450 R, recovery was small at 10 days. Kohn (1955) concluded from his studies, that the dose received by the testis determines the changes observed in that organ. Human testes respond to as little as 15R of localized radiation to the testis indicating that very small direct doses can destroy some cells.

While it now appears that there is very little mediated effect of radiation on the germinal epithelium, a mediated effect is noted on androgen synthesis with at least part of the mediated effect coming from the head. There is also little doubt now that localized radiation to the testes affects androgen biosynthesis. Testicular androgen synthesis also responds to *in vitro* x irradiation. Moreover, the biosynthetic pathway show a differential in response and are dose-dependent. It is also known that ^{226}Ra, ^{239}Pu, ^{228}Th, and ^{228}Ra can inhibit androgen biosynthesis by the testes when they are internally deposited. Likewise, other radionuclides have been shown to destroy the germinal epithelium of the testis; ^{137}Cs; ^{32}P; ^{210}Po; and ^{3}H. Although no gross damage to the germinal epithelium was observed in dog testes after the internal deposition of some radionuclides, it is possible that damage and repair had taken place before the animals were sacrificed. It is generally concluded that the dose or amount of energy deposited in the testes is responsible for the damage to the germinal epithelium.

5

Effect of Light

The processes of life consists ultimately of the synthesis and breakdown of carbon compounds. Because a carbon atom can bind four other atoms to itself at a time and is thereby able to link up with other atoms—especially other carbon atoms—in chains and rings, carbon lends itself to the construction of a virtually endless variety of molecules. These molecules derive their physical characteristics and chemical activity not only from their composition but also from their size and intricacy of structure. The rich variety of life suggests in turn that living cells have gone far in the elaboration of such compounds and the processes that make and unmake them. All these processes depend in the end of a first one. This is the process of photosynthesis, which takes carbon and several other common elements from the environment and builds them into the substances of life. The plant finds most of these elements already bonded to oxygen in oxides such as carbon dioxide (CO_2), water (H_2O), nitrate (NO_3^-) and sulphate (SO_4^-).

Before the plant can bind the elements other than oxygen together as organic compounds, it must remove some of the excess oxygen gas (O_2) and this accomplishment takes a large amount of energy. In the simplest terms photosynthesis is the process by which green plants trap the energy of sunlight by using that energy to break strong bonds between oxygen and other elements, while forming weaker bonds between the other elements and forcing oxygen atoms to pair as oxygen gas. For example, to make the sugar glucose ($C_6H_{12}G_6$) the plant must split out six molecules of oxygen in order to combine the carbon and half the oxygen of six carbon molecules with the hydrogen of six water molecules. The glucose and other organic compounds taken up

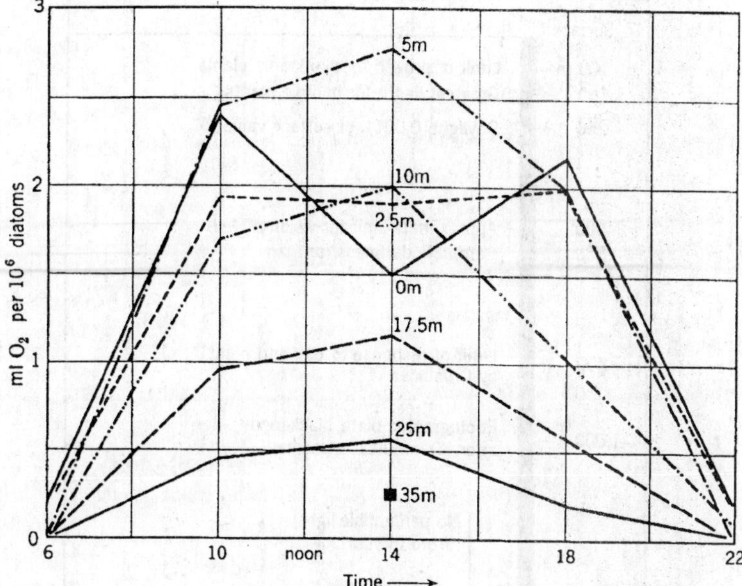

Fig. 5.1. Photosynthesis of a species of phytoplankton (Coscinodiscus) at the indicated depths during the course of the day off.

in the chemical machinery of the plants and the animals that on plants serve both as fuel and as the raw materials for the synthesis of higher organic compounds. That considerable solar energy is bound by photosynthesis becomes apparent when wood or coal is burned.

In living cells the controlled combustion of respiration extracts this energy to power the other processes of life. Both kinds of combustion take oxygen from the air, and breakdown organic compounds to carbon dioxide and water again. In its end result photosynthesis can be defined as the opposite of respiration. Together these complementary processes drive the cycle flow of matter and the noncyclic flow of energy through the living world. From such generalization about the effect and function of photosynthesis in nature it is long step to the explanation of how photosynthesis works. Yet much of the explanation is now complete.

The work has been greatly facilitated by the earlier and more nearly complete resolution of the chemistry of respiration. The two processes, it turns out, are in some ways complementary on the molecular scale, just as they are on the grand scale in the biosphere. Each involves some 20 to 30 discrete reactions and as many intermediate compounds; half a dozen of these reactions and their intermediates are common to both photosynthesis and respiration. Only the first few

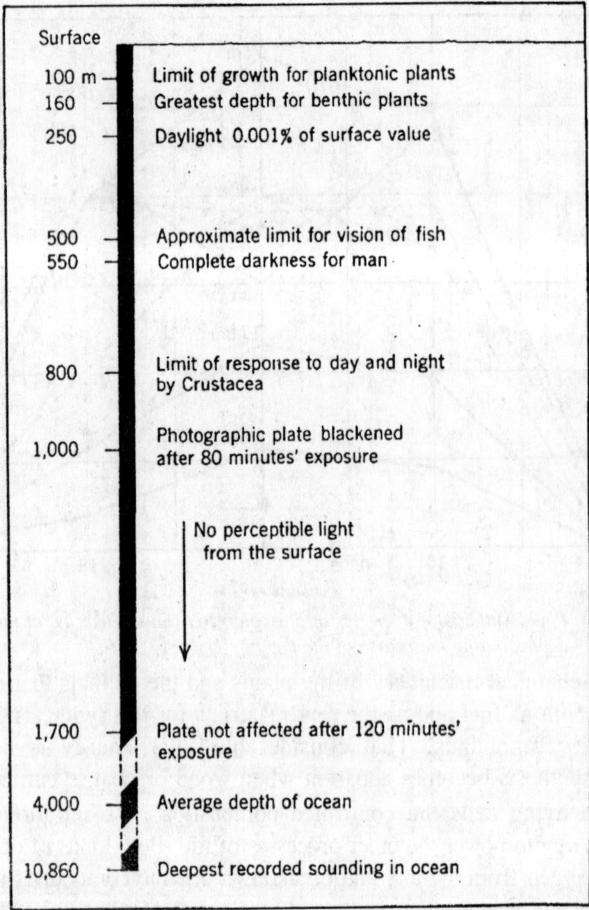

Fig. 5.2. Summary of limitations by light factor in aquatic environment. Values are for clearest water.

steps in photosynthesis are driven directly by light. The energy of light is trapped in the bonds of a few specific compounds. These energy carries deliver the energy in discrete units to the steps of synthesis that follow. The same or closely similar carriers performs corresponding operations in respiration, picking up energy from the stepwise dismemberment of the fuel molecule and delivering it to the energy-consuming processes of the cell.

Although the first, energy-trapping stage in photosynthesis remains to be clarified, it is now possible to trace the path of carbon from the very first step in which a single atom of carbon is captured in the bonds of an evanescent intermediate compound. The term "carbohydrate" recalls the deduction of early 19th-century investigators that photosynthesis made glucose directly by combining atoms of carbon with molecules of water, as the formula for glucose suggests. In line with this idea it was thought that the oxygen transpired by green leaves came from the splitting of carbon dioxide. The progress of chemistry, however, failed to disclose any processes that would accomplish the results so simply. Accumulating evidence to the contrary became convincing some 30 years ago, when C.B. van Niel of Stanford University discovered that certain bacteria produce organic compounds by a process of photosynthesis similar to that in plants but with one important difference. These bacteria use hydrogen sulphide (H_2S) instead of water and liberate elemental sulphur instead of gaseous oxygen. The otherwise complete similarity of the two processes strongly suggested that the oxygen evolved by green plants must come from the splitting of water.

The Capture of Light

Photosynthesis could now be described in terms of familiar chemistry. The splitting of water would be accomplished by the process of oxidation (which means the removal of hydrogen atoms from a molecule), with oxygen gas as the product of the reaction. The free hydrogen atoms would then be available to carry through the equally familiar and opposite process of reduction (which means the addition of hydrogen atoms to a molecule). By the addition of hydrogen atoms (or electrons plus hydrogen ions) the carbon dioxide would be reduced to an organic compound. It is during the first, energy-converting stage of photosynthesis that the water molecule is split.

Initially the energy of light impinging on the plant cell is transformed into the, chemical potential energy of electrons excited from their normal orbits in molecules of the green pigment chlorophyll and other plant pigments. A large part of this energy eventually goes into the splitting of water as electrons and hydrogen ions are transferred from water to the substance triphosphopyridine nucleotide (TPN^+), which is there upon reduced to the form designated TPNH. The TPNH thus becomes not only a carrier of energy but also the bearer of electrons for the subsequent reduction of carbon dioxide. Along with the movement of electrons from water to TPNH, some energy goes to charging the energy-carrying molecule adenosine triphosphate (ATP), specifically by promoting the attachment of a third phosphate group ($—OPO_3^-$) to adenosine diphosphate (ADP), the discharged form of the carrier.

Both ATP and TPNH belong to the family of compounds known as cofactors or coenzymes, which work with enzymes in the catalysis of chemical reactions. ATP, the universal currency of energy transactions in the cell, plays a significant role in respiration as well as in photosynthesis. Needless to say, the manufacture of each of these cofactors involves an intricate cycle of reactions. Although the cycles are not yet fully understood, it is enough for the purpose of the present discussion to know that ATP and TPNH, or closely similar compounds, furnish the energy for the second stage of photosynthesis, during which the carbon atom of carbon dioxide is reduced and joined to a hydrogen atom and a carbon atom in place of an oxygen.

The process of reduction goes forward in small steps. Each reaction brings about some change in a carbon compound until the starting material is at last transformed to the final product. For each reaction there is therefore an intermediate compound. Since every life process involves a more or less extended series of intermediates, cells typically contain a large number of intermediates. Many of them turn up in two or more pathways leading to different end products. The tracing of the path of carbon in photosynthesis required first of all a technique for identifying the intermediates proper to it and for establishing their sequence along the path. Samuel Ruben and Martin D. Kamen, then at the University of California, met this need some 20 years ago by their discovery of the radioactive isotope of carbon with a mass number of 14. This isotope has a conveniently long half life of more than 5,000 years; over the time period of an experiment, therefore, carbon 14 has an effectively constant radioactivity.

Ruben and his colleagues recognized at once the potential usefulness of this isotope as a label for the identification of compounds in biological processes. They prepared carbon dioxide in which the, carbon atoms were carbon 14. When they exposed green plants to an atmosphere containing this gas instead or normal carbon dioxide ($C^{12}O_2$), the plants took up the ($C^{14}O_2$), and made compounds from it. The presence of the carbon 14 in these compounds could be detected by various devices, such as the Geiger-Muller counter, and by radioautography on X-ray film. Unfortunately this work was cut short by the war and by Ruben's death in a laboratory accident. In 1946 Melvin Calvin organized a new group at the Lawrence Radiation Laboratory of the University of California with the primary objective of tracing the path of carbon in photosynthesis with $C^{14}O_2$ as one of its principal tools. Starting as a graduate student in 1947, I had the good fortune to participate in this work with, Calvin, Andrew A. Benson and others.

The early experiments were quite simply contrived. We used leafy plants and often just the leaves of plants. After allowing a Tear to photosynthesize for a given length of time in an atmosphere of $C^{14}O_2$ in a closed chamber, we would bring biochemical activity to a halt by immersing the leaf in alcohol with the enzymes inactivated, the reactions converting one intermediate compound into another would story, and the pattern of labeling would be "frozen" at the point. We soon discovered, however that photosynthesis proceeds too rapidly for completely reliable observation by such a procedure. With few seconds delay in the penetration of alcohol into the cell, for example, the labeling pattern would be disarrayed and no longer representative of the stage at which we tried to halt the photosynthesis. Since rapid and precisely timed killing of the plant is important, we adopted single-celled algae—Chlorella pyrenoidosa and Scenedesmus obliquus—as the subject for many of our experiments. In both species the plant consists of a single cell so small that it can be seen only with a microscope.

Alcohol can quickly penetrate the cell wall and deactivate the enzymes. The algae offer another advantage they can be grown in continuous, cultures, assuring us a supply of material with highly constant properties. An experimental sample is taken from the culture in a thin-walled, transparent closed vessel. Illuminated through the walls of the vessel and supplied with, a stream of ordinary carbon dioxide, which is bubbled through the suspension, the algae photosynthesize at the normal rate. We shut off the supply of carbon dioxide and inject a solution of radioactive bicarbonate ion (carbon dioxide dissolved in our algae culture medium is mostly converted into bicarbonate ion). After a few seconds or minutes the cells are killed. We then extract the soluble radioactive compounds from the plant material and analyze them.

Reduction or CO_2

Calvin and his colleagues soon found that the carbon 14 label was distributed among several classes of biochemical compounds, including not only sugars but also amino acids; the sub units of proteins. As the exposure time was reduced to a few seconds, the first stable intermediate product of photosynthesis was found to be the three-carbon compound 3-phosphoglyceric acid (PGA). The next step was to determine which of the three carbon atoms in the first generation of PGA molecules synthesized in the presence of radioactive carbon dioxide bears the carbon 14 label. We first removed from PGA the phosphate group and then diluted the free glyceric acid with glyceric acid containing the stable carbon 12 isotope in order to have enough material for analysis by ordinary chemical methods. Treatment with reagents that severed the bonds between the carbons produced there different products, one from each carbon atom.

By measuring the radioactivity of each of the products we were able to identify the labeled carbon. In PGA from plants that had been exposed to labeled carbon dioxide for only five seconds we found that virtually all the carbon 14 was located in the carboxyl atom, the carbon at one end of the chain that is bound to two oxygens. This was not surprising because the carboxyl group most nearly resembles carbon dioxide. The carbon is bound to the oxygens by three bonds, however, instead of four; the fourth bond now ties it to the middle carbon in the PGA chain. The transfer of this bond from one of the oxygens to a carbon constitutes the first step in the reduction of the carbon dioxide. This was evidence also that the reduction is accomplished by some sort of carboxylation reaction, a reaction in which carbon dioxide is added to some organic compound.

Ultimately, of course, the two other carbons of PGA must come from carbon dioxide. But it was some time before investigation disclosed the specific compound that picks up the carbon dioxide and the cyclic pathway that makes this carbon dioxide acceptor from PGA. The discovery of the pathway intermediates was made much easier by a then comparatively new technique; two-dimensional paper chromatography, developed by the British chemists A.J.P. Martin and R.L.M. Synge. Closely similar compounds can be distinguished in this procedure by slight differences in their relative solubility in an organic solvent and in water. The extract from the plant is dropped on a sheet of filler paper near one-corner.

An edge of the paper adjacent to the corner is immersed in a trough containing an organic solvent; the paper is held taut by a weight and the whole assembly is placed in a water-saturated atmosphere in a vapour-tight box. The solvent travelling through the paper by capillarity dissolves the compounds and carries them along with it. As they move along in the solvent, however, the compounds tend to distribute themselves between the solvent and the water absorbed by the fibers of the paper. In general the more soluble the compared with the organic solvent, the slower it travels. If the compound is also absorbed to some extent by the cellulose fibers, its movement will be even slower.

As a result the compounds are distributed in a row in one dimension. Depending on the solubility of the compounds and the nature or the solvent used, some compounds may still overlap one another. Repetition of the procedure, with a different solvent travelling at right angles to the direction of the first run, will usually separate these compounds in a second dimension. Since most of the compounds are colourless, special techniques are needed to locate them on the paper. Those that are radioactive will locate themselves, however, if the chromatographic paper is placed in contact with a sheet of X-ray film for a few days. The resulting radioautograph will show as many as 20 or 30 radioactive compounds in the substances, extracted from algae exposed to carbon 14 for only 30 seconds. Clearly the synthetic apparatus of the plant works rapidly.

Chromatographs and Radioautographs

In order to identify these compounds we prepared a chromatographic map by running samples of many known compounds through the same chromatographic system and recording the locations at which we found them on the paper. The locations can be made visible in these cases by spraying the paper with a mist of some chemical that is known to react with the compound to produce a coloured spot. Comparison of the radioautograph of an unknown compound with the map yields a first clue to its identity. This can be corroborated by washing the radioactive compound out of the paper with water and mixing it with a larger sample of the suspected authentic substance. The mixture is applied to a new piece of filter paper and chromatographed. With enough of the authentic material to yield a coloured spot, comparison with a radioautograph of the same paper now shows whether or not the radioactive and the authentic material really coincide.

The possibility that the authentic material and the radioactive material are still not the same can be tested by using different solvent systems in the preparation of the chromatograph and by other means. Over the years these procedures have established the identity of a great many of the intermediate and end products of photo synthesis. Some of the sugar phosphates labeled by carbon 14 proved to be well-known derivatives of triose (three-carbon) and hexose (six-carbon) sugars.

Other were discovered for the first time among the intermediates produced by our algae. Benson showed that among these are a seven-carbon sugar phosphate and also five-carbon phosphates, including in particular ribulose-1, 5-diphosphate. The rapid building of carbon 14 into the more familiar triose and hexose phosphates suggested certain biochemical pathways already established in studies of respiration. It seemed likely that PGA might be linked to these phosphates by the reverse of a sequence of respiratory reactions first mapped many years ago by the German chemists Otto Meyerhof, Gustav Embden and Jakob Parnas.

In the respiratory pathway hexose phosphate is split into two molecules of triose phosphate, with the split occurring between the two carbon atoms in the middle of the chain. The triose phosphate is then oxidized to give PGA. The electrons from this energy-yielding operation are picked up by diphosphopyridine nucleotide (DPN^+), which is thereupon reduced to DPNH. The DPN^+ is a close relative of the TPN^+ that turns up in photosynthesis. In addition this oxidation yields enough energy to make a molecule of ATP from ADP and phosphate ion. In the reverse pathway of these reactions in photosynthesis Calvin concluded, the plant uses the cofactors ATP and TPNH, made earlier by the transformation of the energy of light, to bring about the reduction of PGA to triose phosphate.

In the first step the terminal phosphate group of ATP is transferred to the carboxyl group of PGA to form a "carboxyl phosphate" (really an acyl phosphate). Some of the chemical potential energy that was stored in ATP is now stored in the acyl phosphate, making the new intermediate compound highly reactive. It is now ready for reduction by TPNH. This reducing agent donates two electrons to the reactive intermediate. One carbon-oxygen bond is thereby severed and the oxygen atom, carried off with the phosphate group, is replaced by a hydrogen atom. In this way the carboxyl carbon atom is reduced to an aldehyde carbon atom; that is, it now has two bonds to oxygen instead of three and one bond each to carbon and hydrogen. This is the point in the cycle at which most of the solar energy captured in the first stage of photosynthesis is applied to the reduction of carbon.

Unstable Intermediate

The next development in the plotting of the carbon pathway came from a series of experiments first performed in our laboratory by Peter Massini. He hoped to see which intermediates would be most strongly increased or decreased in concentration by turning off the light and allowing the synthetic process to go on for a while in the dark. In order to establish the concentration of the various intermediates when the reaction proceeds in the light, he bubbled radioactive carbon dioxide through the culture for more than half an hour. At the end of this period every intermediate was as highly radioactive as the incoming carbon dioxide. The radioactivity from each compound therefore gave a measure of the concentration of the compound. He then turned off the light and after a few seconds took another sample of algae in which he measured the relative concentration of compounds by the same technique.

Comparison with the compounds sampled in the light showed that the concentration of PGA was greatly increased. This finding could be readily explained: turning off the light stopped the production of the ATP and TPNH required to reduce PGA to triose phosphate. Of the sugar phosphates present, only one, the five-carbon ribulose diphosphate, was found to have changed significantly; its concentration dropped to zero. Because the PGA had simultaneously increased in concentration, it was apparent that ribulose diphosphate was consumed in the production of PGA. This finding was of great significance because it indicated for the first time that ribulose diphosphate is the intermediate to which carbon dioxide is attached by the carboxylation reaction. For this reaction ribulose diphosphate is prepared by an earlier reaction that goes on in the light and in which ATP donates its terminal phosphate group to ribulose monophosphate.

The more reactive diphosphate molecule now adds one molecule of carbon dioxide by carboxylation. The details of this reaction remain obscure because the resulting six-carbon intermediate is so unstable that we have not been able to detect it by our methods of analysis. As its first stable product this sequence of events yields two three-carbon PGA molecules. Massini's experimental results were confirmed by a parallel experiment devised by Alex Wilson, then a graduate student in our laboratory. Instead of turning out the light Wilson shut off the supply of carbon dioxide. In this situation one might expect to find an increase in the concentration of the compound that is consumed in the carboxylation reaction; ribulose diphosphate showed such an increase. Correspondingly, one would look for a decrease in the concentration of the product of this reaction; PGA did in fact decrease in concentration.

The first steps along the path were thus established. The photosynthesizing plant starts with ribulose monophosphate and converts it to ribulose diphosphate, using chemical potential energy trapped from the light in the terminal phosphate bond of ATP. Carbon dioxide is joined to this compound and the resulting six-carbon intermediate splits to two molecules of PGA. With energy and electrons supplied by ATP and TPNH, PGA is reduced to triose phosphate. In the next step, it was apparent, two triose phosphates must be joined end to end in the reverse of a familiar respiratory pathway to form a hexose phosphate. The pathway from hexose to pentose phosphate remained to be uncovered. We continued the carbon-by-corbon dissection and analysis of these chains by the methods that had earlier shown the carbon 14 in PGA to be located first in the carboxyl carbon.

In the hexose molecules we had found the labeled carbon concentrated in the two middle carbons, just where it should be if two triose molecules mode from PGA were linked together by their labeled ends. We also took apart the seven-carbon and the carbon sugar phosphates to establish the position of the carbon 14 atoms in their chains. As the result of these degradations we were able to show that the over-air economy of the photosynthetic process starts with five three-carbon PGA's, variously transforms them through three-, six-four and seven-carbon phosphates intermediates and returns three five-carbon ribulose diphosphates to the starting point. From carboxylation of these three chains and their immediate bisection, the cycle at last yields six PGA molecules. The net result, therefore, is the conversion of three carbon dioxide molecules to one PGA molecule.

Calvin Cycle

With these steps filled in, the carbon reduction cycle in photosynthesis, called the Calvin cycle, was complete. The intermediates formed in the cycle depart from it on various pathways to be converted to the end products of photosynthesis. From triose phosphate, for example, one sequence of reactions leads to the six-carbon sugar glucose and the large family of carbohydrates. Because the cycle had been established primarily by experiments with algae and the leaves of a few higher plants, it was important to see whether or not the cycle prevailed throughout the plant kingdom. Calvin and Louisa and Richard Norris carried out experiments with a wide variety of photosynthetic organisms. In every case, although they found variation in the amounts of particular intermediates formed, the pattern was qualitatively the same. It also had to be shown that the pathway we had traced out is quantitatively the most important route of carbon reduction in photosynthesis. To this end Martha Kirk and I undertook an intensive study of the kinetics of the flow of carbon in photosynthesis. Our study has helped to solve other general problems, particularly the question of how carbon enters into the pathways leading to the synthesis of proteins and fats.

The biological materials for this work are supplied by an algae culture system under automatic feedback control. In this apparatus we

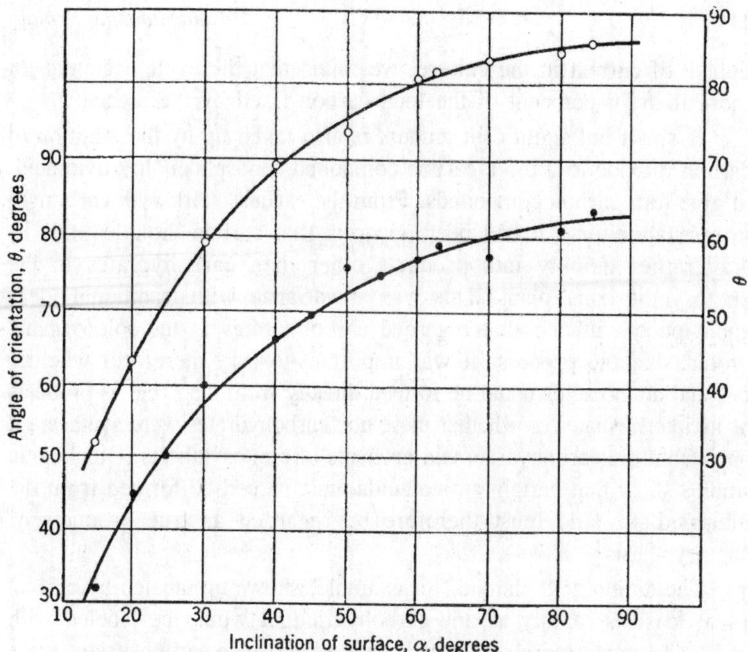

Fig. 5.3. Relation between the angle of inclination of plane (α) and the mean angle of orientation (θ) of caterpillar creeping upon it. Open circles are means for one individual; black circles are means for all individual tested. The curve is that for Δα/Δ log sin α=constant.

are able to maintain the photo synthetic process in a steady state, with nutrients supplied at a constant rate and with temperature, density, salinity and acidity held within narrow limits. At the start of a run we inject radioactive bicarbonate ion into the culture medium along with radioactive carbon dioxide gas and so bring the ratio of carbon 14 to carbon 12 immediately to its final level in both the gas and the liquid phase.

An automatic recorder measures the rate at which carbon is absorbed by the photosynthesizing cells. We take samples every few seconds and kill the cells immediately by immersing them in alcohol. After we have chromatographed the photosynthetic intermediates and measured their radioactivity we then plot the appearance of labeled carbon in each of these compounds as a function of time. By the end of three to five minutes, our records show, all the stable intermediates of the cycles are saturated with carbon 14. Taking the total amount of carbon thus fixed in compounds and comparing it with the rate of uptake of carbon in the culture, we found that the cycle accounts for more than 70 per cent of the total carbon fixed by the algae.

A small but significant amount is also taken up by the addition of carbon dioxide to a three-carbon compound, phosphoenolpyruvic acid, to give four-carbon compounds. From the earliest work with carbon 14 in our laboratory, it had been apparent that carbon dioxide finds its way rather quickly into products other than carbohydrates in the photosynthesizing plant. This was at variance with traditional ideas about photosynthesis that regarded carbohydrates as the sole organic products of the process. It was important to ask, therefore, whether fats and amino acids could be formed directly from the cycle as products of its intermediates or whether these non-carbohydrates were synthesized only from the carbohydrate end products of photosynthesis. Our kinetic studies show that certain amino acids must indeed be formed from the intermediates and must therefore be regarded as true products of photosynthesis.

The amino acid alanine, for example, shows up labeled by carbon 14 at least as rapidly as any carbohydrate; it would be labeled with carbon 14 much more slowly if it were made from carbohydrate, since the carbohydrate would have to be labeled first. We have been able to show that more than 30 per cent or the carbon taken up by the algae in our steady-state system is incorporated directly into amino acids. There is some evidence that fats may also be formed as products of the cycle. The discovery that plants make these other compounds as direct products of photosynthesis lends new interest and importance to the chloroplast, the subcellular compartment of green cell that contains pigments and the rest of the photosynthetic apparatus. It has been known for some time that this highly structured organelle is responsible for the absorption of light, the splitting of water and the formation of the cofactors for carbon reduction. More recent studies have shown that it is the site of the entire carbon-reduction cycle. Now the chloroplast emerges as a complete photosynthetic factory for the production of just about everything necessary to plant's growth and function.

6

Energy Cycle in Ecosystems

The Energy in Ecosystems

Productivity

Productivity is the amount of organic substance acquired by an individual, a population, or a system per unit time. The ecological concept, then, differs from that of agriculture, in which "production" is considered to be only that material utilizable by humans in some form. In the agricultural sense, any roots or above ground parts that cannot be used as food or fodder are ignored. Many misunderstandings in discussions between ecologists and agronomists are based on this discrepancy. A fundamental distinction must be made between the numbers of organisms in a habitat and the amount of organic matter they produce per unit time. A large population by no means implies high production.

Indeed, the suggestion has been widely defended that a large population, a large biomass per unit area, is a typical strategy by which organisms adapt to low nutrient availability; a large population is better able to bridge periods when food is scarce than a small one. The turnover—that is, the production—of such a population is comparatively small. It follows from this hypothesis that small biomass is characteristic of habitats with fairly large amounts of available nutrients, and this small biomass should show a high turnover. Habitats with large biomass and limited available nutrients are thought to contain very many species, whereas those with little biomass and high turnover can house but a few species. We can be quite certain that this hypothesis, in this simple form, is not generally applicable. It is presented here in order to underline the difference between biomass

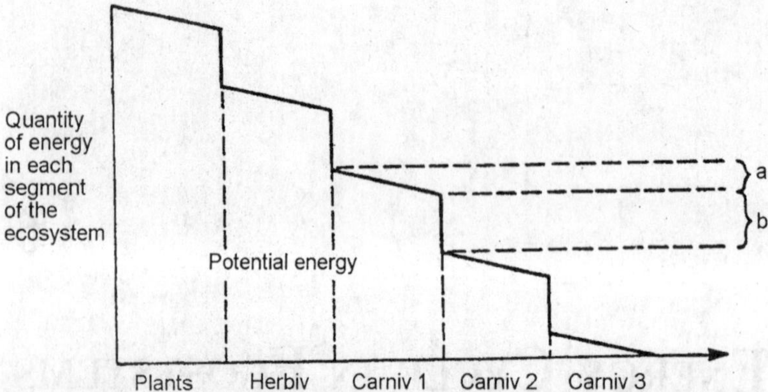

Fig. 6.1. Energy flow through the ecosystem. At each trophic level maintenance energy is lost as heat (a); energy is also lost as heat in each transformation from one trophic level to the next (b).

and production in a habitat, and because it is definitely worth considering, at least, for a number of habitats. In testing its validity, one will have to ask which kind of organism (plant, animal, microorganisms) the large population comprises, and here, again, divide the system into compartments. The amount produced, as a rule, is smaller than the mass of the producing organism. Consider a forest, for example.

The amount of organic matter contained in the stand of plants has, under some circumstances, been developed over centuries; the production in a single year is distinctly lower. But where small organism are concerned the relationships can be different. This is the case with bacteria, for example, and is an especially familiar feature of planktonic algae. The production of new algae per unit time is considerably higher than the stand of algae present in the body of water at any time. Production falls into two major categories.

One is primary production, by green plants and photosynthetically active bacteria, which produce organic matter from solar energy and inorganic matter. The other is secondary production, by organisms not capable of photosynthesis—bacteria fungi and animals, which convert the organic matter of the plants into the substance of their own bodies. Each of these categories can be subdivided into gross and net production. Using light, a plant produces organic from inorganic matter. This is gross production. But part of this matter is respired, used up by the plant's metabolism. Gross production minus respiratory losses is called net production. It is only the net primary production that we can realistically measure; gross primary production is usually simply a

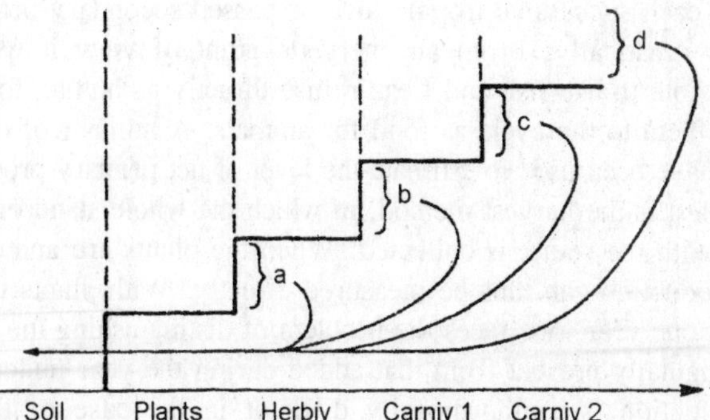

Fig. 6.2. The cycling of a chemical in the ecosystem assuming no erosion. Chemicals in organisms not consumed by the higher trophic level (a) to (d) slowly return to the soil as they are released through decay.

logical construct. Gross production and net production by animals are easier to distinguish. The respiratory losses of animals are relatively accessible to measurement, and the factors determining them are relatively well known.

Net production is of course the most important to ecosystem research, for it is this that can be passed on to the next trophic level. To humans, the animals of especial interest are those with net production not much lower than gross production. The ecological efficiency of animals has therefore been widely studied. For the reason the net primary production called simply primary production in the following discussion) is largely independent of the species of plant. Primary production of a system is affected chiefly by the duration of the growing season and by the water supply. In natural ecosystems it is usually a valid starting assumption that nutrients are present in adequate quantities. The nutrient poverty of many present-day cultivated areas has come about through decades or centuries of overuse.

Fertilization in many cases does no more than restore these areas to their original fertility. Water supply and length of the growing season, then, are a sufficient basis for a first-approximation estimate of primary production in land ecosystems. It is impossible to make direct comparison of this primary production of a system with a the secondary production. Secondary production by a primary consumer—a herbivore—is necessarily higher than that by a secondary consumer that feeds on this herbivore. And production by a tertiary consumer is necessarily still lower. The secondary production of an ecosystem therefore in every case approaches zero—a self-evident fact that need not be discussed further.

The significant point with regard to the human nutritional situation is that as each successive trophic level is passed secondary production decreases drastically. From an energetic point of view it would be more sensible to use fish and flesh refuse directly as human food than to return them to the cycle as food for animals. A number of different methods have been used to estimate the level of net primary production. The simplest is the harvest method, in which the whole stand of plants, together with the roots, is collected. When the plants are annuals, the year's production can thus be measured directly. With plants that live more that one year, one faces the problem of distinguishing the organic matter originally present from that added during the year under study. This distinction is extraordinarily difficult in the case of the root system. For this reason the net primary production given is often that of the parts above-ground alone.

Furthermore, the above-ground net primary production is the most important in general for zoological and ecological purposes. One way of estimating the annual growth is to harvest the stand both at the beginning and at the end of the growing season. More precise data are provided by the gas-exchange measurements. Part of a plant, or the whole plant, is enclosed in a cuvette within which the air is kept at the same humidity and temperature as that of the surroundings. Air is passed through the chamber and the concentration of carbon dioxide in the entering air is compared with that at the exit. The difference in the carbon dioxide that has been incorporated into the plant tissues.

The method is very exact, but because of the accuracy with which the microclimate within the cuvette must be controlled it is technically complicated and laborious. Production in water is measured chiefly by the dark-bottle method. Two bottles are filled with the water to be studied, and they are hung at the desired depth in the water. One of the bottles is made light-tight by covering it with aluminium foil. In the lighted bottle the planktonic algae continue to take up carbon dioxide, whereas in the dark bottle they only respire. The oxygen content of the two bottles is then determined by the Winkler method, and the difference between the two indicates the rate of photosynthesis and thus the amount of carbon dioxide incorporated. In bodies of water with abnormal chemistry the method is not very accurate; the milieu within the bottle changes. Moreover, the endogenous diurnal rhythm of the algae can affect the results.

In many cases, therefore, a more elaborate method using radioactive labelling has been adopted. Here the water is injected with $NaH^{14}CO_3$. Under the assumption that the planktonic algae assimilate $^{14}CO_2$ to the same extent as $^{12}CO_2$ and do not deposit ^{14}C in a special way, measurement of the amount of ^{14}C taken up by the end of the experiment can be used to determine the total amount of carbon assimilated. An objection to this method is, for example, the face that C_4 plants, because of the higher affinity of PEP for carbon dioxide, incorporate ^{14}C to a greater extent than do C_3 plants, so that errors can be introduced. Furthermore, N, P, or Si deficiency can rapidly develop in closed vessels. When this happens protein synthesis is largely blocked, while photosynthesis continues. These cells excrete a considerable percentage of their primary production in the form of soluble compounds; this does not appear in the incorporated ^{14}C activity.

The diversity in the methods of production measurement is necessarily accompanied by a variety of units for production. Sometimes it is expressed in g dry weight, and sometimes in the amount of carbon dioxide taken up. Because inorganic material is often deposited in plants, it is customary to indicate the amount of energy bound in the organic matter (Chart 6.1). Very extensive summaries of the energy content of organic matter are now available. From these data, knowing the general possibilities of error in the methods used, one can derive rules that allow one to omit certain measurements. On the average we can take it that 1g (ash-free) animal substance corresponds to about 5.6 kcal, and 1g of plant substance has about 4.6 kcal. The values for seeds are considerably higher. There is thus no single best method for determining primary production; under different conditions, different methods will be required. All of these are subject to error, and they are not directly comparable. However, the errors have now been reduced to an order of magnitude low enough that estimates of worldwide production is possible.

Although the determination of primary production, risky though it remains, is in principle so well founded methodologically that "recipes" can be given, the same cannot be said of secondary production. Here the difficulties begin as soon as one wishes to transfer the growth rate of a single animal, measured in the laboratory, to field conditions. In the field mortality in the population complicates the issue. Two opposed processes are operating, the growth—the production—of the individuals and the continual decrease in number of individuals by mortality. As has been discussed, the two process do not occur uniformly in time. In insects, for example, production is highest during larval development; the highest mortality occurs in the youngest stages. With species that

Chart 6.1. Data for the Conversion of Various Units in Production Ecology

Production is commonly indicated by a variety of units. So that approximate orders of magnitude can be estimated for purposes of comparison, a number of conversions are listed here. The comparisons themselves, of course, can also be only approximate.

1 g C corresponds to	2.2 g	organic plant dry matter
	3.3 g	dry matter of phytoplankton (i.e., including inorganic components)
	42 g	fresh weight of phytoplankton
	1.7 g	organic animal dry matter
	8.3 g	fresh weight of zooplankton

1 g plant dry matter, ash-free, corresponds to	4,000—5,000 cal
1 g animal dry matter, ash-free	5,000—6,000 cal
1 g carbohydrate	3,700—4,200 cal
1 g fat	9,500 cal
1 g protein	3,900—4,150 cal

1 g oxygen uptake corresponds to 3,280 cal
= 0.345 g fat
= 0.728 g plant matter
= 0.596 g animal matter

1 carbon-dioxide release corresponds to 1,800 cal

1 ml O_2 uptake corresponds to 4,687 cal

1 ml CO_2 release corresponds to 3,558 cal

1 cal=4.186 joule   1 joule=0.23889 cal

have sharply separated generations or clearly distinguishable stages in the life cycle, so that they can be subdivided into cohorts, regular monitoring of the population (concerning the difficulties of measuring a population) allows production to be estimated. But with species having many intermingled generations with ill-defined stages, and within which by no means all individuals proceed to reproduce—as appears to be the case among most mammals and birds, at least—things are considerably more difficult. Special procedures must be adapted to meet each case. An especially problematic task is that of establishing the number of microorganisms. Counts made with different straining methods in many cases give discrepant results.

Furthermore, such counts give no information about the species (and thus the special capabilities) of the organism, nor can one discern whether the cells counted were still alive when the sample was taken. Methods employing vital stains such as acridine orange have also proved

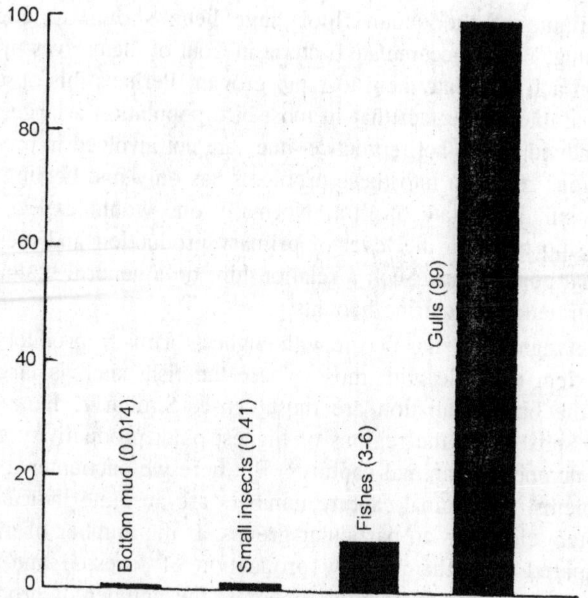

Fig. 6.3. DDT concentration in a Lake Michigan food chain. DDT load in gulls was about 240 times that of small insects.

not to be reliable enough. Recently researchers have turned to a method in which the content of ATP (a compound that breaks down almost immediately when an organism dies) is used to infer at least the biomass of the living microorganisms. This procedure assumes a constant relation between biomass and amount of ATP, which seems to hold only very approximately. In many cases no attempt has been made to estimate biomass; rather, the activity of certain key enzymes is used as the sole indicator of the activity of the microorganisms. Wieser and Zech (1976), for example, showed that there was particularly high dehydrogenase activity at the surface of sand grains on the ocean shore; in the interstitial water the activity of this enzyme, an important part of the electron-transport system, was much lower (only about 10%-20% of the total activity). And yet another problem is generally encountered. This is evident from the fact that, with even a minimal rate of division of the microorganism under the prevailing conditions, the consumption of matter ought to be much greater than it actually is (assuming that the number of bacteria derived by various methods are minimal values.) One can only conclude that the majority of microorganisms in the soil are inactive, whereas others seem to be excessively active.

Animals of the ocean floor have been shown to engage in "gardening." They accumulate bacteria in front of themselves by special secretion and stimulate them to rapid growth. Perhaps this observation is comparable to the fact that in most bird population a large fraction of the individuals do not reproduce—they are not involved in production. But serious research into these problems has only just begun, and the field is still in a state of flux. Normally one would expect a close relationship between the level of primary production and the density of animal populations. Such a relationship, on a general scale, can be demonstrated from marine habitats.

The regions of the ocean with highest primary production to a large extent coincide with those where the fish catch is largest and where the bird population are most dense. Similarly, Hinz showed that on Spitsbergen the regions of highest plant productivity gave the largest numbers of animal captures. But here we encounter a number of problems. No animal-capture numbers are absolute indicators, for they refer selectively to particular groups. If the number of mammals is compared with the primary production of forested and treeless regions, no sensible correlation emerges; the number of productivity of mammals in forests are considerably lower than in any region lacking trees, from the tundra to the steppes at tropical latitudes. In part this simply results from the facts that large mammals cannot move freely in the dense jungle and that most of the production of a primeval forest occurs in the crown stratum, out of reach of the large mammals. (Table 6.1) This interpretation accounts for the fact that a poor arctic tundra sometimes supports a hundred times as many mammals as a primeval forest in central Europe.

At the moment it is impossible to tell whether taking all animal groups into account would give appreciably different results. There is another reason why the level of primary production cannot be closely correlated with animal productivity; from one year to the next there can be extreme differences in the amount of accumulated plant substance, depending on the differences in the weather conditions from year to year. The number of animals present is not necessarily controlled by the same factors and in the same way as the productivity of plants. Thus there are differences in the rate of utilization of the primary production in different years. In central Europe a year very favourable to grasshoppers is not noticeable as such until the following year, when the new generation emerges from the abundantly produced eggs and begins to grow. Plants, by contrast, tend to respond immediately

Table 6.1. Degree to which primary production is utilized by warm-blooded berbivores

Type	*Harvestable primary roduction per ha per yr. kg dry matter*	*No. sheep per ha per year this energy can support*	*No. of animals per ha*	*Degree of utilization in % (rough estimate)*
Tundra Spitsbergen	(270-500 kg ha)	1	Reindeer 0.006 Wild geese (90 days) Ptarmigans Musk oxen	1-2
Steppe, Africa (Serengeti)	7,500 kg/ha ($30x10^6$ kcal)	20	0.5 animals of many species ($2.2x10^6$ kcal/ha/yr)	8
European mountains	9,000 kg/ha	24	Red deer 0.006 Roe deer 0.02 Hares	0.02
European mountains, herb and shrub strata only	500 kg/ha	1.4	Red deer 0.006 Roe deer 0.02 Hares	1.0
Hirta St. Kilda, maritime mountain meadow	2,000 kg/ha	5	Soay sheep 1.72	31
Carpathian mountain meadow	$6x10^6$ kcal/ha/yr	4	15 Muridae	1

to the prevailing conditions. The situation becomes quite difficult when comparisons over large geographical regions are attempted. Because there are no herbivorous ectothermic animals in the far north, these regions cannot be directly compared with stepes with respect to the ectothermic fauna. Nor can marine and terrestrial regions be compared.

On land the primary production is used by animals of direct interest to man (plant-eating birds and mammals). In the ocean, on the other hand, the fish interesting to humans are all secondary, tertiary, or quaternary consumers. They feed on large clams or fish; only very few species are able to live on planktonic crustaceans—that is, primary consumers. For this reason the production of animals important to man in the ocean must be well below that on land, even for a given level of primary production. Finally, special features may be overlooked when large-scale comparisons are made. Though primary production is about the same in the two bodies of water, the western Baltic Sea provides far higher yields of food fish than does the Skagerrak. The reason is that the two differ in salinity. The Skagerrak has normal marine salinity, with the normal marine fauna.

In the western part of the Baltic the salt concentration is reduced, and a great many animal forms have disappeared. These—irregular sea urchins, sea stars, and large gastropods—are competitors of other marine animals, especially clams. Clams (Cyprina islandica above all) are excellent food for the cod, whereas the animals that have vanished from the Baltic would not have been eaten by the fish in any case. Furthermore, there is the following consideration. Two bodies of water may exhibit identical levels of primary production, although in one the producers are very small planktonic algae and in the other very large species.

The very small algae in the first habitat can be utilized only by very small planktonic animals, which in their turn are the food of the larger planktonic animals on which fish feed. By contrast, the larger planktonic algae in the second habitat can be eaten directly by large planktonic animals; in this case the food chain is shorter by one link. At a rough estimate, this shortening increases fish production by a factor of 10. Where vascular plants on land are concerned, it is self-evident that only part of the production is normally used by animals, or that different parts of the plant tissue produced are used by different animals—some eat roots, others leaves, other seeds, and still others the nectar of flowers. Here the central issue is the part of the plant by which production is represented; the overall level of production is not the first consideration. For all these reasons, it is inappropriate to try to find a direct relationship between the level of primary production and the number of animals—especially if one wants anything but a very rough indication, and if one is interested primarily in the animals important as food for humans.

The differences in production within a system can be well illustrated by a food pyramid, in which the individual trophic levels are set one above another. But this form of presentation conceals awkward problems. Should the numbers of individual organisms be indicated, or their biomass? A great many Orchestes fagi can live on a single beech tree, but if the biomasses of the green leaves and of the beetles are compared, the picture looks quite different. And if production is judged entirely on the basis of physiological experiments, still another picture emerges. For example, a certain amount of hay can feed a certain number of rabbits for a certain period of time.

In theory, a certain number of martens can feed on this number of rabbits for the same length of time, and a certain number of wolves can live on these martens for the some period. Finally, it has unfortunately become accepted practice to limit such a pyramid to systematic groups; a certain number of plant-eating birds is associated with a certain number of carnivorous birds, and these with a certain number of birds that prey on the carnivores. This approach is unreasonable, as the example of the birds—hardly any of which are plant-eaters-emphasizes. The only kind of pyramid that should be published is one in which the net production at each trophic level is indicated. Such a "Production pyramid" is considerably more difficult to construct than one based on number of individuals or biomass, but has the advantage of being really unambiguous.

The "upside-down" food pyramid for open water should be dispensed with entirely. This structure has been derived from the fact that the biomass of the consumers in open water is greater than that of the primary producers, the planktonic algae. But the productivity of the planktonic algae—their rate of cell division—is so enormous that the situation, as expressed by a production pyramid, appears entirely normal.

Measurements of Stand

When a stand of plant or population of animals is to be quantitatively evaluated, a distinction is made between number of species and number of individuals. In general species number is relatively easy (though by no means simple) to determine. Models derived from the island theory of MacArthur actually allow one to estimate the number of species present fairly reliable on the basis of the area covered by region of interest, if the number of species in comparable regions is known as area increases the number of species is greater, even if the larger area contains no habitats. The determination of number of individuals—an indispensable item of information in ecosystem research—is far more difficult, especially in the case of land plants. Estimates of root mass, in particular, are frequently challenged. Only very recently have reliable data been published, chiefly by Kummerow.

In general rough approximations are offered, and these are not dependable; under different conditions a given plant can have quite different ratios of root-system size to phytomass above ground. There is therefore also a high probability of error in constructing a simple energy pyramid based on numbers of individuals per unit area (or their biomass). So far, there is no method for the quantitative determination of animals on land; enormous errors can occur. But we must know precisely the number of animals in an area if we want to say anything quantitative about the ecosystem. We also need to know the number of offspring that the animals in the system produce per unit time, and the mortality of the progeny (when bird's nests are deserted or robbed, for example, or caterpillars are infested by parasites), but this information is almost never available.

Indeed, it is practically impossible to obtain a simple count of the adult population of large, relatively conspicuous animals. A discouraging example is the recent attempt to quantify the roe deer population in the cultivated areas of central Europe. Danish studies, in which the area investigated was fenced off, showed that the real population was at least three times as large as all the other counts by experts had indicated. This fundamental discovery—that the true number is three to five times greater than the best estimates—may well apply to all the roe deer populations in central Europe. Even enclosed populations in which every individual is marked can never be completely counted by the professionals in charge of them. There is an annual cycle in "observability." If it is so difficult to obtain a reliable count of the roe deer in central Europe, a region relatively easy to inspect, one can imagine how far wrong the number given for other animals and birds may be—apart from animals so extremely rate that they deserve protection, such as the peregrine falcon, golden eagle, and eagle owl. The counts of the various mouse species per unit area so far available are exceedingly unsatisfactory. Almost all data on mass reproduction of mice are based on relative studies; traps are set up at the same spots year after year, according to a maintained pattern. The differences in number of animals caught are used as indicators of fluctuations in population density.

A method that ought in theory to work is to catch as many members of a species as possible, mark them, and then let them go. Then there is a second trapping session, and one can compute the population sizc from the ratio of marked and unmarked animals. But not animals are caught so easily a second time, not all can be permanently marked, and marking can put the animals at additional risk. And special difficulties arise where the animals are clustered, territorial, or randomly distributed—that is, everywhere. None of the techniques for counting insects gives anything like a satisfactory result. The most reliable counts seem to be obtained for terrestrial insects, those of the soil as well as in the herbaceous, shrub, or canopy strata, by collection by hand from a test area. It is necessary to distinguish precisely among the different species, and in many cases to distinguish the young and adult stages of a single species. Remember that animals of different size can vary greatly in metabolic rate and thus in the amount of food eaten per unit time and in productivity.

It is therefore pointless from the outset to lump animals at the same trophic level in a single biomass. Bearing all this in mind, we must regard the data in Table 6.1 as very rough indicators. There have been an especially large number of studies on the density of bird populations. But here, too, the variations are huge and there is a very high probability of error. Nevertheless, at present these bird counts are probably the most reliable data of all on the density of terrestrial animals. Because birds have been counted over a very long time in very many parts of the world, and because there are large numbers of extremely knowledgeable amateurs in the field of ornithology, estimates of bird population size have attained a special significance. Birds seem to be useful as "bioindicators;" routine monitoring of the bird population can provide evidence about the state of the environment. For this reason programmes have been instituted in several countries (the "Common Bird Census" in the British Isles, for example) in which quantitative determinations of the bird population are regularly carried out.

Despite the likelihood of error this method still gives us the best information about changes occurring on our planet. The situation in water seems to be simpler. Here the level of primary production evidently depends chiefly on the availability of nutrients and light. When larger amounts of nutrients are provided, productivity can be decisively increased. The number of algae in a body of water rises sharply, but at the same time the transmission of light through the water is reduced. At the deeper levels, where photosynthesis had been possible when minerals were scarcer, autotrophic algae can no longer exist. Now oxygen is used up at these depths, but no more is liberated. As a result, in the deeper parts of nutrient-rich waters anoxia kills off both animals and plants. The eutrophication of lakes and regions of oceans is dangerous because of the associated reduction in available oxygen. On land, where oxygen is always available, overfertilization does not have such consequences. The process in water can be followed closely by filtering a certain amount of water through a simple plankton net.

The number of individual planktonic algae and animals in a whole body of water can be readily derived from these results. Even planktonic microorganisms are now relatively easy to count. Bottom-dwelling organisms are also accessible; grabs can be used to bring to the surface a piece of bottom material of known area, and with enough samples of this sort the number of animals colonizing the bottom can be determined. As always in biology, the identification of species can be difficult, but here again the problems in the aquatic realm are slight compared to those encountered on land. For these reasons there have been many more studies of aquatic populations and ecosystems than of terrestrial ones. In many cases, the results have been applied directly to land communities; but it has turned out that such application is almost always incorrect. The situation on land is not only much more difficult methodologically, but the general functional principles of populations and ecosystems on land are more complicated than in the water.

Food Chains and Food Webs

The distributions of plants and animals derived from such measurement procedures can be arranged in terms of function; the ordering most commonly used is based on the flow of food. We can distinguish food chains, food webs, and—associated with these-trophic levels. Food chains are found primarily in habitats with very few species. In the eastern African Lake Nakuru, a soda lake, one species of blue-green algae [Oscillatoria (Spirulina) platensis] is the producer for two consumers, a copepod and flamingo. When the systems are more complicated, the simple food chain tends more and more to become a food web. This tendency can be followed very well in ponds of different size. In very complex habitats the networks become so complex that they cannot be traced in detail. Whereas the food chains consist of readily distinguishable links—primary producer, consumer I (which feeds on the primary producer), consumer II (which feeds on consumer I), and sometime still higher "trophic levels" —such distinctions are no longer possible in food webs. One can make at most an approximate subdivision into different trophic levels. Such food webs have been presented as descriptions of the "community nexus" in a wide variety of aquatic and terrestrial habitats. They have profound implications for the stability of habitats.

It is a priori clear that a system comprising a network has greater stability than one composed of independent chains, However, no such web has so far indisputably been shown to exist. Proof of its existence would require that the quantitative relationships be known. A hypothetical food web as it is classically presented, and with it—hypothetically!—the precisely quantified food relationships. The quantification reveals that in actually we are dealing not with a web, but with chains. In reality, the netlike arrangement plays no role at all. The frequently drawn conclusion that web structure implies stability, then, is not reliable in the absence of quantification. Because no

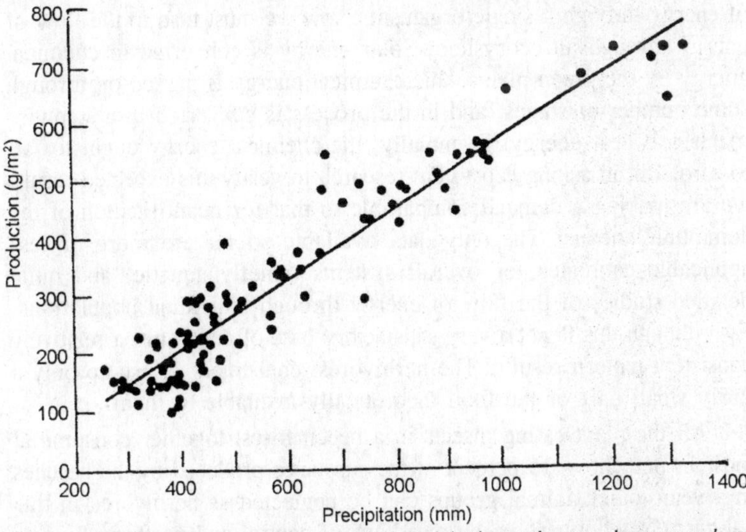

Fig. 6.4. Relationship between mean annual precipitation and mean above-ground net primary production for 100 major land resource areas across the central grassland region.

ecosystem on earth has been described in really quantitative terms, there is as yet no confirmation of the web structure. Before the question of stability can be argued further, we must apply ourselves to quantification. Such studies as are available do show that stands of cultivated plants are less vulnerable to mass outbreaks of "pests" when they are not monocultures, but comprise two or more species (for example, the maize and sweet-potato plantations in Costa Rica).

The leaf beetle Phyllotreta cruciferae, which is a severe pest, on cabbages in the USA, did considerably less damage and occurred in considerably less dense population when the cabbage fields also contained other plants, such as tomatoes, tobacco or Ambrosia ambrosia artemisiifolia). The reason was not a decrease in the number of predators or parasites, but the fact that the insects find their food plants by following the chemical stimuli the plants release; these are not as easily recognizable in the polycultures. At the same time, this example shows that associations based on food relationships and thus on energy flow are only one facet of the complex ecosystem; in this case food relationships are drastically altered by parameters of sensory physiology.

Energy Flux

The ecological significance of food supply, we considered the flow of energy through a single organism. Now we must turn to the flow of energy through an ecosystem. Solar energy is converted to chemical energy by the green plants; this chemical energy is passed on through some number of stages, and in the process is converted in a stepwise manner to heat energy. Eventually, the chemical energy ought to fall to zero. But in asking ecosystem research to verify this expected result, we are making a demand comparable to that for quantification of the community nexus. The only data available so far are more or less applicable estimates for overall systems (chiefly aquatic) and quite detailed studies of the flow of energy through individual populations. Even though this is not a very satisfactory base of departure, a relatively consistent pattern results. The herbivores, consumers 1, use up only a fairly small part of the food theoretically available to them.

All the plant-eating insects in a beech forest together consume at most 10 percent — 15 percent of the substance produced by the beeches in a year; other animal groups can be neglected as herbivores in this habitat. Grasshoppers in the meadows of central and northern Europe consume no more than 10 percent of the plant tissue formed. The warm-blooded herbivores of the tundra—reindeer and musk ox above

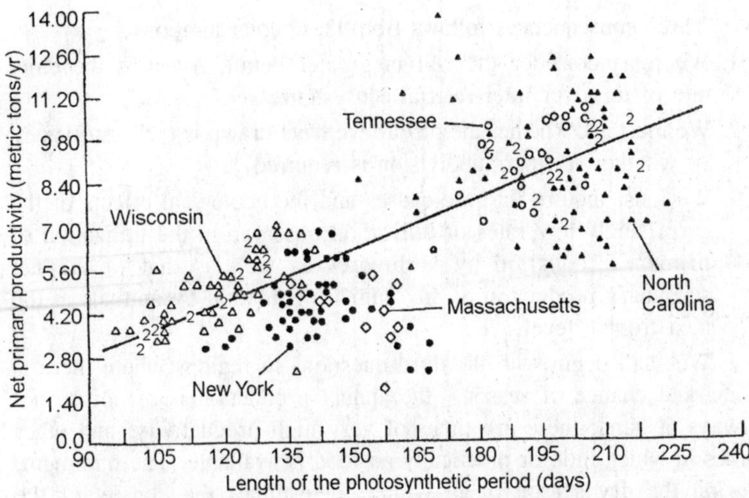

Fig. 6.5. Relationship between net primary production and the length of the growing season for stations in the eastern deciduous forest region of North America.

all—utilize only 5 percent—10 percent, and the large grazing animals on the African steppes barely exceed 10 percent utilization. Still less of the plant matter in the forest is consumed by mammals, for the main productive stratum, the canopy, is out of their reach. Small mammals in general consume barely 1 percent of the net primary production; only in a very few cases can this figure rise to as much as 6 percent. However, particular parts of the plants, such as the seeds, can be consumed in higher proportions -30 percent, 75 percent or even 90 percent.

It is generally true that on land a maximum of 10 percent–20 percent of the substance formed by plants is directly consumed by herbivores. All the rest eventually (very much later, in the case of trees) enters the food chain responsible for the decomposition of dead matter. By far the greatest energy turnover on land, therefore, as chemical energy is used up, occurs on the soil layer. One must bear in mind the excrement, exuvia, and other things derived from the bodies of the first consumers (hair, feathers, shed antlers) are also passed on directly to the food chain in the soil. Now the question arises, how large a percentage of the net production of the first consumers enters the secondary consumers—predators and parasites? Here, again, the value seems in general to vary between 5 percent and 15 percent. The greater part of the substance produced by the first consumers, then, also enters the soil stratum, from the dead bodies.

Three consequences follows from these considerations.

1. We must consider the soil in greater detail, for it is the chief site of turnover inter-restrial ecosystems.
2. We must ask whether the picture we have drawn is really sufficient, or whether, a finer subdivision is required.
3. We must inquire into the causes and the ecological effects of the surprisingly low rates of utilization observed-in the utilization of primary production by herbivores as well as that of the net secondary production of any animal population by animals at the next trophic level.

We shall begin with the third question. In regions where there is a marked change of seasons, the annual production is of course not always at hand; there are times of very high productivity, and other times in which little or practically no food is available. The minimum, during the dry season or in winter, determines the density of the large-mammal population during the main production season, and thus the degree of utilization. The longer or more severe the winter or dry season, the smaller the percentage of production likely to be utilized. This tendency is to seem extent compensated by migrations, like those which used to bring red and roe deer down from the mountains in winter. In the tropical rainforests there are no seasons. But here the productive zone, the crowns of the trees, is largely inaccessible to the herbiverous mammals.

In treeless regions where persons are absent or not well marked, it may be that the proportion utilized by mammals is much higher. This could happen in temperate-zone heaths near the coast, where because of the prevailing winds trees cannot grow. The situation is similar on high-altitude steppes near the equator; here the very small territories of the vicuna suggest that their rate of utilization is very high. Furthermore the small mammals and insects with high rates of reproduction appear unable to "keep up" with the rapid new growth of vegetation in the spring. They are directly dependent on plant food, and cannot survive warm periods without food. Their lives would be endangered if they were to emerge before the leaves sprout in the spring. If they appear after leafing-out has begun—even immediately after—as a rule the plants have gained an advantage in development that cannot be overcome. But if leafing-out and the appearance of the insects are simultaneous, the trees can easily be stripped bare; this happens, for example, during mass outbreaks of June beetles. A yet we cannot specify a percentage utilization in such cases. All the trees so damaged in temperate zone can put out new leaves and exhibit full production after the June beetles have dies out. However, another example shows that mass outbreaks in natural regions of vegetation can have extremely dangerous results.

In the Finnish and Swedish birch forests, mass outbreaks of the moth Oporinia autumnata cause widespread destruction; the forest boundaries in many cases are pushed back by 30 km. Ultimately we must admit that the causes of the low rate of utilization of the permanently available food reserves in the tropical rainforest by herbivorous insects are not known. Though low utilization is the rule, we must always be prepared for surprises. In the two examples that follow, plant-eating animals use up nearly 100 percent of the primary production. The first of these is the only native herbivore of central Iceland, the pink-footed goose. The contemporary populations of this bird here are so dense that they consume nearly all of the vegetation of the few oases during the breeding season. This is possible only because the geese can migrate so rapidly; the arrive only after the vegetation has sprouted and leave after they have eaten everything up.

Another likely contributing factor is the very favourable conditions under which this population spends the winter. The losses that probably were once routinely suffered in the overwintering region no longer occur. The second example is provided by the studies of Reichholf on the lakes produced by damming the lower reaches of the Inn River. He showed that the large flocks of ducks that overwinter there consume almost all the aquatic plants. This happens only when the animals are not disturbed during the winter, so that they can stay in the region all day and night. Here, again, the high rate of utilization is possible only because the birds migrate, arriving after the vegetation has grown tall and leaving the lakes when the water plants have been grazed off. The consequences are important ecologically. The energy contained in the plants is respired above the water by the ducks. If it were not for the ducks, the plants would sink to the bottom in the winter and rot there, withdrawing oxygen from the water in the process. Without the ducks anoxic conditions would develop; with them, they do not. In this sense the ducks are responsible for the large stocks of fish in the lakes.

In evaluating energy flow, one encounters a special problem with regard to the food not used by the animals. Here, we do not mean the part that leaves the gut undigested—although this element is frequently underestimated. For example, the dipteran larvae feeding in acid beech forests chew up about 30 percent of the litter from the stand, whereas they actually utilize only a few percent. What is meant by "unused food" is the amount destroyed, but not consumed, by animals during feeding. The larvae of the alder-leaf beetle eat only the photosynthetically active layer of the leaf; most of the leaf, with all the veins, remains behind and turns brown. The amount of food eaten is thus considerably less than the amount of plant matter destroyed. In this respect there are large differences among species, so that it is impossible to present a generally valid picture. Grasshoppers and hawkmoth caterpillars, for example, all begin to eat a leaf at its tip. These animals consume all the plant matter they destroy.

Let us turn to the second question, whether these results really suffice to qualify the biocenotic interactions. The answer is an unequivocal "no". A much more refined subdivision of the system is necessary for such quantification. We cannot speak of "the primary production" of which a certain percentage is utilized. It is possible, for example, that certain species, or certain parts of a plant, are 100 percent utilized, whereas others are essentially ignored by the herbivores. Oaks at the edge of a forest are exploited to a considerably greater extent by all sorts of herbivores than beeches similarly located. Willows, poplars, and black-thorns are relatively frequently and obviously attacked by plant-eating insects, whereas pines and spruces are heavily infested by insects only in exceptional cases. But these are qualitative observations, which have not been quantified. Among the large ungulates of Africa, as we have seen, the different species exhibit strict specialization to the different strata of vegetation. But even if we distinguish among species we have gained relatively little.

The individual plants are also differentially vulnerable to herbivores. These differences are in part genetically based, and in part associated with the habitat. Alders or spruces in dry habitats (Or in dry years) tend to be attacked by herbivores more extensively than they are under moist conditions. And finally, the separate parts of a plant are used as food by herbivores to widely differing degrees. So far we have been discussing the parts of plants that grow above ground, disregarding the fact that there is about the same phytomass underground. We know nothing about the degree to which this part is utilized. Moreover, different above-ground organs are formed in different years. The irregular production of seeds by forest trees is an example of this effect. In some years the accumulated production of many years is evidently channelled into seed formation. As yet we cannot say whether more plant matter is consumed by herbivores in a seed year than in those when no seeds are produced.

Presumably consumption is higher in a seed year, but it would make sense to give a percentage figure only if a mean value for many years could be cited; voluminous acorn production is possible only because reserves have accumulated in the oak tree during many previous year. Most of the organic matter formed by the plants is not consumed by herbivores, but eventually reaches the ground. Normally, these substances are continuously decomposed and worked into the soil. In terrestrial habitats, therefore, by far the highest energy turnover occurs in the soil stratum. In spruce and pine forests the amount of litter falling per year has been estimated as 1,500—3,000 kg per hectare, and in good mixed deciduous forests of central Europe as 4,000 kg/ha or more. In a balanced system, then, this amount must be decomposed each year; otherwise organic matter would accumulate on the forest floor. This is not to say that a particular beech leaf is decomposed in the course of a year.

In general, a fallen leaf in an acid beech forest requires about 5 years before it is no longer visually identifiable. This decomposition is not a rapid and complete breakdown into CO_2, H_2O, and minerals. The process is slow and involves complicated resynthesis, in the course of which humus is produced. Humus substances, because of their great stability, are of very special significance in the preservation of soil structure. As far as the participation of organisms in the breakdown of litter is concerned, there are both qualitative and quantitative considerations. In view of the enormous number of soil-dwelling animals it is a priori likely that they play a central role in decomposing plant matter.

Generally, for practical purposes, macrofauna are distinguished from microfauna; the former is understood to include chiefly the earthworms, enchytraeids, isopods, diplopods, and large insect larvae that in Europe are especially important in the breakdown of litter. Evidently the litter is first eaten by the larger animals, which can attack even the less digestible elements with the aid of microbes. Isopods, snails, and dipteran larvae are equally important. The significance of these primary decomposers is evident; in their absence, the rate of decomposition is almost halved. During this first passage through a digestive system, little energy is withdrawn from the material. All animals take up very large amounts of fallen leaves and produce very large amounts of excrement; they utilize the litter relatively poorly. Nevertheless, this first passage is of fundamental importance to the system, as follows:

1. The water-binding capacity is increased. The excreta of the primary decomposers holds moisture more uniformly than the original litter.
2. Having been comminuted, the material becomes accessible to a large number of so-called secondary decomposer—primarily springtails, mites, and nematodes. The excreta of some primary decomposers are more readily consumed by the secondary decomposers than those of others, so that if the process of decomposition is to proceed evenly there must be an undisturbed, well-balanced community of animals in the soil.
3. The excreta of both primary and secondary decomposers provide particularly good conditions for the development of microorganisms, above all actinomycetes and bacteria. Their presence can be inferred by comparison of the metabolic activity (CO_2 release) in fallen leaves with that in excrement.

This first passage through the gut is followed by repeated passage, through the same or other animals. Wieser (1965) showed that isopods flourish particularly well when they can eat their own excreta over and over again. The microorganisms break down those plant parts that are hard to digest, and the animals are nourished primarily by these microorganisms. Would decomposition proceed more rapidly if only the microorganisms were present? After all, the activity of the animals appears to cause a considerable reduction in the numbers of microorganisms! This is the sort of false conclusion that is so commonly encountered in analysis of predator-prey systems.

If the animals were not present, the microorganisms would very rapidly reach an extremely high population density and then almost stop reproducing; their cultures would be in the stationary phase. By virtue of the fact that their population density is kept low, they maintain maximal production-and thus optimal performance—over the long term. The importance of the comminution of litter is also shown by certain experiments done in Russia, and others by Herlitzius and Herlitzius (1977). Completely intact beech foliage on acid forest soil is very resistant to attack and hard to break down. The excrement of leaf-eating beetles that falls down from the canopy and the dead bodies of these beetles provide nuclei in which microorganisms can begin to grow. During the repeated passage through animal bodies, easily decomposable substances are withdrawn from the substrate.

The carbohydrates are predominantly broken down into carbon dioxide and water. The percentage of the most important plant nutrients (nitrogen, phosphorus, potassium) rises, and lignin and tannins—products of the polymetization of phenols—accumulate in the excreta. In this neutral or weakly alkaline milieu these substances come onto close contact with nitrogenous metabolites produced by the animal decomposers. The conditions thus favour the microbial breakdown of the lignins to phenols, at a rate that depends on the amount of nitrogen available to the microorganisms. The analysis of humus fractions from different soils, together with model studies, shows that in the milieu described, when nitrogen and sufficient oxygen are present, phenolic substances are oxidized to oxyquinones, which polymerize to form larger molecules.

The process is accelerated by phenoloxidases of the microorganisms. The end result is the production of humic acids—polymers of complicated structure which are extraordinarily stable. Breakdown of humic acids in the soul occurs very slowly; it is likely that the chief agents of their decomposition are mycorrhiza fungi. Humic acids, then, are an extremely stable element of the soil, especially because they can form clay-humus complexes by combining with aluminium and iron hydroxides. These complexes ultimately determine the fine structure of the soil, the friability required for good ventilation and water retention. The proportion of clay-humus complexes in the soil is thus responsible for soil fertility—at least in the majority of soils. The quantitative division of the labour of remineralization, between herbivores and soil organisms, has probably resulted from a many-faceted coevolution that has led to optimization of the system. When the decomposition of litter proceeds as described, the result is the ideal form of humus, mull. But it does not take the same course in all habitats; there are wide differences in rate of decomposition in different types of forest.

The quality of the humus, the rate of breakdown of leaf litter, and the number of litter consuming animals in a habitat are closely related. Comparison of three forest habitats with decreasing humus quality (mull on lime soil, mull on acid soil, and mor on acid soil—the last is a very unfavourable form of humus) showed that the rate of decomposition decreased in this sequence, and the number of animals was related about as 3:1-5:1. Soil pH and the organisms supported are frequently correlated, and even though the dominant plant species are in many cases the same (beech on chalk and variegated sandstone;

pines on sand and chalk, birch forest on quite different geological substrata in Greenland, Iceland, and Scandinavia) the differences are striking—particularly with respect to the density of the animal species present. A number of other factors are associated with pH, which together make acid soils generally poorer in nutrients. Humus, as an organic substance, does not continue to exist indefinitely. It in turn is eventually broken down by microorganisms. Naturally, this process occurs especially rapidly under hot, humid conditions (in a tropical rainforest). Although humus is not a plant food, in most soils it is essential to plant nutrition. The humic acids can absorb molecules that otherwise would soon be leached out by the rain. Moreover, humus substances can retain large amounts of water.

Humus therefore guarantees the plants a steady supply of minerals and water. The destruction and disappearance of humus in many soils is responsible for a decline in primary production. In principle, the same relationships are found in water; only a relatively small fraction of he producing planktonic algae is consumed as living cells by herbivores. Most of the algae die, sink toward the bottom, and are eaten in deeper water by consumers in the detritus food chain. On land as in water, then, we have what has been called a Y-shaped model of energy flow. It is biocenotic interrelationships which, while unquantified, appeared to be a very complex food web, and found that quantification produced two nearly independent food chains; although the latter was a purely hypothetical model, the above considerations show it to be quite realistic. Most of the production in water is by phytoplankton rather than by sessile plants.

It is therefore restricted to water levels near the surface. The more transparent the water, the deeper the producing layer can be. The higher the nutrient content, the more dense the phytoplankton , so that light penetrates less deeply and the producing layer is shallower. In this sense a large body of water can be compared with a forest, where production is also restricted to the upper surface; in the shaded region near the ground the chemical energy formed is used up. In deep bodies of water where oxygen is abundant the rain of detritus is essentially all consumed before reaching the bottom. This is generally true in the oceans; in the Baltic Sea, for example, it has been documented by analyses of protein content. In such waters the bottom contains very little organic matter, and as a result the fauna is relatively diverse and in many cases consists of fairly large animals which presumably develop very slowly. In waters with little oxygen at depth consumption occurs anaerobically or not at all; the bottom becomes covered with decaying mud in which H_2S is formed.

Oxygen deficiency occurs in the depths when the water becomes opaque because the nutrient content is too great, and/ or when the deeper water of an ocean or take is never exchanged with the more superficial water because autumn and spring storms fail to churn the water sufficiently. Under such conditions the water becomes increasingly uninhabitable down to a certain depth. At great depths more and more organic material accumulates. These large-scale relationships probably can serve as models of the origin of petroleum. Conditions are like this in the Black Sea, and it is feared that they will become so in the Baltic. But the situation in the Baltic is complicated by irregular inputs of oxygen-rich, high salinity water from the North Sea; these masses of water, being more saline and thus heavier that the Baltic water, move into the Baltic along the bottom, displacing the water that is low in oxygen and salts. So far this process has repeatedly brought about complete decomposition of the deposited organic matter. In shallow bodies of water the organic matter produced reaches the bottom before it is decomposed.

The abundance of this food source is evident in the large numbers of filter-feeding animals that inhabit fresh water and the shallow parts of the oceans—barnacles (Balanus), mussels, bryozoans, a variety of polychaetes, and many others. Let us consider these relationships in detail. Primary production is entirely performed by unicellular algae of several sorts—peridinaeans, diatoms, and the very small flagellates of the nannoplankton. In cooler regions or seasons the diatoms predominate, whereas under warmer conditions they are less in evidence. In warm oceans primary production depends heavily on the nannoplanktonic flagellates. The significance of this fact is enormous, even though the level of primary production may be identical in the two cases.

The organism of the nannoplankton are so small that they escape not only ordinary plankton nets but also most of the animals that eat plankton; only highly specialized animals, usually very small, are capable of utilizing the nannoplankton. The best known example is the mantle of the planktonic tunicate Oikopleura, an extremely complicated structure within which the animal encloses itself. The mantle presents a very fine-meshed lattice to the flow of water generated by the tunicate, and the nannoplankton that accumulate on it are used by the animal as food. In fact, it was this observation that first led to the discovery of the nannoplankton, which until that time had been overlooked. The nannoplankton, then, serve as food for small to very small planktonic animals, and these in turn are eaten by larger zooplankton. But if the primary production is by relatively large phytoplankton, the large planktonic animals can utilize it directly. By the rule of thumb that only about 1/10 of the energy is retained at each transition from one trophic level to another, it is clear that this extension of the food chain when the nannoplankton are the primary producer must have a drastic effect on secondary production by larger inhabitants of the open water; the necessary consequence is a paucity of fish.

Recent studies indicates that the poverty of tropical oceans is not necessarily all a matter of mineral deficiency; to some extent it results entirely form the fact that the nannoplankton are responsible for primary production. (Evidently this nannoplankton population is extraordinarily resistant to a great variety of influences, so that if it were to replace the primary producers in other parts of the ocean fish production, and thus the world's fisheries, would be endangered). For this reason the precise analyses of the requirements and reproductive rates of phytoplankton currently underway should have high priority. The ocean's planktonic animals comprise two fundamentally different groups. The animals in one group spent their entire lives in the open water, whereas those in the other are the larval forms of bottom-dwellers. The latter are naturally seldom encountered in the large oceans.

Planktonic larvae are characteristic of the smaller seas—North Sea, Baltic Sea, and Mediterranean, for example-but are almost nonexistent in fresh water. The planktonic animals in lakes spend their whole lives in the plankton, with at most certain temporary stages (as happens among the water fleas and their relatives) resting on the bottom. There are probably great differences in the food requirements of the zooplankton, as has been suggested by the studies of Lampert. But we know almost nothing about them. The next trophic level is represented by small or some what larger plankton-eating fishes, such as herring (Clupea) and whitefish (Coregonus), and these in natural systems are eaten by large predatory fishes, mammals and birds. Here, again, we may take it that in general only about 10 percent of the algae are directly ingested and digested by animals (ingested algae, if undamaged, are often passed out again in the excreta and then simply continue to grow).

Yet another factor is involved. Planktonic algae "lose" a large part of their photosynthates (though there is no reliable, quantitative estimate of how large it is) into the water, where they drift about as relatively short-chain carbohydrates, as fatty acids, and as amino acids. These losses are greater if, while photosynthesis is proceeding rapidly under strong light, certain substrate materials fall below a critical concentration; then subdivision can no longer occur. If the nitrogen concentration is below the critical level no more amino acids and protein molecules can be synthesized, but carbohydrates continue to be produced, and these must be eliminated.

As a result, the water accumulates large amounts of free organic molecules. The same thing happens when planktonic organisms die or are injured. For this reason the amount of dissolved organic compounds in the ocean is considerably higher than that of all organic compounds bound in plants and animals. The compounds dissolved is fresh water are all taken up, incorporated and broken down by bacteria; in the ocean there are many organisms that are either obligatory or facultative feeders on such dissolved organic compounds. In water, then, the formation of a soil such as occurs on land is neither necessary nor possible. The principle of energy flow in water is based on complete breakdown of organic matter. There is no question of humus formation. If organic matter is deposited on the bottom and not decomposed, the first result is a decaying sludge; then firm deposits of organic matter are formed which eventually can become fossil fuels (peat, coal, petroleum). These regions are quite unlike humus soils-they are extremely hostile to life, and practically devoid of animals. Bacteria and protozoans are the sole inhabitants of such reducing sediments.

Significance of Animals in an Ecosystem

Animals are much more conspicuous in water than on land. On land they are surrounded by higher plants and are barely apparent, whereas on the bottoms of bodies of water they play a leading role. This is especially true of the ocean; the subdivision of marine bottom-dwelling animals has been done according to principles very like those used to subdivide terrestrial regions on the basis of plant communities. The large, well-known associations in the ocean are the pure animal communities of the coral reef; the benthic animal and plant communities hardly enter the macroscopic picture. On land, by contrast, animals in general contribute little to the flux of energy and the cycling of materials. Rarely do herbivores consume more that 10 percent of the energy flux by the plants. For this reason, many ecosystem analyses have given only cursory treatment to the animals. Not a few researchers—particularly botanists interested in productivity and system analysts interested in the system as such—are inclined to deny the animals any appreciable role in the ecosystem. But nothing could be further from reality.

An animal like the rhoe deer, which preferentially consumes the buds of trees, can dramatically reduce the production of a forest; the rabbit, an equally selective feeder, in population of normal density can keep the stands of many special plants down the extent that they are no longer ordinarily encountered. After myxomatosis had drastically reduced the rabbit population in the British Isles certain wildflowers appeared in all sorts of places where people had been looking in vain for years. Beavers in a ecosystem can back up large lakes behind their dams; if the dams break extensive meadows appear in otherwise continuous woodland. There is not much quantitative data available, and that has been taken from only a few animals.

Indeed, effects such as those described can hardly be expressed quantitatively or included in treatments of energy flow or of the cycling of matter. Cycling and energy flow are but two facets in the intricate pattern of an ecosystem. We shall now present a few examples that illustrate the complex involvement of animals on ecosystems. These must be anecdotal and in some cases have not been completely worked out, but they reveal the difficulties inherent in this line of research.

Leaf-eating beetles evidently cause no decline in production in a deciduous forest. Because the beetles eat holes in the upper leaves more light reaches those below, so that these can proceed with photosynthesis at a greater rate. Here loss and profit are in balance. In fact, there is an added benefit. The excrement of the beetles, which contains finely ground up leaves and in many cases is full of minerals, as well as the dead beetles, laden with nitrogen and phosphorus, fall to the ground and act like nuclei of crystallization—centres from which the decomposition of the leaf litter proceeds must more rapidly and efficiently than it would without these highly nutritive particles. The leaves shed from the trees contain very few nutrients even, for bacteria and fungi, and are excessively hard to digest. Only by way of the excreta and bodies of animals are they vulnerable to attack. With these aids, decomposition can occur rapidly; without them it would take so long that natural rejuvenation of the woods would be nearly impossible, for a thick layer of undecomposed leaves would pile up. The beetle is thus a fundamental element in the system. But its influence can be detached only by laborious and extremely meticulous research; without such research its effects might be preserved only when it was 80 or 100 years too late. Moreover, the effect varies on different types of soil.

On chalky soils, where the decomposition of organic matter proceeds relatively rapidly, the contribution of the beetle corpses is less important than on acid sand and silicate soils. Because plant-eating insects preferentially infest trees in which the sap flow is not completely intact, aging trees with lowered productivity, but which are still highly competitive, tend to be invaded and killed. Competition is thus reduced for the younger trees, and in the long term productivity of the system is increased.

Aphids withdraw from their host plants large amounts of fluid containing sugar. These sucking insects also require nitrogen, present in much smaller concentrations in plant sap; by the time they have acquired enough nitrogen they have taken in far too much sugar, and this is excreted in the form of "honeydew". Are the plants seriously damaged by this loss? Experiments have shown that nitrogen-fixing bacteria can be brought to peak performance by aqueous solutions of sugar; after application of aqueous sugar solution the fixation of atmospheric nitrogen increases sharply. It is possible that the following system has developed by coevolution. The plants pass on the surplus carbohydrates they have formed to the sapsuckers, which pass most of these carbohydrates on to the soil and thus activate the bacteria to produce more of an element that is ordinarily a limiting factor for plant development. That is, the plant releases a surplus product in order to obtain adequate amounts of a factor in minimum. In such a case we would be dealing not with a plant and a pest, but with a system which-as a system-has been optimized.

Spittlebugs and their relatives can occupy tropical forests in such crowds that a continual fine rain drips from them onto the ground. They withdraw huge amounts of water from the plants. With what results? The soil of tropical rainforests contain almost no nutrients. Only the uppermost layer, the leaf litter, contains minerals that can be used by the plants, but in general this layer is too dry for the plant roots to penetrate.

Because of the continual sprinkling by spittlebugs, some roots can invade even this top parts of the soil and thus provide the plant with nutrients, whereas the greater part of the root mass, deeper in the soil, is responsible exclusively for the provision of water. In this case, too, plant and sucking insect together would represent an optimized system. Studies of soil animals and litter decomposers have shown that they feed chiefly on bacteria, which carry out the acute decomposition. Without these animals the bacteria would be much more numerous. It would appear, then, that the soil animals inhibit decomposition of the litter. But the opposite is actually true. By their constant eating of bacteria, as has clearly been shown for Orchestia, these animals keep the population of bacteria in the exponential phase of growth. If the soil animals were not present the bacteria would rapidly increase to capacity and enter the stationary phase, in which they become practically inactive.

By keeping the population density low the animals ensure the highest productivity of the bacteria—that is, the greatest effectiveness in breaking down the litter. In the last analysis this example is simply another expression of the predator-prey relationship we found for the red grouse. In that case, too, the activity of the predator kept the rate of grouse reproduction as high as possible; without predation, the grouse reproduction rate falls off dramatically.

The lemmings in the arctic tundra show oscillation in population size, with mass reproduction bringing a peak every 3-5 years. The plants over large areas of tundra can be killed when the animals cut through their roots. But without the turnover of the soil that occurs at regular intervals by the large numbers of burrowing lemmings, breakdown of the ground litter in these arctic regions would take an exceedingly long time. The outbreaks of lemmings are necessary if organic material is not to accumulate to excess. The dung produced by the cattle introduced into the Australian steppes could not be naturally re-mineralized there. It accumulated on the ground, often drying into a hard crust, and destroyed the vegetation.

Moreover, it provided a haven for stinging flies that had also been introduced by accident and proved severely annoying, and it gave the parasitic worms passed out by the cattle in their dung ample opportunity to re-infect them. In 1963 the Australian government began a massive effort to get rid of this nuisance. Many dung beetles were subjected to the most through tests, and finally—from April of 1967 on—about 275,000 beetles of four species were released. One species (Onthophagus gazella) reproduced and spread at a spectacular rate. Wherever this beetle settled in well, the cattle dung was worked into the soil in about two days (there the beetle larvae continued to feed on it). Two other African species have reproduced to a similar extent in other parts of Australia, with similar effects.

The danger of parasites is very much less, and the troublesome flies have almost disappeared. But at the moment the additional release of mites is under consideration. All dung beetles normally carry mites with them; these move into the dung, reproduce there, and then attach to the next beetle to be carried to the next pile of dung. The European burying beetle Nicrophorus is also associated with mites. Its larvae as a rule are subordinate in competition with fly maggots, but the mites of the beetle attack the fly's eggs—the competitors of the beetle larvae are eliminated. The relationship between the dung beetles and their mites is thought to be similar, and therefore experiments are currently underway to provide the dung beetles with normal mites. No one pays much heed to dung beetles encountered in the countryside; but they can be vital to the functioning of the system, as the Australian example shows. For reasons obvious in these examples, Mattson and Addy (1975) do not regard the animals of an ecosystem in isolation., but rather consider the whole as a system for long-term optimization, with uniform operation over centuries. The example presented, in which the ducks consumed the primary production almost completely and thus prevented a lake from reverting to land, is an argument in this direction.

But the principle of optimization becomes especially apparent when we consider the implications of animals as flower pollinators and seed-dispersing agents. Here they determine the structure of the plant community, altogether their function could not be adequately comprehended in the context of energy flow or the cycling of matter. Remarkably, these functions of animals—despite their outstanding significance to plant communities—are often neglected in botanically oriented analyses of ecosystems. Plants growing in stands comprising a few species, or in natural monocultures, usually are wind-pollinated.

The greater the number of species in a stand, and the more nearly random the distribution of the individuals of each species, the more the species are forced to adopt a better-aimed method of pollination, by animals. A logical consequence of pollination by animals is seed dispersal by animals. This connection has rarely been noted; Regal (1977), in particular, has called attention to it. When many species are distributed in a mosaic throughout a system, individuals have a good chance of survival only if the seeds can be transported over considerable distances. Thus, in quite general terms, we find that wind pollination prevails in the northern tundras and in the forests of northern and temperate latitudes, where the number of species are small, as well as in windswept steppes. Even in the temperate-zone forests pollination by insects and seed transport by birds gains in importance, and in subtropical and tropical forest regions animals are indispensable to most of the plants.

It is true that many plants, whether pollinated by wind or animals, can exist for quite long times (sometimes for several generations, by parthenogenesis or self-pollination) without normal pollination. But in the long run an exchange of genes is necessary, as it is in animals. Just how necessary is evident in the complicated coevolution that has occurred between plant and insect. Examples have been presented in many research reports and handbooks. The enormous energy expenditure by the plants is understandable only in this context. Similar instances of coevolution to ensure seed transport are not known to such a great extent, and studies of this aspect have been fewer. Evidently real specialization, so common where pollination is concerned, is rare in the case of seed disposal. But the fact that many plant seeds (mistleote, tomato) germinate with difficulty or not at all unless they have passed through the body of an animal points in the direction of coevolution. Under heavy grazing pressure most dicotyledonous plants disappear and are replaced by grasses and their relatives. When cattle are allowed into a wood to feed, it becomes overgrown with grass.

It is quite certain that grasses, with their ability to grow continually, become so widespread on as a result of the activity of grazing mammals. Grazers can convert a plant community containing many species into a quite different, much less diverse community. A good example is given by the meadows in coastal regions which are sometimes flooded by the sea—the North Sea, for instance. Here many species grow naturally, but when sheep are kept on the meadows most of them disappear. Only meadow-grass (Puccinnellia), with its persistent growth, can withstand the pressure of grazing. The very dense grass vegetation that is produced under the influence of the sheep holds the soil together much more firmly than did the original salt-meadow vegetation. For this reason sheep are a valuable aid in converting these coastal fields into land that can be used for agriculture. The natural vegetation allows channels to open in the ground from time to time, so that conversion to dry land is much less likely; the system is maintained.

Another outstanding way in which plant-eating animals can affect ecosystems is by transmitting plant diseases (fungus and virus diseases). Bark beetles, long-horned beetles, and sap-sucking insects are disease carriers. Previously this fact has been viewed entirely as an injurious aspects of insects, so that at present nothing can be said about its significance in a balanced ecosystem. An ecologist would infer that such diseases do the most damage to already weakened plants, thus making room for other species that find conditions more favourable at that location. The spread of such pathogens would then be regarded as a mechanism by which turnover in the ecosystem is accelerated. But this is pure speculation, unsupported by any data. The examples discussed so far have illustrated the significance of animals to the ecosystem as a whole. The importance of many animal species to the existence of other animals has been known for much longer, and indeed is self-evident.

Woodpeckers that dig out holes provide breeding places for stock doves, owls, hoopoes, rollers, and tits, as well as for the dormouse and many insects. By their burrowing activities foxes in the forest, marmots and prairie dogs in the steppe, and rabbits in the savanna and at the edges of forests provide a continual supply of fresh soil; this is colonized by pioneer plants that otherwise could not exist in such mature stands. The rich fauna in the Florida Everglades results from the presence of alligators. These animals dig deep pits for themselves, where all sorts of other water animals–fish, turtles, insects-can shelter during the dry season, while at their margins even sensitive water plants can survive hot dry periods. If It were not for the alligators the water birds would find no food during the dry seasons, and the aquatic animals for which the Everglades are famous would die out.

In the mud flats of the North Sea the factor restricting colonization by sessile marine animals is lack of an adequate substrate. Only a few species are capable of living directly on the mud. One of these is the oyster. Where oysters appear, they are followed by a great number of sessile species which can now attach to the oyster shells. It was the observation of this relationship that led to the discovery by Mobius of the phenomenon of the biocenosis, and thus to the actual beginning of the science of ecology. These examples should illustrate the point sufficiently, but let us mention just a few more of the many others that could be cited.

The big Chilaces typhae lives in cattails, but can enter the reeds only if the caterpillars of a moth have bored passages into them. Some parasitic wasps cannot find their host by themselves; they follow the trails of other species that prefer the some host and, having found their victims in this way, proceed to lay their eggs. Their larvae develop more rapidly than those of the first wasps, and thus win the competition against the actual discoverer of the host. Given all these possibilities, it is quite inappropriate to limit an ecosystem analysis to energy flow and the cycling of matter. To do so would be like making a physiological study of an animal in which nothing but feeding and digestion were considered.

The functions of animals described here determine the composition of the ecosystem, the level of primary production (of which they than consume 10%), and the form in which this primary production is delivered (i.e., by which species and which parts of the plant). The action of animals in an ecosystem can be compared with that of switches and amplifiers in an electronic system, or with the sense organs and nervous system of an individual organism. But these effects are very difficult to represent numerically. It is just as futile to try to express in numbers the way a system is affected by the removal of a pollinator or seed transporter as it would be to find a numerical expression for the loss of an individual's vision, hearing, or sense of smell. On land, too, animals can determine the entire structure of their system.

Particularly striking examples are the islands colonized by sea birds, which build up large deposits of guano in arid regions and can thus be an economic factor of major importance. Seabird colonies situated in trees can kill the trees with their droppings; cormorants have been known to do this. In the water, animals dominate the structure of the habitat as the higher plants ordinarily do on land.

7

Community Dynamics

The previous chapter stressed the fact that a community is a functional entity: this complexity of structure and organisation lacks a dimension, until it is seen in operation for it is a system which only exists in motion. Clearly the operation of any community is a function of the interaction of the various processes affecting its component organisms: competition; parasitism predation and so on, and these individual processes will each be examined in detail in later chapters. But to stick first with our 'overview' let us consider the operation of the community as a whole: the net result of all these separate interactions combined.

This is, of course, somewhat more easily said than done, and for many years ecologists lacked this overview. They had, founded their science on meticulous observation and field recording; with such an approach, a picture of community function only could arise as the product of the interactions of its component species. It was not possible to study the dynamics of the whole, for suitable techniques were simply not available; even within the community it was often not possible to form more than a qualitative, subjective, estimate of the roles of the various organisms, and there, were many problems that could not be tackled at all, or others which, through their complexity, could only be partially resolved. In short, both in terms of understanding relationships within the community, and in understanding the dynamics of the whole, ecologists were looking for a more rigorous unifying approach, a method of quantification with respect to some very basic common theme. And they took their cue from the theoretical physicists, by' turning to a study of energy flux.

Community as a System of Energy Transformations

It is indeed valid to consider any biological system-individual population or community-as a system for the transfer, storage and dissipation of energy. It is, what is more, a system that obeys all the basic laws of thermodynamics: that energy can neither be created nor destroyed, and that each transfer, or transformation of energy cannot be 100 per cent efficient: energy will be lost and there will be an increase in entropy at each stage. Animals and plants during their lifetimes act as stores of potential energy. In feeding or in being fed upon, that energy is transferred to and from other organisms. Not all the energy is retained as potential energy within the body: the life processes of maintenance-basal metabolism, or whatever you like require energy expenditure; still more energy is released in 'work'. Thus some of the original energy is dissipated from the system in respiration.

Within an ecological system, of course this 'energy' is not free energy but is the chemical energy incorporated into the structure of complex organic materials-the energy of molecular bonds. Animals, and certain heterotrophic plants (fungi and many bacteria) take in their energy ready-stored within the organic molecules of their food, but what is the ultimate source of energy for the ecological system? For this we must look back to the Producer level: to those autotrophic organisms that are able to synthesise organic compounds from inorganic precursors, using free energy in the process. Certain bacteria use energy derived from the low potential chemical energy of is the single most important chemical substance in the physiology of the ecosphere.

Each year the oceans evaporate a quantity of water equivalent to an average depth of one meter. The total evaporation from land and bodies of fresh water is one sixth of the evaporation from the sea, and at least one fifth of this evaporation is from the transpiration of plants growing on land. The grand total of water evaporated annually is roughly 100,000 cubic miles, and this must be roughly the annual precipitation. The precipitation on land exceeds the evaporation by slightly over 9,000 cubic rules, which therefore represents the annual runoff of water from land to sea. It is astonishing to me to note that more than one tenth of this total runoff is carried to the sea by just two rivers—the Amazon and the Congo.

Precipitation supplies nonmarine organisms with the water which they require in large quantities. Protoplasm averages at least 75 per cent water, all plants require something like 450 grams of water to produce one gram of dry organic matter. The water moving from land to sea also erodes the land surface and dissolves soluble mineral matter. It brings to the plants the chemical nutrients that they require and it tends to level the land surface and deposit the minerals in the sea. At present the continents are being worm down at an average world-wide rate of one centimeter per century. The levelling process, however, apparently ha; never gone on to completion on the earth. Geological uplift of the land always intervenes and brings marine sediments above sea level, where the cycle can begin again.

The rivers of the world are now washing into the seas some four billion tons of dissolved inorganic matter a year, about 400 million tons of dissolved organic matter and about five times as much undissolved matter. The undissolved matter represents destruction of the land where organisms live, but the dissolved materials is of greater interest, because it includes such important chemicals as 3.5 million tons of phosphorus, 100 million tons of potassium and 10 million tons of fixed nitrogen. In order to say what these losses may mean to the biosphere we must review a few facts about the chemical composition of the earth and of organisms.

Every organism seems to require at least 20 chemical elements and probably several others in trace amounts. Some of the organisms requirements are rather surprising. Penicillium is said to need traces of tungsten, and the common duckweed demands manganese and the rare earth gallium. There is a European pansy which needs high concentrations of zinc in the soil, and several plants in different parts of the world are so hungry for copper that they help prospectors to find the mineral. Many organisms have fantastic abilities to concentrate the necessary elements from dilute media. The sea-squirts have vanadium in their blood, and the liver of the edible scallop contains on a dry-weight basis one tenth of 1 per cent of cadmium, although the amount of this element in sea water is so small that it cannot be detected by chemical tests.

But the exotic chemical tastes of organisms are compratively unimportant. Their main needs can be summed up in just five words-oxygen. carbon, hydrogen. nitrogen and phosphorus, which account for more than 95 per cent of the mass of all protoplasm. Oxygen is the most abundant chemical element on earth, so we probably do not need to be concerned about any absolute deficiency of oxygen. But nitrogen is a different matter. Whereas protein, the main stuff of life, is 18 per cent nitrogen, the relative abundance of this element on the earth is only one 10,000th of the earth's mass. It is apparent that our land forms of life could not long tolerate a net annual loss of 10 million tons of fixed nitrogen to the sea. Fortunately this nitrogen loss from land is reversible, so that we can speak of a "nitrogen cycle." Organisms in the sea convert the fixed nitrogen into ammonia, a gas which can return to land via the atmosphere.

Carbon also is not in too abundant supply, for it amounts to less than three parts in 10,000 of the total mass of the earth's matter. But once again the biosphere profits from the fact that carbon can escape from the oceans as a carbon dioxide. This gas goes through a complex circulation in the atmosphere, being released from the oceans in tropical regions and absorbed by the ocean waters in polar regions. Because some carbon is deposited in ocean sediments as carbonates, there is a net loss of carbon from the ecosphere. But there seems to be no danger that a shortage of this element will restrict life. The atmosphere contains 2,400 billion tons of carbon dioxide, and at least 30 times that much is dissolved in the oceans, waiting to be, released if the atmosphere should become depleted. Volcanoes discharge carbon dioxide, and man is burning fossil fuels at such a rate that he has been accused of increasing the average carbon dioxide content of the atmosphere by some 10 per cent in the last 50 years, In addition, lots of limestone, which is more than 4 per cent carbon dioxide, has been pushed up from ancient seas by uplifts of the earth.

The story of phosphorus appears somewhat more alarming, This element accounts for a bit more than one tenth of 1 per cent of the mass of terrestrial matter, is enriched to about twice this level in plant protoplasm and is greatly enriched in animals" accounting for more than 1 per cent on the weight of the human body, As a constituent of nucleic acids it is indispensable for all types of life known to us. But many agricultural lands already suffer a deficiency of phosphorus, and a corn crop of 60 bushels per acre removes 10 per cent of the phosphorus in the upper six inches of fertile soil. Each year 3.5 million tons of phosphorus are washed from the land and precipitated in the seas. And unfortunately phosphorus does not escape from the sea as a gas. Its only important recovery from the sea is in the guano produced by sea birds, but less than 3 per cent of the phosphorus annually lost from the land is returned in this way.

It must agree with agriculturalists who say that phosphorus is the critical, limiting resource for the functioning of the ecosphere. The supply is at least shrinking (if dwindling is too strong a word) end there seems to be no practical way of improving the situation short of waiting for the next geological cycle of uplift to bring phosphate rock above sea level. Perhaps we should also worry about other essential elements, such as calcium, potassium, magnesium and iron, which behave much like phosphorus in the metabolism of the ecosphere, but the evidence clearly indicates that if present trends continue phosphorus will be the first to run out.

This brings me to the close of a very superficial summary of dome of the physiological processes of the ecosphere. There are srastic oversimplifications in this treatment; the importance of some processes may be overestimated and others (e.g., dumping sewage in rivers and oceans) may not have received enough attention. The figures for the total quantity of energy received by the earth, for total annual precipitation and for the total supply of some chemical elements may overlook the very irregular distribution of these resources in time and space. Much solar energy falls on deserts and fields of snow and ice where it cannot be used by plants, and much precipitation arrives at unfavourable seasons or in such torrents that it does more harm than good to organisms.

Our survey suggests that man may be justified in feeling some real concern about the problem of erosion. It should also make us aware of the important role played by organisms that we might otherwise ignore or even regard as pests. The dung beetles, the various scavengers and the termites and other decomposers all play important bit parts in this great production. At least six diverse groups of bacteria are absolutely essential for the proper physiological functioning of the nitrogencycle alone. Man in his carelessness would probably neither notice nor care if by some unlikely chance his radioactive fallout or one of his chemical sprays or fumes should exterminate all of the microorganisms that are capable of decomposing chitin. Yet, as we have seen, such a tragedy would eventually mean an end to life on earth.

Finally, it is interesting to ask how large a role man plays in the physiology of the ecosphere. The Statistical Office of the United Nations estimates the present human population of the earth at 2.7 billion persons. Each of these is supposed to consume at least 2,200 metabolizable kilocalories per day. This makes a total food requirement of 22×10^{14} kilocalories per year. I have estimated that all of the plant growth in the world amounts to an annual net of all 5×10^{17} kilocalories, of which not more than 50 per cent is metabolizable by any primary consumer. Thus if man were to feed exclusively on plants he would require: almost exactly 1 per cent of the total productivity of the earth. To me this is a very impressive figure. There are more than: one million species of animals, and when just one of these million species can cover 1 per cent of the total food resources, this form is truly in a position of overwhelming dominance. The figure becomes even more impressive when we reflect that 70 per cent of the total plant production takes place in the oceans, and that our figure for productivity includes inedible materials such as straw and lumber.

8

AIR POLLUTION

The atmosphere forms an insulating blanket around the earth. Without it the temperature at the equator would rise to 180 during the day and drop as low as—220°F at night. It burns up meteors that would bombard the surface of the earth from space. Without the atmosphere there would be no sound and no flight. There would be no conventional long-distance radio communication, for this is dependent on the electrons in the upper atmosphere. Without air there would be no lightning, no clouds, no wind, no rain, no snow, and no fire. The surface of the earth would be as bleak and sterile as the moon. The atmosphere

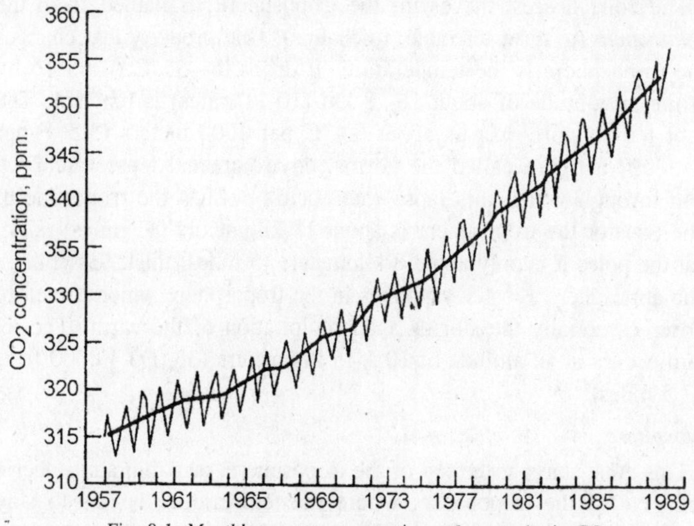

Fig. 8.1. Monthly mean concentrations of atmospheric CO_2.

shields of the earth from lethal concentrations of ultraviolet radiation. It selectively filters the sun's rays so that only two small segments of the electromagnetic spectrum penetrate to the earth's surface in appreciable amounts.

One segment is the *optical window*, consisting essentially of the visible spectrum of light, from near-ultraviolet to near-infrared. The other is the *radio window*, consisting of radio waves from about 1 centimeter to 40 meters in length. It is impossible to define the limits of the atmosphere because the atmosphere becomes progressively tenuous with increasing distance from the earth. There is no boundary between the atmosphere and the void of outer space. However, 75 percent of the earth's atmosphere lies within 10 miles or the surface and 99 percent of the atmosphere lies below an altitude of 30 kilometers (19 mites). The total mass has been estimated at 5,500 trillion tons.

The Zones of the Atmosphere

Based somewhat on the way in which the temperature of the atmosphere changes with increasing altitude, scientists have defined zones of the atmosphere. Each zone is more or less a spherical shell of the atmosphere. The thickness of a zone varies with latitudes and with the season of the year.

Troposphere

The zone nearest the earth, the troposphere, is named from the *Greek tropein* (to turn, to rotate, to change). One property that changes in the troposphere is the temperature. It drops to—55°C (—65°F) by the time an altitude of about 16-18 km (10-11 miles) is reached. The rate of temperature drop is about 6.4°C per 1000 meters (3.5°F per 1000 feet), a figure called the normal environmental lapse rate. The region through which this lapse rate occurs defines the troposphere. At the equator the troposphere is about 18 kilometers (11 miles) thick, but at the poles it is only about 8 kilometers (5 miles) thick. Essentially all the atmosphere's water vapour is in the troposphere which therefore encloses essentially the storms and precipitation of the earth. The jet stream occurs at an altitude of 10.5-15 kilometers (35,000-40,000 feet; 6.5-7.5 miles).

Stratosphere

This zone starts at the top of the troposphere, at a thin atmospheric "shell" called the tropopause, where the temperature begins to stay fairly constant, increasing slightly, in going to an altitude of 50 kilometers (31 miles). The stratosphere includes much of the *ozone layer*. The warming trend in the stratosphere is caused by a cycle of chemical changes in the ozone layer which converts all the incoming high energy ultraviolet rays to heat (see Ozone). If these rays reached the earth's surface life as we know it would be impossible. The *sratopause* marks the narrow zone at the top of the stratosphere where the temperature begins to fall again with increasing altitude as the mesosphere is entered.

Mesosphere

The mesosphere, on top of the stratosphere, extends roughly to 80 kilometers (50 miles), and its temperature drops to between—80 and—90°C at the mesopause, a thin zone where the temperature stabilizes and soon starts to rise again as the *thermosphere* is ascended.

Thermosphere

This zone is above 80 kilometers (50 mites) altitude. The temperature becomes very high but the air has such a low density that it can hold very little heat. Any, denser object in the thermosphere will be extremely hot in sunlight but very cold at night. The zones just discussed may be defined by changes of temperature with altitude. Changes in chemical composition or activity, also occur with altitude. Up to about 80 kilometers (50 miles) the proportions of components given in Table 8.1. remain nearly constant, and the long zone up to that altitude is sometimes called the *homosphere*.

Table 8.1. The Gaseous Composition of Unpolluted Air (Dry Basis)

	ppm (vol)	*μG/m³*
Nitrogen	780,900	8.95×10^8
Oxygen	209,400	2.74×10^8
Water	—	—
Argon	9,300	1.52×10^7
Carbon dioxide	315	5.67×10^5
Neon	18	1.49×10^4
Helium	5.2	8.50×10^2
Methane	1.0-1.2	$6.56\text{-}7.87 \times 10^2$
Krypton	1.0	3.43×10^3
Nitrous oxide	0.5	9.00×10^2
Hydrogen	0.5	4.13×10^1
Xenon	0.08	4.29×10^2
Organic vapours	ca. 0.02	—

Within it is a region called the *chemosphere* extending from the upper troposphere where, a huge number of chemical changes occur powered by solar radiation. Ozone production is one such activity. Above the homosphere lies the *heterosphere*. Fot its first 40 kilometers (25 miles), the ionosphere, the very thin concentration of atoms and molecules exists largely as electrically charged particles called ions. This globe-enveloping band of ions reflects outgoing radio waves back to earth, making radio transmission possible well beyond the horizon. Television waves are not reflected, and a TV transmitter's range does not extent beyond the horizon Meteorites entering the earth's atmosphere generally out in the ionosphere.

The stratosphere has been of interest to aeronautical scientists because it is traversed by airplanes; to communication scientists because of radio and television communications; and to air pollution scientists because global transport of pollution, particularly the debris of above ground atomic bomb tests and of volcanic eruptions, occurs in the stratosphere and because absorption and scattering of solar energy also occurs in the, stratosphere. The troposphere has been the region in which we live and is the region to which this book is primarily devoted.

Atmospheric Pressure (Barometric Pressure)

A column of air whose base is one square inch and which extends outward beyond the thermosphere weighs at sea level 14.7 pounds on the average, when the surface air temperature is 0°C.

Standard Atmosphere

1. An arbitrarily agreed-to vertical distribution of atmospheric pressure, temperature, and density used in altimeter design and, ballistic calculations 2. A pressure unit with a value of 1013.2 millibars, 29.9213 inches of mercury (750 millimeters of mercury) the standard atmospheric pressure at 0°C and under the standard value of the gravitational constant, 980.665 centimeters per second.

Unpolluted Air

The gaseous composition of unpolluted tropospheric air is given in Table 8.2. Unpolluted air is a concept, i.e., what the composition of the air would be if man and his works were not on earth. We will never know the precise composition of unpolluted air because by the time we had the means and the desire to determine its composition, man had been polluting the air for thousands of years. Now even at the most remote locations, at sea, at the poles, in the deserts, and mountains, the air may be best described as dilute polluted air. It closely approximates unpolluted air, but differs from it to the extent that it contains vestiges of diffused and aged man-made pollution.

Table 8.2. The Gaseous Composition of Unpolluted Air (Wet Basis)

	ppm (vol)	*μG/m³*
Nitrogen	756,500	8.67×10^8
Oxygen	202,900	2.65×10^8
Water	31,200	2.30×10^7
Argon	9,000	1.47×10^7
Carbon dioxide	305	5.49×10^5
Neon	17.4	1.44×10^4
Helium	5.0	8.25×10^2
Methane	0.97-1.16	$6.35\text{-}7.63 \times 10^2$
Krypton	0.97	3.32×10^3
Nitrous oxide	0.49	8.73×10^3
Hydrogen	0.49	4.00×10^1
Xenon	0.08	4.17×10^2
Organic vapours	ca. 0.02	—

The real atmosphere has been more than a dry mixture of permanent gases. It has other constituents-vapour of both water and organic liquids; and particulate matter held in suspension, Above their temperature of condensation, the vapour molecules act just as permanent gas molecules in the air. The predominant vapour in the air has been water vapour. Below its condensation temperature, if the air is saturated, it changes from vapour to liquid. We are all familiar with this phenomenon since it appears as fog -or mist in the air, and as condensed liquid water on windows and other cold surfaces exposed, to the air. The quantity of water vapour in the air gets varied greatly from almost infinite number of possible organic compounds that may be in the air have been identified. The major constituents of air, nitrogen (78%), oxygen (20.94%) and argon (0.93%), do not react with one another under normal circumstances.

Similarly, the trace components helium neon, krypton, xenon, hydrogen and nitrous oxide have little or no interaction with other molecules. A number of other gases also present in trace quantities have been not inert chemically but interact with the biosphere, the hydrosphere and each other and so have a limited residence time in the atmosphere and characteristically variable concentrations (Table 8.3). It is this reaction group of gases which are considered pollutant

Table 8.3. The Composition of Dry Air in the Lower Troposphere (Free of Water Vapour)

	Chemical Symbol	*Concentration (a) %*	*Calculated Residence Time*
Principal Gases			
Nitrogen	N_2	73.0	Continuous
Oxygen	O_2	20.9	Continuous
Argon	A	0.93	Continuous
Carbon Dioxide	CO_2	0.032(b)	20 years (c)
Trace Gases			
(a) Permanent Gases (non-reactive)		Ppm	
Helium	He	5.2	Continuous
Neon	Ne	18.0	Continuous
Krypton	Kr	1.1	Continuous
Xenon	Xe	0.086	Continuous
Hydrogen	H_2	0.5	?
Nitrous Oxide	N_2O	0.25	8-10 years
(b) *Reactive Gases*			
Carbon Monoxide	CO	0.1	0.2-0.3 year
Methane	CH_4	1.4	< 2 years
Non-Methane Hydrocarbons	HC	0.02	?
Nitric Oxide	NO	0.2 to 2.0×10^{-3}	2.8 days
Nitrogen Dioxide	NO_2	0.5 to 4.0×10^{-3}	
Ammonia	NH_3	6 to 20×10^{-3}	1-4 days
Sulphur Dioxide	SO_2	0.03 to 1.2×10^{-2}	1-6 days
Ozone	O_2	0 to 0.05	?

Notes:

(a) This is the atmospheric background concentration, and not the concentrations found in polluted areas. When a range of concentrations is given, it indicates that these have been measured by different workers at different places.

(b) Minimum concentration of CO_2 measured away from centres of population. In population centres, CO_2 concentrations vary from about 0.034 to 0.035 percent.

(c) For photosynthesis. Turnover time with the deep ocean is in the order or centuries

? Indicates that little is known about the residence time of the gas.

when they are produced by, man in sufficient quantities for the background concentrations of Table 8.3 to be significantly exceeded. The most important gases in this group have been those which are universally present in the air of the world's cities, namely, sulphur dioxide (SO_2), nitrogen oxides (NO and NO_2, carbon monoxide (CO), and non-methane hydrocarbons. Other reactive gases can also bring about pollution problems at elevated concentrations, for example, the halogen gases chlorine and fluorine and their acid derivatives hydrochloric and hydrofluoric acid, but these problems tend to be local rather than universal, and their background concentrations an order of magnitude or more less than the gases reported in Table 8.3.

Polluted Air

Man's activities take place, for the most part, on the earth's surface within the first 2 km of the atmosphere. The pollutants produced by these activities get injected directly into the troposphere where they are mixed and transported. The background concentrations of the reactive gases have remained, to the best or our knowledge, constant with time.

It implies that the sources and sinks (as the formation and removal processes are generally called) are in balance, and also, for gases

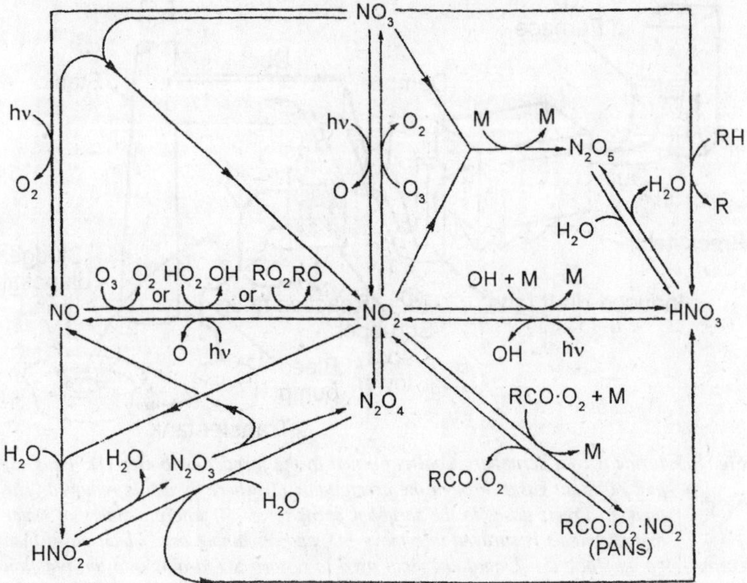

Fig. 8.2. The principal tropospheric chemical transformations of NO_x.

with a high pollutant contribution, that the sinks are able to cope with the additional burden from man. It could be partly explained by reactivity and partly by the fact that these gases natural source greatly exceeds the pollutant one in the majority of cases (Table 8.4), It can be seen from Table 8.4 that nature produces over 10 times as much hydrogen sulphide (H_2S), at least an equivalent amount of nitrogen oxides (NO and NO_3 written as NO), and over 100 times as much ammonia (NH_3) as is produced by man. Sulphur dioxide (SO_2) looks like an exception to the rule, however. Hydrogen sulphide (H_2S) has been ultimately converted to sulphur dioxide (SO_3) in the atmosphere and is, therefore, a source of SO_3. When it is taken into account, together with the different relative molecular masses of H_2S and SO_2, it can be demonstrated that the natural and man-made sources have been equivalent. Pollution problems associated with the gasses of Table 8.4 (CO_2 excepted) arise not because of the magnitude of the man made (anthropogenic) emission but because this emission gets concentrated in the areas where people live and work, and more specifically in the cities of the industrial world.

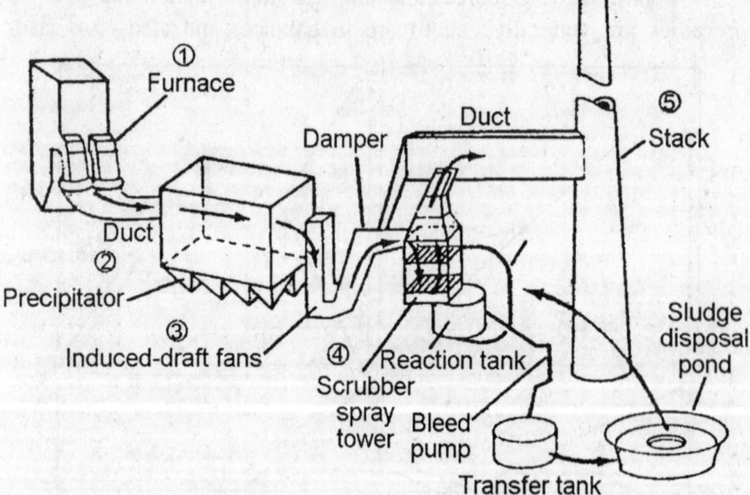

Fig. 8.3. Inside a coal scrubber. Coal is burned in the furnace or boiler (1). Fans (3) pull resultant gases through the precipitator (2) where fly ash is removed. The damper directs gases to the scrubber spray tower (4) where a slurry of water and chemicals is sprayed to remove SO_2 and remaining ash. Clean gases then go up stack (5). Liquid chemical used to absorb SO_2 drains into the reaction tank where sulfur is removed through a chemical process. The bleed pump routes it to the transfer tank from which is drains into the sludge disposal pond.

Table 8.4. Source of air pollutants

Gas	Source Major Pollutant Sources (Anthropogenic Sources)	Natural Sources	Pollution	Quantity ($X 10^4$ toms per annum) Natural
1	2	3	4	
Sulphur Dioxide (SO_2)	Combustion of coal and oil, roasting of sulphide ores	Volcanoes	146	6-12
Hydrogen Sulphide (H_2S)	Chemical processes, sewage treatement	Volcanoes, biological action in swamps	3	30-100
Carbon Monoxide (CO)	Combustion, principally motor car exhausts	Forests fires terpene reaction	300	>3000
Nitrogen Oxides (NO_2)	Combustion	Bacterial action in soils	50*	60-280*
Ammonia (NH_2)	Waste treatment	Biological decay	4	100-200
Nitrous Oxide (N_2O)	Indirectly from use of nitrogenous fertilizers	Biological action in soil	>17	100-450
Hydrocarbons	Combustion, exhausts, chemical processes	Biological processes	88	CH_4 : 300-1600 Terpenses : 200
Carbon Dioxide (CO_2)	Combustion	Biological decay, ocean release	1.5×10^4	15×10^4

*Expressed as tons NO_2

Further, most of the world's industry has been concentrated in the northern hemisphere (over 90%), and the great majority of this between the latitude of 30°N and 60°N. In this region, the anthropogenic emission has been considerably more important than the natural one, and the potential effects have been more serious because of the concentration of the receptor of most vital interest to us—man himself. Natural emissions have been only rarely concentrated in limited areas in this way and when they are, as in the case of volcanoes, their intermittent nature and scattered location have been found to minimize the effects of their reactive gas emissions. With the exception of volcanoes, natural emissions do not get fluctuate significantly from year to year, although they may fluctuate considerably within a year. Man's emissions, on the other hand, have been steadily increasing as populations and industry expand.

The estimated worldwide production of the past century of one such pollutant-sulphur dioxide was about 5 million tonnes, mainly from coal combustion. Today, about 190 million tonnes get produced, of which about 100 million is from coal (which generally contains about 0.5 to 4 percent sulphur), 50 million tonnes from the refining and burning of oil, and the remainder chiefly from the smelting of copper, lead and zinc ores. Similar curves could be drawn for all other anthropogenic gases, despite the increased use of pollution, control measures. In the long term our most serious pollution problems have been not arising from the reactive gases but from unreactive emissions, like carbon dioxide which are not having harmful interaction, with living systems.

Like the reactive pollutant gases the emission of carbon dioxide is increasing from year to year; this increase, in contrast with the

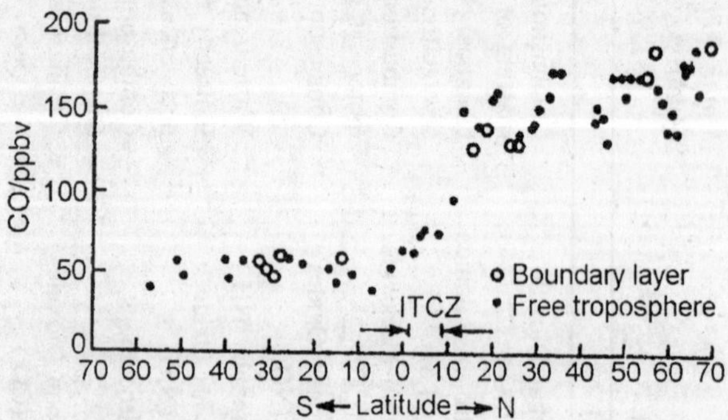

Fig. 8.4. Latitudinal variation in carbon monoxide concentrations.

reactive gases, is accompanied by an increase in concentration on the pollutant. Carbon dioxide plays a vital role in the earth's radiation balance and, therefore, in the earth's climate. The change in the CO_2 concentration, it is predicted, will be able to produce an associated change in the climate of the earth. Any inert pollutant will tend to increase in concentration with time if there have been no sinks available. However, this is rarely the case in practice. For CO_2 the problem has been the lack of balance between the sources and sinks over a time scale that is relevant to the atmospheric concentration (Table 8.3).

Achievement of equilibrium with the oceans would need centuries if all anthropogenic emission ceased. For that other group, of inert pollutants, and chlorofluoromethanes, whose concentration is also increasing, it has been the nature of the final removal process in the stratosphere which is the cause of concern. Although the chlorofluoromethanes appear to have no tropospheric sink, they get decomposed in the stratosphere, giving products which react with ozone. Because this removal process exists, the concentration of chlorofluoromethanes at a steady state of usage would eventually reach and equilibrium concentration in the troposphere. However, it would take decades rather than years. The real atmosphere is also having particulates and vapours, both of which are having natural cycles in the atmosphere which get affected by human activities. Man produces particles directly because of his agricultural and industrial activity, and indirectly because of atmospheric reactions of anthropogenic gas emissions.

Although the total particulates from natural sources, these particles get concentrated (like the pollutant gases which are their major source) in the industrial regions of high population density. Particles are having an effect on amenity by producing soiling, and on health by acting as carriers into the lung of trace substances like lead or the polyaromatic hydrocarbons. In the atmosphere, particles act as reaction centres, facilitating gas reactions and also adversely affecting visibility and sunlight penetration. The major vapour phase atmospheric species has been water. This persists as a vapour until supersaturation occurs, when it condenses, the vapour concentration for condensation being a function of the temperature. In practice, the water content of air has been very variable and ranges from 0 to about 4 percent by weight.

Many industrial activities emit water vapour directly—thermal power generation for example—and others effect the water cycle indirectly by release of heat to the atmosphere or to large bodies of

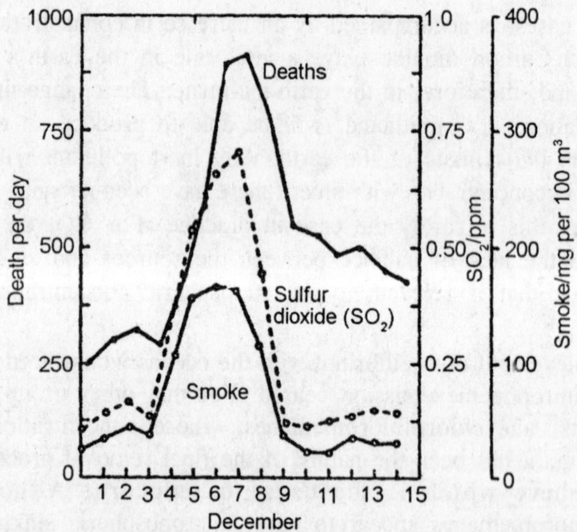

Fig. 8.5. Death rates and atmospheric pollution during the London smog of 1952.

water. Other substances also exist in the vapour phase in the atmosphere, in particular the non-methane hydrocarbons from both pollutant and natural sources, and sulphuric acid, the product of sulphur dioxide oxidation. Both vapours get associated with the incidence of smog in 'favourable' circumstances. For gases and particles to be air pollutants it is known that they must exceed their normal background concentrations to a significant extent.

It can be put in another way, substances in the air are called pollutant when their concentrations are sufficient to have adverse effects on man and his environment. Indirect effects arise because of changes in the physical properties of the earth's atmosphere system—the infra red radiation balance in the case of CO_2, and the amount of incoming ultra violet radiation in the case of ozone depletion caused by chlorofluoromethanes. The potential effects have been global in their implication and as such require international rather than national cooperation for their study and, if necessary, for their solution. Direct effects take place because of the interaction between a pollutant and a receptor such effects results from the higher than, background concentrations pertaining near the source of pollutants.

Direct effects can be immediate if the concentration has been high enough (acute effects), or can develop over the long term as a result of continuous exposure to higher than background levels (chronic effects). The extent of area affected distant from the source or sources has been a function of the distance the pollutant is transported before the concentration is reduced, by reaction or removal at the earth's surface, to a level at which no direct effects take place. To date, it is the direct form of air pollution of which we have been most aware, particularly in its most acute manifestation—the London smogs of the 1950s. It was direct pollutian, by hydrochloric acid gas from the alkali industry, which led to the first Alkali Act in the UK in 1863 and most control legislation since that date.

Where the pollutant emission has been typical of the process, as in the case of the alkali industry, and there has been only one plant carrying out the process in the area, then an association between effects and pollutant is readily made and control measures can be taken where necessary. Other examples of unique pollutants from particular processes would be hydrogen fluoride (HF) from fertilizer plants and the aluminium industry, and odourous reduced sulphides (hydrogen sulphide and mercaptans) from the wood pulping industry. The products of fuel combustion like particulates, sulphur dioxide, nitrogen oxides, carbon monoxide and hydrocarbons are common to most industries and, in addition, are having domestic and vehicular sources, making it more difficult to lay blame for observed effects individually.

In cities, pollution comes from a multiplicity of sources, not least of which is the motor car, which contributes more than 50 percent of the hydrocarbons and nitrogen oxides to most urban airsheds and 90 percent of the carbon monoxide. These pollutants influence the city producing them most seriously under low wind speeds and stable atmospheric conditions, which prevent the dispersal of pollutants and maximize their opportunities for reaction. Reactants and products have been forming a complex mixture which can have an additive effect on a receptor or result in a significant enhancement of effect over and above what would be expected from the sum of the individual pollutants present.

9

Agrochemicals

Human beings have learned to line and compete with the insect world, even though insects appeared long before humans did. According to recent discoveries humanoids have exited on earth more than 3 million years, while insects are known to have existed for 250 million years. We can guess that the first materials used by our primitive ancestors that could be classed an insecticides (repellents) in the crudest definition of the word were mud and dust spread over their skin to repel biting and tickling insects, a practice that resembles the habits of water buffalo, pigs and elephants. History doesn't tell us very much about chemicals used against insects.

The earliest records of insecticides pertain to the burning of "brimstone" (sulfur) as a fumigant. Pliny the Elder (AD 23-79) recorded most of the earlier insecticide uses in his Natural History. Gall from a green lizard to protect apples from worms and rot were also included in it a variety of materials have been used with doubtful results: extracts of pepper and tobacco, hot water, soapy water, whitewash, vinegar, turpentine, fish oil, brine, lye and many others. Even as recently as 1940, our insecticide supply was limited to several arsenicals, petroleum oils, nicotine, pyrethrum, rotenone, sulfur, hydrogen cyanide gas, and cryolite. World War II opened the Chemical Era with the introduction of a totally new concept of insect control chemical–synthetic organic insecticides, the first of which was DDT.

Organochlorines

As the names indicates the *organochlorines* are insecticides that contain carbon chlorine and hydrogen. They are also referred to by other names: *chlorinated hydrocarbons*, *chlorinated organics*, *chlorinated insecticides*, *chlorinated synthetics*.

DDT and Related Insecticides

DDT affect human health agriculture and the environment. It can be considered the pesticide of greater of historical important Since EPA canceled all uses of DDT effective from January 1, The story of its rise of stardom, carrying with it the Nobel Prize, and decline to infamy is rather sensational and should be briefly narrated for the uninitiated. DDT which is more than 100 years old is probably the best known and most notorious chemical of this century. It is the most fascinating, and remains to be acknowledged as the most useful insecticide developed. It was first synthesized in 1873, by a German graduate student who had no idea of its tremendous insecticidal value, and after synthesis it was put on the shelf and forgotten. In 1939 a Swiss entomologist, Dr. Paul Miller, rediscovered DDT while searching for a long-lasting insecticide against the clothes moth. DDT proved to be extremely effective against flies and mosquitoes, ultimately bringing to Dr. Mller the Nobel Prize in medicine in 1948 for his lifesaving discovery. We should bear in mind that its most beneficial use was in public health, for malaria control, and in many nations it still is so used.

More that 1.8 billion kg of DDT have been used throughout the world for insect control since 1940, and 80 percent of that amount was used in agriculture. Production reached its maximum in the United States in 1961, when 73 million kg were manufactured. The greatest agricultural benefits from DDT have been in the control of the Colorado potato beetle and several other potato insect, the codling moth on apples, corn earworm, cotton bollworm, tobacco budworm, pink bollworm or cotton, and the worm complex on vegetables. It has been most useful against the gypsy moth and the spruce budworm in forest. From the point of view of human medicine, DDT has been most successful against mosquitoes that transmit malaria and yellow fever, against body lice that can be carry typhus, and against fleas that are vectors of plague.

One of the most surprising features of DDT was its low cost. Most of that sold to the World Health Organization went for less than 22 cents pound. Without question, it was the most economical insecticide ever sold. A federal ban on the use of DDT, declared by the EPA on January 1, 1973, named DDT an environmental hazard due to its long residual life and to its accumulation, along with the metabolite DDE, in food chains, where it proved to be detrimental to certain forms of wildlife. It is no longer available to the grower or home-owner.

DDT belongs to the chemical class of diphenyl aliphatics, and as the name indicates it consist of an aliphatic, or straight carbon chain, with two (di-) phenyl rings attached, as in the illustrations. DDT was first known chemically as *dichloro diphenyl trichloroethane*, hence DDT. Five members of DDT are of importance because they all had an early role in pest control : TDE (or DDD), methoxychlor, ethylan, dicofol, and chlorobenzilate. The latter two are not really insecticides, but rather are acarocodes (miticides).

How does DDT Kill? The mode of action is not very clear. In some complex manner it destroys the delicate balance of sodium and potassium within the neuron, thereby preventing it from conducting impulses normally. In order to understand some of the well-documented evils attributed to it let us consider a few important pents concerning DDT. The first point is DDT's chemical stability, DDT and TDE are *persistent*, that is, their chemical stability gives the products long lives in soil and aquatic environments and in animal and plant tissues. They are not readily broken down by microorganisms, enzymes, heat, or ultraviolet light. The remaining DDT relatives are considered nonpersistent. Second, DDT has been reported in chemical literature is the most insoluble compound ever made sure that DDT'S solubility in water is only about six parts per billion part (ppb) of water. However, it is quite soluble in fatty tissue, and, as a consequence of its resistance to metabolism, it is readily stored in fatty tissue of any animal ingesting DDT along or DDT dissolved in the food it eats, even when it is part of another animal.

DDT gets accumulated in every animal that preys on the other animals because of the fait eat it is not readily metabolized and thus not exerted. It also accumulates in animals thateat plant tissue bearing even traces of DDT. For example, the dairy cow excretes (or secretes) a large share of the ingested DDT in its milk fat. Humans drink milk and eat the fatted calf, and thereby ingest DDT. The same story is repeated in food chains ending in the osprey, falcon, golden eagle, seagull, pelican, and so on. The principle of those food chain oddities is this : Just as DDT, any chemical that possesses the characteristics of stability and fat solubility will follow the same biological magnification (*biomagnification*). The polychlorinated biphenyls (PCBs), a group of chemicals that have no insecticidal properties, are stable and fat soluble and have climbed the food chain just as DDT has.

Other insecticides incriminated to some extent in biomanginification, belonging to the organochlorine group, are TDE, DDE (a major metabolite of DDT), dieldrin, aldrin, several isomers of HCH, endrin, heptachlor and mirex.

The rise and fall of DDT contains a lesson, and this is perhaps as good a place as any to moralize. Because of DDT's great success in World War II against body lice in Neples, during the typhus outbreak, and in the Pacific, against mosquitoes known to carry malaria, after the war it was rapidly adopted for agricultural use with inadequate basic knowledge. And, because of its effectiveness against a host of agricultural insect pests and its ridiculously low cost, it was overused, and abused, land then—when it caused problems-was banned in rather a panic. DDT moved to be highly effective against body lice in Neples, during world war II in typhus our break, and in the pacific against mesquitves which carried malaria.

Due to the above mentioned reasons DDT was readily adopted for agricultural use without knowing much about its disadvantages. Only when it started to cause problems, it was banned, rather in a panic. A lesson that should be learnt is that no matter how effective it is against a host of agricultural insect pests, a how cheap it is, there is on absolute need for an informect caution and according to available knowledge correct way of employing a specific chemical for pest control. We need basic research performed in autonomous institutions not subjected to competition in the economic marketplace. And because this research it basic, as opposed to applied, it will of necessity be slow, expensive, long-term, and not immediately applicable. But our social policy must support such research, to enable us to cope with—if not to avoid-those chemicals that prove to pose dangers to our environment as well as our health.

Hexachlorocyclohexane (HCH)

Hexachlorocyclohexane (HCH)-previously erroneously called benzenhaxachloride (BHC)-was first discovered in 1825. But, like DDT, it was not known to have insecticidal properties until 1940, when French and British entomologists found in the material to be active against all insects tested. It is made by chlorinating berizence, which results in a product made up of several isomers, that is molecules containing the same kinds and numbers of atoms but differing in the internal arrangement of those atoms.

HCH, for instance, has five isomers, named, after the Greek letters, *alpha, beta, gamma, delta* and *epsilon*. After much laboratory work in isolating and identifying these isomers, the chemists found to their great surprise that only the gamma isomer had insecticidal properties. In a normal mixture of HCH, the gamma isomer makes up only about 12 per cent of the total, leaving the other four isomer as inert material or insecticidally inactive ingredients. Since the gamma isomer is the only active ingredient, methods were developed to manufacture lindane, a product containing 99 per cent gamma isomer, which is effective against most insect, because of its cest being high it is impractical for crop use. Technical grade HCH has one highly undesirable characteristic, a prominent musty order and flavour.

The order is form the inert isomers, which are more persistent than the orderless gamma isomer in animal and plant tissues as well as in soil. As a result, root and tuber crops planted in soils previously treated with HCH retain its order and are usually unsalable. The same problem is reported with leafy vegtables, poultry, eggs, and milk that directly or indirectly come in contact with HCH residues. The effects of HCH on insects and mammals superficially resemble those of DDT. Lindane is a neurotoxicant whose effects are normally seen within hours and result in increased activity, tremors, and convulsions leading to prostration. Lindane is orderless and has a high degree of volatility. It quickly became popular as a household fumigant sold as pellets to be attached to light bulbs or to small, decorative electric wall vaporizers. These were later found to be hazardous to humans and house pets and were removed from the market.

Cyclodienes

The cyclodienes, also known as the *diene-organochlorine insecticides*, are of more recent origin than DDT (1939) and HCH (1940) which were developed after World War II and are therefore of more recent origin than DDT (1939) and HCH (1940). The eight compounds listed here were first described in the scientific literature or patented in the year indicated : chlordane, 1945; aldrin and dieldrin, 1948; heptachlor, 1949; endrin, 1951; endosulfan, 1956; and chlordecone (Kepone) 1958. Other cyclodienes developed in the United States and Germany are of minor importance in a general survey. These include isodrin, adodan, bromodan and telodrin. Cyclodienes, which are stable in soil & relatively stable to ultraviolet action of sunlight are persistent insecticides. Consequently, they have been used in greatest quantity as soil insecticides (especially chlordane, heptachlor, aldrin, and dieldrin), for the control of termites and soil-borne insects whose immature stages (larvae) feed on the roots of plant. Because of their persistence, the use of cyclodienes on crops was restricted; undesirable residues remained beyond the time for harvest.

To understand the effectiveness of cylodienes as termite control agents, consider in the year of their development are still protected from damage more than thirty years later. These insecticides are the most effective long-lasting, economical, and safest termite control agents known. However, several other soil insects became resistant to these materials in agriculture, resulting in a rapid decline in their use. In 1975 EPA banned most of the agricultural uses of cyclodienes. The most valuable, and produced in the greatest quantity, were chlordane and dieldrin. Structures of the common cyclodienes are presented to illustrate their similarity and complexity. The nomenclature and chemistry of the cyclodienes are quite complicated.

The cyclodienes have three-dimensional structures and thus possess stereoisomers; that is, forms that have the same kinds and numbers of atoms, but their atoms differ in their spatial arrangement. For instance endrin is a stereoisomer of deildrin. The toxic effects of not very clear. Cyclodienes on insects mammals and birds are of equal intensity but more on fishes perhaps because when the compound is introduced into water the fish continually respire and ingest any toxic compound contained in their aquatic environment. The modes of action of the cyclodienes are not clearly under-stood. It is known that they are neurotoxicants, that have effects similar to those of DDT and HCH. They appear to affect all animals in generally the same way, first with nervous activity followed by termors, convulsions, and prostration. The cyclodienes undoubtedly disturb the delicate balance of sodium and potassium within the neuron but in a way differing from that of DDT and HCH.

Polychlorotepene Insecticides

Toxapliene (1947) and strobane (1951) are the only two polychloroterpene material. Neither have ever been considered urban insecticides. Toxaphene is manufactured by the chlorination of camphene, a pine tree derivative. Toxaphene had by far the greatest use of any single insecticide in agriculture. It was used on cotton, first in combination with DDT, for alone it has a low order of toxicity to insects. In 1965, after a number of cotton insects became resistant of DDT, toxaphene was formulated in combination with methyl parathion, an organophosphate insecticide discussed later in this chapter. As late as 1976 some 11.8 million kg of toxaphene was used on cotton, or 41 per cent of all insecticides on cotton that year.

Toxaphene is an extraordinary mixture of more than 177 polychlorinated derivatives, which are 10-carbon compounds including CI_6, CI_7, CI_8, CI_9 and CI_{10} constituents. Most are probably isomeric CI_7., CI_8, and CI_9. bornanes. No single component makes up more than a small percentage of the technical mixture. The toxaphene components of greater concern are those most toxic to mammals and fish. One of these is toxicant A, shown in its three-dimensional form. Toxicant A makes up only 3 per cent of technical toxaphene, but it is 18 times more toxic to mice, 6 times more toxic to houseflies, and 36 times more toxic to goldfish than technical toxaphene. These materials are persistent in the soil, though not as persistant as the cyclodienes, and narish in three to four weeks from the surfaces of mosts plant tissues. This disappearance is attributed more to volatility than to actual metabolism or photolysis (disintegration from the effects of ultraviolet light in sunlight). They are fairly easily metabolized by mammals and birds, are not stored in body fat to any great extent, as are DDT, HCH, or the cyclodienes. Despite low toxicity to insects, mammals, and brids fish the highly susceptible to toxaphene poisoning, in the same order of magnitude as to the cyclodienes.

The modes of action for toxaphene and strobane are similar to the cyclodiene insecticides, acting on the neurons and causing an imbalance in sodium and potassium ions.

Organophosphates

The organophosphate compounds have been replaced by the chemically unstable organophosphate (OP) insecticides, especially with regard to use around the home and garden. *Organophosphate* is usually used as a generic term to include all the insecticides containing phosphorous. The OPs have several other commonly used names, some of which are organic phosphates, phosphorus insecticides, nerve gas relatives, phosphates, phosphate insecticides, and phosphours esters or phosphoric acid esters. They are all derived from phosphoric acid and are generally the most toxic of all pesticides to vertebrate animals. Because of their chemical structures and mode of action, they are related to the "nerve gases." Their insecticidal action was observed in Germany during World War II in the study of materials closely related to the nerve gases sarin, soman and tabun.

Initially the discovery was made is search of substitutes for nicotine, which was in critically low supply is Germany. The OPs have two distinctive features. First, they are generally much more toxic to vertebrates than are the organichlorine insecticides, and, second, they are chemically unstable or nonpersistent. It is this latter quality that brings them into agricultural use as substitutes for the persistent organochlorines, particularly DDT. The OPs exert their toxic action by inhibiting certain important enzymes of the nervous system, cholinesterases (ChE). This leads to the accumulation of acetycholine (ACh), which interferes with the neuromuscular junction producing rapid twitching of voluntary muscles and finally paralysis.

OPs that have attached to their phosphorous atoms combinations of different alcohols and different phosphours acids are termed esters. Esters of phospherous have varying combinations of oxygen, sulfur & nitrogen. The nuclie of the six subclasses of OPs are shown to help explain some of the seemingly odd chemical names gives to these insecticides. The Ops are divided into three groups—the aliphatic, phenyl and heterocyclic derivatives.

Aliphatic Derivatives

The term *aliphatic* literally means "carbon chain," and the linear arrangement of carbon atoms differentiates them from ring or cyclic structures. All the aliphatic OPs are simple phosphoric acid derivatives bearing short carbon chains. TEPP was the first OP introduced into agriculture is the only useful pyrophosphate and is probably the most toxic. It was never available for home use. Because TEPP is very unstable in water, it hydrolyzes (breaks down) quickly after spraying on crops and disappears with 12 to 24 hours.

The oldest and most heavily used aliphatic OP is Malathion introduced it was quickly adopted by agriculture for use on most vegetables, fruits, and forage crops for control of an extensive range of insect pests. Malathion is so safe that it is suitable for home use aho since it is safe to use around pets, fast acting and can controlled practically every kind insect including garden and household aphids and cockroaches. It is so safe that it is perscribed by physicians for use on humans for the control of head, body, and crab lice. It commonly appears in flea powders for dogs, cats, and other domestic animals, and is used in dips for the control of mange mites.

In 1981, and again in 1982, malathion became the insecticide of choice in the control of the Mediterranean fruit fly, which had invaded the rich fruit-growing areas of California. Malathion was mixed with a protein bait made of molasses and yeast and sprayed from ground equipment and by helicopter over the infested and surrounding areas. Both male and female and female fruit flies are attracted to the bait and die a few hours after feeding on the tasty morsels. The bait formulation applied by helicopter is the most selective and inoffensive of all forms and methods of malathion use. With this technique malathion is appled at the astonishingly low rate of 171 g/ha mixed with 684 g of the bait. When applied by ground equipment it is used at the rate of 2 kg of active ingredient in 378 L of spray per hectare as a foliar or vegetation spray. This malathion-bait mixture was used successfully in the eradication of the Medfly from Florida in 1956-1957, and again in 1962-1963, from Texas in 1966, and from Los Angeles in 1975-1976. It was again placed in service in the brief 1981 Florida outbreak of the Medfly, because of its exceptionally low acute toxicity to human and other warm-blooded animals.

Trichlorfon is a chlorinated OP, which has been useful for crop pest control and fly control around barns and others farm buildings. Monocroptophos is an aliphatic OP containing nitrogen. It is a plant-systemic insecaticide, but it has had limited use in agriculture because of its high mammalian toxicity. It is not available to the home gardener. Systemic insecticides are those that are taken into the roots of plants and translocated to the above-ground parts, where they are toxic to any sucking insects feeding on the plant juices. Normally caterpillars and other plan tissue-feeding insects are not controlled, because they do not ingest enough of the systemic-containing juices to be affected.

Contained among the alophatic derivatives are several plant systemics, dimethoate, dicrotophos, oxydemetonmethyl, and disulfoton, and of which can be used safely by the homeowner. Dichlorvos is an aliphatic OP with a very high vapour pressure, giving it strong fumigant qualities. It has been incorporated into polychlorovinly resin pet collars and pest strips, from which it is released slowly. It lasts several months and is useful for insect control in the home and other closed areas.

Mevinphos is a highly toxic OP. Because of its short insecticidal life, Mevinphos can be used in commerical vegetable production. It can be applied up to one day before harvest for insect control, yet it leaves no residues on the crop to be eaten by the consumer. Two of the recent arrivals in the aliphatic organophosphate structures are methamidophos and acephate. Both have proved highly useful in agriculture, especially for vegetable insect control. In summary, the aliphatic organophosphate insecticides are the simplest in structure of the organophosphate molecules. They have wide range of toxicites, and several possess a relatively high water solubility, giving them plant-systemic qualities several of which are important around the home.

Phenyl Derivatives

When the Benzene ring is attached to other groups it is referred to as *phenyl*. The phenyl OPs contain a benzene ring with one of the ring hydrogens replaced by attachment of the phosphorus moiety and others frequently replaced by Cl, NO_2, CH_3, CN, or S. The phenyl OPs are generally more stable than the aliphatic OPs; consequently their residues are longer latsting. Parathion is the most known of the phenyl OPs, being 1947, the second phosphate insecticide introduced into agriculture. The first, TEPP, was introduced in 1946. As a result of it's age and utility, parathion's total usage is greater than that of many of the less useful materials combined. Ethyl parathion was the first phenyl derivative used commerically and, because of its harmful effects has not been available for home use.

Methyl parathion became available in 1949 and proved to be more useful than (ethyl) parathion because of its lower toxicity to humans and domestic animals and borader range of insect control. Its shorter residual life also makes it more desirable in certain instances. This material is also not used by the layperson. Systemic insecticides are also found in the phenyl OPs. They are, however, usually animals systemics used for the control of the cattle grub; ronnel and crufomate are examples. Stirofos is a home-safe OP much like malathion in its overall usefulness against home and livestock pests. Profenophos and Sulprofos are two of the more recently registered phenyl derivatives that have a wide of insecticidal activity and are used only on field crops today. Isofenphos is used as a soil insecticide in field crops and vegetables, against corn rootworm and onion maggot, and also for white grubs, chinch, bugs, and sod webworms in turf.

Heterocyclic Derivatives

The term *heterocyclic* means that the ring structures are composed of different or unlike atoms. In heterocyclic compound, for example, one or more of the carbon atoms is replaced by oxygen, nitrogen or sulfur, and the ring may have three, five or six atoms. The first insecticide made available in this group was probably diazinon, in 1952. Note that the six-membered ring contains two nitrogen atoms, very likely the source of its proprietary name, since one of the constituents used its manufacture is pyrimidine, a diazine. Diazinon is a comparatively relatively safe OP that has an surpirsingly good track record around the home. It has been effective for practically every conceivable use : insects in the home, lawn, garden, ornamentals, around pets, and for fly control in stables and pet quarters. The second oldest of this group is Azinphosmethly (1954) and is used in U.S. agriculture. It serves both as an insecticide and acaricide in cotton production and is not available to the layperson.

Chlorphyrifos has become the most frequently used insecticided by pest control operators in homes and restaurants for controlling cockroaches and other household insects. Methidathion is not particularly new, but it has in the last few years acquired registrations for forage and field crops, true fruits, and crops for an extraordinary wide variety of insect and mite pests. Phosmet has a set of registration credentials similar to methidathion, including the infamous boll weevil and the plum curculio, two closely related weevil pests. Dialifor was first introduced in the mid-1960s as were methidathion and phosmet. It uses are somewhat more limited, however, to apples, grapes, pecans and citrus.

The heterocyclic organophosphates have generally longer lasting residnes than many of the aliphatic not simple molecules. Also, because of the complexity of their molecular structure, their breakdown products (metabolites) are frequently many, making their residues sometimes difficult to measure in the laboratory. Consequently, their use by growers on food crops is somewhat less than either of the other two groups of phosphorus- containing insecticides.

Organosulfurs

The organosulfurs, as the name suggests, have sulfur as their central atom. They resemble the DDT structure in that most have two phenyl rings. In hot weather dusting sulfer alone is a good Acaricide (miticide).

The organosulfurs, however, are far superior,, requiring much less material to achieve control because sulfur in combination with phenyl rings is particularly toxic to mites. Of greater interest, however, is that the organosulfurs have very low toxicity to insects. As a result, they are used for selective mite control. This group has one other valuable property. They are usually ovicidal as well as being toxic to the young and adult mites. Tetradifon is one of the acaricides and typically bears the sulfur and twin phenyl rings, as do most of the organosulfurs. No doubt the oldest of this group is Aramite, introduced in 1951. Notice that Aramite has only one phenyl ring and is, therefore, an exception to the general rule that organosulfurs have two phenyl rings.

Carbamates

Since the organphsphate insectides are derivatives of phosphoric acid, the carbamates must be derivatives of carbamic acid HO—C—

NH_2, And like the organophosphates, the mode of action of the carbamates is that of inhibiting the vital enzyme, chloinesterase (ChE). In 1951 the carbamate insecticides were introduced by the eigy Chemical Company in Switzerland. They fell by the wayside because the first ones were not very effective, while being quite costly. The early carbamates are shown in the margin at the right.

At the time it was not known that the N,N-dimethyl carbamates, as shown in these structures, were generally less toxic to insects than the N-methyl carbamates, which were developed later and which make up the bulk on the currently used materials. In 1956, Carbasyl which was the first successful carbanate, was introduced. More of it has been worldwide than all the remaining carbamates combined. Two distinct qualities have made it the most popular material : very low mammalian oral and dermal toxicity and a rather broad spectrum of insect control. This has led to its wide use as a lawn and garden insecticide. Notice that carbarly is an *N-methyl carbamate*. Most of the carbamates are plant systemics, indicating that they have a high water solubility, which allows them to be taken into the roots or leaves. They are also not readily metabolized by the plants.

Methomyl, oxamly, aldicarb, and carbofuran have distinct systemic characteristics, making them useful also as nematicidies. Of these, only aldicarb and carbofuram are used as soil insecticides and nematicides. Under rare circumstances, adlicarb has been detected in shallow groundwater following certain uses. Methomyl has proved especially effective for worm control on vegetables. Bufencarb is used in agriculture exclusively as a soil insecticide, becoming a repalcement for the long- residual organo-chlorine insecticides, aldrin, dieldrin and heptachlor. Methiocarb, aminocarb, and promecarb and effective againt foliage-and fruit-eating insects. Methiocarb and aminocarb are both excellent molluscicides, used for slug and snail control in flower gardens and orgnamentals. Methiocarb is also registered as a bird repellent for cherries and blueberries and as a seed dressing. For cockroaches and other household insects that develop resistance to the organochlorines & organophosphates, propoxure is highly effective.

It is used by most structural pest control operators for cockroaches and other household insects in restaurants, kitchens and homes. For home use it is formulated in bottled sprays. Similarly, bendiocarb has found its greatest use in the United States as residual household insecticide. In summary, the carbamates are inhibitors of cholinesterase, are plant systemics in several instance, and are, for the most part, wide ranged in effectiveness, being used as insecticides, miticides, and molluscidides.

Formamidines

The formamidines consist a new, small, promising group of insecticides. Three examples are chloridimeform, formetanate, and amitraz. They are effective against the eggs and very young caterpillars of several moths of agricultural importance, and are also effective against most stages, of mites and ticks. Thus, they are classed as ovicides, insecticides and acaricides. Late in 1976, chlordimeform was removed from the market by its manufactures, Ciba-Geigy Corporation, land Nor-Am Agricultural Products, Inc., because it proved to be carcinogenic to a cancer-prone strain of laboratory mice during high-level, lifetime feeding studies. In 1978 it was returned for use on cotton, but under very strict application restrictions. Their present value lies in the control of organo-phosphate and carbamates-resistant pests. Poisoning symptoms are distinctly deferent from other materials. It has been proposed that one possible machanism of action is the inhibition of the enzyme monoamine oxidase. This results in the accumulation of compounds termed *biogenic amines*. Thus formamidines introduce a new mode of action for the insecticides and acaricides. This fact alone makes them extremely useful, for we are slowly losing ground in the battle or insect resistance to the modes of action of the older insecticide groups.

Thiocyanates

Thicyanates have easily recognized structural formulas. Remembering the *theion* of the Greek word for "sulfur" and that the cyanides or cyanates end in –CN we have molecules that bear- SCN, or thiocynate endings. These insecticides have very distinct, creoste like odours, comparatively safe to use around humans and animals, and the astonishingly quick knockdown of flying insects. Their mode of action is somewhat complex and can be simply to interfere with cellular respiration and metabolism. These materials may be found in aerosols to be used around horses and other farm animals. However, with the appearance of the synthetic pyrethroids, the thiocyanates have deceased demand and will probably sood disappear.

Dinitrophenols

They act by uncoupling oxidative phosphorylaction or basically by preventing the utilization of nutritional energy. In the 1930s, certain dinitrophenols were give by ininformed physicians to their overweight patients ot induce rapid weight loss. They were extremely effective, but quite toxic, and their use resulted in several widely publicized deaths. The older of this group is DNOC (3,5-dinitro-o-cresol), introduced as an insecticixe in 1982. DNOC has also used as an ovicide, herbicide, fungicide, and blossom-thinning agent. Its use has declined today to herbicidal applications in which all plants are to be killed. Dinoseb is used as a dormant fruit spray for control of many insects and mites. Another a caricide which was introduced in 1960 in Binapacryl Dinocap in 1949) is still another as well as fungicide and is one of the rare materials made up of several related molecular structures, only one of which is shown. Dinocap is particularly effective against powdery mildew fungi. Owing to its safety to green plants, it has often replaced the phytotoxic sulfur that is so effective against powdery mildews. In summary, the dinitrophenols have been used as pesticide in practically all classification: ovicides, insecticides, acaricides, herbicides, fungicides, and blossom-thinning agents.

Organotins

The organotins are a relatively new group of acaricides, which double as fungicides. Of particular interest here in cyhenxatin, one of the most selective acaricides presently known introduced in 1967. Proved to be most effective against mites on deciduous fruits, citrus, greenhouse crops and ornamentals is fenbutatinoxide which was introduced somewhat later. The mode of action of this group is not completely known but is believed to be the inhibition of oxidative phosphorylation at the site of dinitrophenol uncoupling (the production of energy in the form of adenosine triphosphate, ATP). These trial tins also inhibit photophosphorylation in chlorplasts (the chlorophyll-bearing subcellular units) and can thus serve as algicides.

Botanicals

Botanical insecticides are of great interest to many, because they are "natural" insecticides, toxicants derived from plants. Historically, the plant materials have been in use longer than any other group, with the possible exception of sulfur. Tobacco, pyrethrum, derris, hellebore, quassia, and turpentine were some of the more important plant products is use before the organized search for insecticides began. Some of the widely used insecticides come from plants. The flowers leaves, and roots are tinely ground and used in this form, or the toxic ingredients are extracted and used alone or in mixtures with other toxicants.

There are five natural or botanically derived insecticides that are of interest to gardeners in general, but especially to the organic gardener : pyrethrum, rotenone, sabadilla, ryania and nicotine. All except nicotine are exempt from the requirement of a tolerance when applied to growing fruit and vegetables. This is, they can be eaten anytime after application, but these batanical insecticides must be used according to label directions. Batanical insecticide use reached its maximum in the United States in 1966 and has declined steadily since.

Pyrethrum is now the only botanical of significance in use, typically in rapid knowdown sprays in combination with synergists and one or more synthetic organic insecticides formulated for use in the home and garden. Botanical insecticides are naturally occuring chemicals, synthesized by plants. They are really no safer than most of the currently available synthetic insecticides, at least as compared to those available to the layman. These chemicals would be unaffordable as insecticides if they were synthesized in the laboratory, which is also possible, as Botanicals are expensive to extract from plant tissues.

Nicotine

Smoking tobacco was introduced to England in 1585 by Sir Walter Releigh. As early as 1690, water extracts of tobacco were reported as being used to kill sucking insects on garden plants. As early as about 1890, the active principle in tobacco extracts was known to be nicotine, and, from that time on, extracts were sold as commercial insecticides, for home, farm and orchard. Today organic gardeners may soak a cigar or two in water overnight and spray insect-infested plants with the extract, achieving some success. "Black Leaf 40," which has long been a favourite garden spray, is a concentrate containing 40 per cent nicotine sulfate. Today steam distillation or solvent extraction are the two processes for the commercial extraction of Nicotine from Tobacco.

Nicotine which is a physiological compounds is an alkaids & contains nitrogen. It shows prominent physiological properties. Other well-known alkaids, which are not insecticides, are caffeine (found in tea and coffee), quinine (from cinchona bark), morphine (from the opiumpoppy), concaine (from coca leaves), ricinine (a poison in castor oil beans), *Strychnos nux vomica*), confine (from spotted hemlock, the posion that killed Socrates), and, finally, LSD (a haflucinogenic derived from the ergot fungus attacking grain). As its mode action, nicotine mimics acetylocholine (ACh) at the neuromuscular (nerve-muscle) junction in mammals, the results in twitching, convulsions, and death, all in rapid order. In insects, the same action is observed, but only in the gangila of their central nervous systems.

Nicotine sulfur, as it is commonly marketed, is highly toxic to all warm-blooded animals, as well as insects; for example, the LD_{50} (dose that proves lethal for 50 per cent to the test population) for rats is 50-60 mig/kg. This makes it the most hazardous of the botanical insecticides of home gardeners. It has been used with great success since before the turn of the century. Dusts are very much toxic to humans hence are not available for garden use.

Nicotine sulfate us used primarily for piercing-sucking insects such as leafhoppers, aphids, scales, trips and whiteflies, but it can kill all insects and spider mites on which it is sprayed directly. Many caterpillar pests, however, are very resistant to nicotine. It is more effective during warm weather, but degrades quickly. It is registered for use on flowering plants, and ornamental shrubs and trees to control aphids, merlybugs, scales and thrips, lace bugs, leafminers, leafhoppers, rose slugs, and spider mites. Its use for most greenhouse pests is also acceptable. Nicotine sulfate is registered for use on a variety of vegetables and fruit trees, but it cannot be used in most cases within seven days of harvest, as required by EPA. There are several ornamental plants that are sensitive to nicotine, such as roses. The label should identify those sensitive plants. Nicotine is also registered for out-of-door use as a dog and can repellent as well as for furniture in the home. Additionally, tobacco dust in acceptable as a dog and rabbit repellent out-of-doors.

Rotenone

Rotenoids, the retenone-related materials, have been used as crop insecticides since 1848, when they were applied to plants to control leaf-eating caterpillars. However, they have been used for centuries (at least since 1649) in South America to paralyze fish, causing them to surface. Roteniods are produced in the roots of two genera of the legume (bean) family : *Derris*, grown in Malaya and the East Indies, and *Lonchcarpus* (also called cubed or cube), grown in south America. Rotenone has an oral LD_{50} of approximately 350 mg/kg (in rats) and has been used for generations as the ideal general garden insecticides. It is harmless to plant, highly toxic to fish kind many insects, especially caterpillars, moderately toxic to warm-blooded animals, and leaves no harmful residues on vegetables. There is no waiting interval between application and harvest of a food corp.

Rotenone is marketed as spray concentrates and ready to use dust, it is both a contact & a stomach poison to insects. It kills insects slowly, but causes them to stop their feeding almost immediately.

Like all the other botanical insecticides its life in the sun is short, one to three days. It is useful against caterpillars, aphids, beetles, true bugs, leafhoppers, thrips, spider mites, ants, rose slugs, whiteflies, sawflies, bagworms, armyworms, cutworms, leafrolles, midges and a host of other pets. Next to pyrethrum, rotenone is probably second in the number of approved use, exceeding 1000. Rotenone is the most useful piscicide available for reclaiming lakes for game fishing. It eliminates all fish, closing the lake of reintroduction of rough species. After treatment, the lake can be restrocked with the desired species. Rotenone is a selective piscicide in that it kills all fish at dosages that are relatively nontoxic to fish foiod organisms. It also breaks down quickly leaving no residues harmful to the fish used for restocking. The recommended rate is 0.5 part of rotenone to one million of water (PPM), or 5.1 kg per hectarameter of water (1.36 pounds per acre-foot).

Sabadilla

Sabadilla is extracted from the seeds of a member of the lily lfamily. Its oral $LD_{50}$is approximately 5000 mg/kg, making it the least toxic to warm blooded animals of the five botanical insecticides discussed. It acts as both contact stomach poison for insect. Cevatridine ($C_{32}H_{49}NO_9$) and veratridine ($C_{36}H_{51}NO_{11}$) are the two known active alkaloids that are included in Sabadilla. Neither chemical structure has been established. It is irritating to human eyes and causes violent sneezing in some sensitive individuals. It deteriorates rapidly in sunlight and can be used safely on food crops with no waiting interval required by the EPA.

Sabadilla is probably the most difficult of the five botanical insecticides to purchase, simply because there was hardly any demand for it for about 15 years. Sabadilla is registered for most commonly grown vegetables and will control caterpillars, grasshoppers, beetles, leafhoppers, thrips chinch bugs, stink bugs, larlequin and squash bugs, other true bugs, and patato psyllids. It is not very useful against and will not control spider mites.

Ryania

Ryania is a safe insecticide, quite safe for humans & domestic animals so safe that no waiting is required between the time of application to food crops and harvest, as there is for most other insecticides. Ryania is made from the ground roots of the ryania shrub grown in Trinidad and, like nicotine, belongs to the chemical class of alkaloids. It has an oral LD_{50} of approximately 750 mg/kg. It is a slow-acting insecticides, requiring as long as 24 hours to kill. Insects exposed to ryania usually stop their feeding almost immediately, making it particularly useful for caterpillars.

Ryanodine ($C_{25}H_{35}No_9$), is the active principle of ryania, and its chemical structure is still not determined. Ryanodine affects insect muscles directly by preventing contraction, resembling the effects of strychnine in mammals. The preferred uses for ryania are against fruit-and foliage-eating caterpillars on fruit trees, especially the codling moth on apple trees. However, it is useful against almost all plant-feeding insects, making it an ideal material for small orchards of deciduous fruits. It is not effective against spider mites.

Ryania is exempt from a waiting period between application and harvest and is registered by the EPA for the control of a host of insect pests on a wide variety of plants, shrubs and trees. Vegetable garden pests include aphids, cabbage loopers, Colorado patato beetles, corn borers, cucumber beetles, diamond back moths, flea beeles, leafhoppers, Mexican bean beetle, spittle bugs, and tomato horn-worms.

Ryania is registered for deciduous fruit trees of control aphids (except the woodly aphids), codling moth, Japanese beetles, and cherry fruit fly. It has long been used to control citrus thrips on all citrus. Though not very effective, ryania can be used in the home to control ants, silverfish, cockroaches, spiders and crickets. On ornamentals it is registered for aphids and lace bugs. It can used on roses against aphids, Japanese beetles, thrips, and whiteflies; on brambles, for aphids, raspberry fruitworms, and sawflies; and on grapes for aphids (except the woolly aphid) kand the Japanese beetles. Ryania is difficult to obtain since its importation into the United States has been stopped by its major distributor.

Pyrethrum

Pyrethrum is extracted from the flowers of a chrysanthemum grown in Kenya, Africa and Ecuador, South America. It has an oral LD_{50} of approximately 1500 mg/kg and is one of the oldest household insecticides available. In the early 19th century during the Napoleonic wars, ground dried flower heads were used to control body lice. Pyrethrum acts on insects with phenomenal speed causing immediate paralysis, thus its popularity in fast knockdown household aerosol spray. However, unless it is formulated with one of the synergists, most of the paralyzed insects recover to once again become pests. Pyrethrum is formulated as household sprays and aerosols and is available as spray concentrates and dusts for use on vegetables, fruit trees, ornamental shrubs, and flowering plants at any stage of growth. Vegetables and fruit sprayed of dusted with pyrethrum may be harvested or eaten immediately; there is no waiting interval required between application and harvest of the food crop. Pyrethrum has been approved by EPA, and has more uses than any other insecticides, numbering in thousand because of its general sagety to humans & domestic animals & its effectiveness against practically every known crawling and flying insect pest.

It is a mixture of four compounds: pyrethrins I and II and cinerin I and II. Their structures can be assembled by attaching the R1 and R2 in their proper positions on the large ester structure to the left.

Synthetic Pyrethroids

The natural insecticide is quite unstable in sunlight and aho it is very costly, therefore, it is seldom used for the agricultural purposes. Recently however, several synthetic pyrthrin-like materials have become available and are referred to as synthetic pyrethroids These meterials are very stable in sunlight and are generally effective against most agricultural kpests when used at the low rate of 0.11 to 0.32 kig/ha. Examples are permethrin (Ambuhs® or Pounce®) and fenvalerate (Pydrin®). The synthetic pyrethroids, or more correctly *pyrethroids*, have a rather long and successful history.

For ease of classification, they are placed in four categories, or generations. The first generation contains but one pyrethroids, allethrin (Pynamin). Allethrin was commercially available in 1949. It consists of 22 chemical reactions to produce the final insecticide, in this sense it marked the beginning of an era of complex syntheses. Allethrin in merely a synthetic duplicate of cinerin I (a component of pyrethrum), with a slightly more stable side chaïn, and it is more persistent than pyrethrum. Equally effective against houseflies and mosquitoes, but less so against cockroaches and other insects it was readily synergized by the common pyrethrum synergists.

The second generation includes tetramethrin (Neo-Pynamin), which appeared in 1965. It gives stronger knockdown of flying insects than allethrin and is readily synergized. Resemthrin (NRDC-104, SBP-1382), and FMC-17370) appeared in 1967, is approximatley 20 times more effective than pyrthrum is housefly knockdown, and is not synergized to any appreciable extent with pyrethrum synergists. Bioresmethrin (NRDC-107, FMC-18739, and RU-11484), also described in 1967, is 50 times more effective than pyrethrum against normal (suceptible to insecticides) houseflies, and also not synergized with pyrthrum synergists. Both resmethrin and bioresme-thrin and more stable than pyrethrum, and decompose fairly rapidly on exposure to air and sunlight, which explains why they were never developed for agricultural use. (Resemthrim has become the most used of the second generation pyrethroids for sprays and aerosols to control flying and crawling insects indoors.) In 1969 Bioallethrin (d-trans-allethrin) was introduced & it is more potent than allethrin and readily synergized, but it is not as effective as resmelthrin. The last of this period was phenothrin (Sumithrin®), introduced in 1973. It, too, is intermediate in quality and slightly enhanced by synergists. The third generation includes fenvalerate (Pydrin®) and permethrin (Ambush, Pounce, and Pramex), which appeared in 1972 and 1973 respectively. These became the first agricultural pyrethroids because of their exceptional insecticidal activity (0.11 kgAI/ha) and their *photostability*. Photostability means they are unaffected even by ultraviolet in sunlight, lasting four to seven days on crop foliage as effective residues.

The contemporary forth generation, still being developed and regsitered, is truly exciting, for their rates of application are again reduced to one-tenth of the previous generation, or to the order of 0.11 to 0.06 kg AI/ha required of the phenomenal compared to the rate of 1.1 to 2.3 kg) AI/ha. This is truly phenomenal compared to the rate of 1.1 to 2.3 kg) AI/ha required of the organophosphate, carbamate, and organochlorine insecticides. Emerging in this truly revolutionary form of chemical insect control are the pyrthroids; cypermethrin (Ammo® Cymbush®, and Ripcord®), fenpropathrin, flucythrinate (Pay-Off®), fluvalinate (marvil®), and decamethrin (Decis®). All these insecticides are photostable, providing long residual effectiveness in the field, and are not significantly improved with the addition of pyrethrin synergists.

Synergists or Activators

Activators or Synergists as the name indicates are not in themselves considered toxic or insecticidal, but are materials used with insecticides to synergize or enhance the activity of the insecticides. Synergists are added to certain insecticides in the ratio of 8 : 1 or 10 : 1. The first synergist was introduced in 1940 to increase the effectiveness of pyrethrum. Since then many materials have been introduced, but only a few are still marketed beccause of cost and ineffectiveness. Synergists are found in practically all the "bug-bomb" aerosols to enchance the action of the fast knockdown insecticides pyrethrum, allethrin, and sometimes resemthrin against flying insects. Although initially developed for use with pyrethrum, they have since been observed to synergize some, but not all, organophosphates, organochlorines, carbamates, as well as a few of the botanical, or plant-derived, insecticides.

The mode of action of the synergists is to inhibit mixed-function oxidases, enzymes that metabolize foreign com- pounds, which in this instance would be the preythrum. The most popular synergists belong to the only two molecular groups, or moieties. The first is the methylendioxyphenyl moeity. The R_1 and R_2 are simple or oxygenated dcarbon chains or other groups of varying combinations. The second synergistic moiety does have a single name but is characterized by either of the following structures. Notice that all three moieties involve a five-membered ring associated with two oxygens. Because their mode of action k is the inhibition of insecticide-metabolizing enzymes, it is likely that this steric three-dimensional structure is generally the most effective in enzyme binding.

The synergists are usually used in sprays prepared for the home and garden, stored grain, and on livestock, particularly in dairy kbarns. Synergists are quite expensive, thus seldom if ever used on crops. Sesamin was the name given to a material containing the methyl endioxpenhyl groups, first is sesame oil. As mentioned earlier, many compounds having this moeity are synergistic, but the structures of piperonyl butoxide, mostly in livestock and animal shelter sprays. In summary, the synergists are used in many of the insecticide mixtures for home, garden and barn. Their mode of action of their binding to oxidative enzymes the would otherwise degrade the insecticide.

Inorganics

These insecticides which do not contain carbon are called Inorganic insecticides. Usually they are white and crystalline, resembling the salts. They are stable chemicals, do not evaporate, and are frequently soluble in water. They are mentioned here for their historical siignificance. Sulfur, 1, is very likely the oldest known *effective insecticide*. Sulfur and sulfur candles were burned by our great-grandparents for every conceivable purpose, from bedbug fumigation to the cleansing of a house just removed from medical quarntine of smallpox.

Sulfur is a highly useful; pesticide in integrated pest management programmes where target pest specificity is important. Sulfur dusts are especially toxic to mites of every variety, such chiggers and spider mites, thrips, newly-hatched scale insects, and as a stomach poison for some caterpillars. Sulfur dusts and sprays are also fungicidal particularly against powdery mildews. Several other inorganic materials have been used insecticides. These include compounds of mercury, boron, thallium, arsenic, antimony, selenium and fluoride. The only of these used extensively today is arsenic, which is used in two forms, the arsenites (salts of arsenious acid) and the arsenates (salts of arsenic acid).

The first commonly used arsenical was Paris green (green because of its copper content). It's a water soluble assenite. Next was lead arsenate. Finally, the third and last of these arsenicals was calcium arsenate, which was used for a time on vegetables in the 1930s and on cotton in the 1930s and 1940s. *Arsenicals* are truly stomach poisons, exerting their toxic action following ingestion by the insects. Their action is attributed to the arsenate or arsenate ion. The arsenicals have a rather complex mechanism of action. First, they uncouple oxidative phosphorylation (bysubstitution of the arsenite ion for phosphourus), a major enter-producing step of the cell. Second, the arsenate ion inhibits certain enzymes that contain enzymes that contain sulfhydryl (-SH) groups. And, finally both the arrsenite and arsenate ions coagulate protein by causing the shape or configuration of proteins to change.

Arsenical insecticides were very useful agricultural tools form 1930 until 1956, as we were making the transition form the simple to the complex synthetic molecules. They were, in fact, responsible for the initiation of large-scale insecticide applications eventually leading to the intensive use of fungicides and herbicides in modern agriculture.

Fluorine insecticides also included organic fluorine compounds, but these were of little importance and seldom used. The inorganic fluorides were sodium fluoride, used for cockroach and ant control around the home, and barium fluosilicate, sodium silicofluoride, and cryolite (NaF, $BaSiF_6$, and Na_3ALF_6, respectively). The last three were used for a time in plant protection. The fluoride ion inhibits many enzymes that contain iron, calcium and magnesium. Several of these enzymes are involved in energy production in cells, as in the case of phosphatases and phosphorylases.

Boric acid (H_3BO_4), used as an insecticide against cockroaches and other crawling household pests in the 1930s and 1940s, has returned in the 1980s. Although manufacturers claim it is "safe, does not evaporate, and continued to kill for years," in fact, it is only moderately effective, acting as both a stomach poison and adsorber of insect cuticle wax. The last group of inorganics are the silica gels or silica aergels. These are light, white fluffy silicates, used for household insect control.

The silica aerogels kill insects by adsorbing waxes from the insect cuticle, permitting the continuous loss of water from the insect body. The insects then gradually become desiccated and die from dehydration. These include Dir-Die®, Drianone®, and Drione®. The latter two are fortified with pyrethrum and synergists, which enchance their effectiveness.

Fumigants

The fumigants are small, volatile, organic molecules that become gases at temperatures above 5°C. They are usually heavier than air and commonly contain one or more of the halogens (CL, Br, or F). Most are highly penetrating, reaching through large masses of materials. They are used to kill insects eggs, and certain microorganims in buildings, warehouses, grain elevators, soils, and green-houses and in packaged products such as dried fruits, Ibeans, grain, land breakfast cereals. Fumigants, as a group, are narcotics.

The mechanism of action of the fumigants are more physical than chemical. The fumigants are liposoluble (fat soluble); they have common symptomogoly; their effects are reversi- ble; and their activity is altered very little by structural changes in their molecules. Narcotics induce narosis, sleep, or unconsciouness, which is effect in their action on insects Liposolubility appears to be an important factor in the action of fumigants, since these narcotics lodge in lipid-containing tissues, which are found in the nervous system.

Microbials

Certain microbes or micro-organisms are and used for controlling the insects, hence are called microbial insecticides like mammals insects are susceptible to diseases caused by fungi, bacteria, and viruses. In several instaces, these have been isolated, cultured, and mass-produced for use as pesticiedes. The insect disease-causing microorganisms do not harm other animals or plants. The reverse of this is also true. This method of insect control is ideal in that the diseases are usually rather specific. Undoubtedly the future holds many such materials in the arsenal of insecticides, since several new insect pathogens are identified each year. However, at the persent only a few are produced commercially and approved by the EPA for us on food and feed crops. There is still some concern regarding the every remote chance of human susceptibility to these diseases of invertebrate animals, hence the slow advances into this relatively new field exceptional precautionary testing. The microorganism *Bacillus thuringiensis* is a disease-causing bacterium whose spore are necessary induction. These spores produce compounds that injure the gut of the insect larve in such a way that invasion of the body cavity follows. This organism produces four substances toxic to insect.

The first, and most important, is a crytalline protein whose ingestion by caterpillars results in paralysis of their gut. The second is a toxin, a water, a water-soluble nucleotide derivative, that passes unchanged in the manure. The commerical preparation of Bacillus caused the loss of remaining two substance which are enymes. Only one insect virus has been registered for agricultural use by the EPA. Two of the most destructive pest caterpillars is agriculture, known as the *Heliothis* nuclear polyhedrosis virus, it is specific for *Heliothis zea* (corn earworm, cotton bollworm) and *Heliothis virescens* (tobacco budworm).

Recently registration has been extended not only for cotton but also for all crops attacked by *Heliothis* species including beams, corn, lettuce, okra, peppers, sorghum, soyabeans, tobacco and tomatoes. Proprietary names for this knaturally occuring viral pathogen are Elcar® and Biotrol VHZ® . Several other insect viruses are in the development stage, all of which are aimed at caterpillars and are only experimentally available. Viruses are highly specific and have modes of action that may not be identical throughout. Generally, the viruses result in crystalline proteins that are eaten by the larva and begin their activity in the gut. The virus units then pass through the gut wall and into the blood. There are units multiply rapidly and take over complete genetic control of the cells, causing thier death. A recent and innovative development in the agricultural use of microbial insecticides is the addition of feeding or gustatory stimulants.

The feeding stimulants apparently attract the caterpillars to treated foliage, which increases their consumption of the microbial. Two successfully marketed products are Coax® and Gustol® , both of which are wettable powders and used at 1.1 to 2.3 kg/ha. Abbott Lboratories developed a mycoacaricide, mycar miticide which is a new biological control. The microorganism is *Hirsutella thompsonii*, a parasitic fungus that infects and kills the citrus rust mite. Under optimum conditions *H. thompsonni* can infect spider mites and other nontarget mites.

It is, however, consistenty effective only against the citrus rust mite; this it is selective miticide. When sprayed on plants the spores grow into colonies that attach to the mite. In the presence of ample free moisture the spores germinates and infects the mite. The mite dies in about three days, and the fungus spread, continuing to propagate itself. In this manner it offers potentially long residual effectiveness.

Because the active agent is a fungus, all commercial fungicides, copper chemicals, and certain other metal salts such as zinc, lead, and manganese are detrimental to *H.thompsonii*. Because of its specificity *H. thompsonii* should be highly compatible with other suppression techniques used in the integrated past managemnt of citrus crops. For the control in grasshoppers, Sandoz Inc. developed a new biological control, *Nosema Locustae*. Marketed under the name NOLOC® and Trojan 10®, the microorganism is a protozoan. Depending on the method of application, climatic conditions, and grasshopper density, the protozoa kill up 50 per cent of the insects and sterilize up to 30 per cent of the survivors. Maximum effect occurs over a two to four-week period. Through subsequent generation by transmission through the eggs upto three or four years the residual effect of a single treatment continues to control grasshoppers.

Insect Growth Regulators

The first generation insecticides were the stomach poisons, such as the arsenicals, heavy metals, and fluorine compounds. The second generation includes the familiar contact insecticides roganochlorines, organophosphates, carbamates and formamidines. The third generation of insecticides are the *biorationals*. These chemicals are environmetally sound, closely resembling or identical to chemicals produce by insects and plants. Among these are the insect growth regulators (IGRs), a relatively new group of chemical compounds that alter growth and developments in insects. Their effects have been observed on embryonic and larval and nymphal development on metamor-phosis, on reproduction in both males and females, on behaviour, and on several forms of diapause. They include ecdysone (the molting hormone), juvenile sghormone (JH), JH mimic, JH analog (JHA), and their broader synonyms, juvenoids and juvegens.

More recently another growth effect, chitin inhibition, has been identified and is discussed later, with the EPA-registered IGRs. The IGRs are effective when applied in very minute quantities and apparently have no undersirable effects on humans and wildlife. They are, however, nonspecific, since they affect not only the target species, but most other arthropods as well. IGRs may play an important role in future insect control, consequently, when used with precision. Several hormone producing glands in insects perform the principal functions of which are the control of reproductive processes, motting land metamorphosis.

Here we are interested only in the hormone ecdysone, which is responsible for molting, and JH, which inhibits or prevents metamorphosis. When insects are treated with ecdysones, they usually die in all stages of growth, making ecdysones similar to second-generation insecticides. One attractive feature of ecdysones are potential tools is their widespread distribution in plants,, and these may play as yet unrecognized roles in insect-plant relationships. Keen interest has been directed toward JHs. These are not in the usual sense, toxic to insects. Instead of killing directly, they interfere in the normal mechanisms of development and cause the insects to die before reaching the adult stage.

One JH is the classical juvabione, found in the wood of balsam fir. Its effect was discovered quite by accident when paper towels made from this source were used to line insect-rearing containers, and the insects' development was suppressed. Some of these plant-derived substances actually serve to inhibit the development of insects feeding thereon, thus protecting the host plant. These are referred to broadly as antijuvenile hormones, or more accurately, *antiallatotropins*. Recently, antillatotropins, called *precocenes*, have been discovered. These are considered a potential fourth generation of insect control agents.

Incidentally, the name precocene results from the precocious metamorphosis stimulated by compounds having the chromene nucleus. It is known that precocenes depress the level of juvenile hormone below that normally found in immature insects, although the mode of action is still unclear. Dramatic results with JHs have been obtained in the laboratory, the most promising effects being on mosquito larvae, caterpillars, and hemipterans (bugs), although effects have been observed on practically all insect orders. Most insect species respond to treatment with JHs by producing extra larval, nymphal, or pupal forms that vary from giant, almost perfect, forms, to intermediates of all sorts between the immatures and the adults. For the most part, the periods of greatest sentivity for metamorphic inhibition are the last larval or nymphal stages and the pupa, in those having complete metamorphosis. One recognizable problem is the precision of timing applications to achieve meximum damaging effect on the uncoming life stage of a particular insect.

For practical purposes. IGRs could be used on crops to suppress damaging insect numbers. They would be applied with the purpose lof preventing pupal development or adult emergence, thus keeping the insects in the immature stages, resulting eventually in their deaths. To date, only three IGRs are registred by the EPA. The first, methoprena (Altosid®), manufactured by the Zoecon Corporation, was registered early in 1975 as a mosquito growth regulator, for use against second-through fourth-larval-stage floodwater mosquitoes at 0.11 to 0.14 kg/ha to prevent adult emergence. Larvae exposed to methoprene continue their development to the pupal stage, when they die. When applied to pupae or adult mosquitoes. Additionally methoprene has been formulated as PrecorØ for indoor control of dogs and cats fleas. It acts by interrupting the flea's life cycle at the larval stage, preventing emergence of adult fleas for lup to 75 days. Its use in combination with a conventional insecticides is usually necessary to control adult fleas not affected by the growth regulator.

Methoprene is also registered for use on tobacco to control cigarette beetles (KabatØ), on cattle feed to control horn flies, and on mushrom-growing media to control hungus gnats (ApexØ). In order to control mosquitoes in drinking water it has been approved by the World Health Organization for its use. The second IGR is diflubenzuron (Dimilin®), manufactured by Thompson-Hayward Chemical Company. Currently registered for gypsy moth caterpillars in forest and for the boll weevil in cotton, it is awaiting registration for a wide variety of pests.

Experimentally, it has controlled insect pests in forest, woody ornamentals, fruit, vegetables, cotton, soyabeans, citrus, larvae of flies, midge, gnats, and mosquitoes : on mosquitoes, a dose as small as 1.5g/ha is effective. Diflubenzuron is not truly a growth regulator of the juvenoid class but rather another insecticide with a different mode of action. However, it is tentatively classed with the IGRs. It acts on the larval stages of most insects by inhibiting or blocking the synthesis of chitin, a vital and almost indestructible part of the hard outer covering of insects, the exoskeleton. The third, and most recent, IGR is kinoporene (EnstarØ),also developed the Zoecon Corporation. It is effective against aphids, whiteflies, mealybugs, and scales (both soft and armored) on ornamental plants and vegetable seed crops grown in greenhouses and shadehouses. It is specific for insects in the order Homoptera, and results in a gradual reduction rathar than an immediate kill, by inhibiting development, reduction egg-laying, killing eggs already laid, and sterilizing mature whiteflies and aphids.

It is virtually nontoxic to humans and other warm blodded animals because it is specific for insects like all other IGRs. Cans IGRs become successful pest control agents? Certainly they can in time. They will, however, have to meet general criteria for other pests control agents; thus they must be effective in reducing insect populations below economic damage levels, be competitive with second-generation insecticides in cost, and have no undersirable side effects. In summary, IGRs hold intriguing possibilities for future use in practical insect control. It should be kept in mind that IGRs are insect-controlling chemicals and thus fall within the same legal confines as other insecticides. Being Toxic to populations of insects rather than to individuals they are however distinguished from other. Likewise the ability to control the fecundity (reproductive capacity) of the pest to the cause of the ultimate success of any pest control agent.

Insect Repellents

Historically, repellents have included smokes, plants huge in dwellings or rubbed on the skin as the fresh plant or brews, oils, pitches, tars, and various earths applied to the body. Camel urine sprinkled on the clothing has been of value, although questionable, in certain locales. Camphor crystals sprinkled among woolens have been used for decades to repel clothes moths. Before humans developed a more edified approach to insect olfaction and behaviour, it was assumed that if a substance was repugnant to humans it would likewise be repellent to annoying insects. Prior to World War II, there were only four principal repellents: (1) oil of citronella, discovered in 1901, used also as a hair-dressing fragrance by certain Eastern cultures, probably to control lice and other head ectoparasites; (2) dimethyl, phthalate, discovered in 1929, (3) Indalone® introduced in 1937, and (4) Rutgers 612, which became available in 1939. It became essential to find new repellents that would survive both time and dilution by perspiration, with the onslaught of World War II and the introduction of American military personnel into new environments, particularly the tropics.

The ideal repellent would be nontoxic and nonirritating to humans, nonplasticizing, and longlasting (12 hours) against mosquitoes, biting flies, ticks, fleas and chiggers. Unfortunately, the ideal repellent has still not been found. Some repellents have unpleasant odours, require massive dosages, are oily or effective only for a short time, irritate the skin, or soften paint and plastics. Insect repellents come in every conceivable formulation - undiluted, diluted in cosmetic solvent with added fragrance, aerosols, creams, lotions, treated cloths to be rubbed on the skin, grease sticks, powders, suntan oils, and clothes-impregnating laundry emulsions.

Regardless of the formulation, the periods of protection they offer vary with the chemical, individual, the general environment, insect species, and avidity of the insect. What happens to repellents once applied ? Why aren't they effective longer than one or two hours? No single answer is satisfactory–but generally it is because they evaporate, are absorbed by the skin, are lost by abrasion of clothing or other surfaces, and are diluted by perspiration. The following chemical structures are those of the most commonly found in today's repellents. Of these, diethyl toluamide (Delphene®) is by far superior to all others against biting and mosquitoes.

Insecticide Selectivity

We must supplement insecticides with other forms of insect control as there are no main tool or insect control at least or the foreseeable future. Selective insecticides must be used and insecticides must be used in selective way, lest the effect of beneficial insects and natural enemies be minimizeds in any programme requiring multiple insecticide applications. Insecticides affect various insects differently. An insecticide that is lethal against one group may have little, if any, effect on another. Generally, however, most insecticides kill many kinds of insects; broad-spectrum insecticides are those that kill more kinds than others. A selective pesticide is one that is toxic to some pests, but has little or no effect on other similar species. For example, a selective herbicide is one that kills certain undersirable species of weeds while harming the crop plants little or not at all. Certain fungicides are selective to the point that they control only powdery mildews and no other fungi.

A selective insecticide kills selected insects, but sprares many or most of the other insects, including beneficial species, either through different toxic action or the manner in which the insecticide us used. Though desirable the development of such a product would be uneconomical for the manufactures since not enough of the spectes specific insecticide could be sold to recoup research development and marketing costs. Thus, the chemical industry seek rather to develop a general purpose, broad-spectrum chemical that will control not only elm leaf beetles, but boll weevils on cotton, caterpillars on vegetables, the gypsy moth in deciduous forest, green bugs in small grains, alfalfa weevis, the codling moth in apples, and scales on citrus.

General-purpose insecticides offer the best possible incentives in the economic market : satisfied growers and contented stock-holders. A selective insecticide, in the narrowest use of term, simply means that the active ingredient is inherently more toxic to one group of insects than to others; in other words most groups of insects are physiologically more tolerant to the chemical than are the few readily killed by it. More commonly, selective is used to designate an insecticide that does not harm a beneficial species of insect while killing pest species. This physiological selectivity results from differences between target and nontarget species in (1) penetration rate of the toxicant, (2) tissue binding and loss of the toxicant, (3) the speed of toxicant and metabolite excretion, (4) metabolic alteration or detoxification, (5) target sites of action or biochemical lesion, or (6) polyfactorial selectivity where more than one of the above processes is involved. (For a clear, detailed discussion of physiological selectivity, see Hollingworth, 1976.)

Physiologically nonselective insecticides may achieve use selectivity if they are applied in such a way that certain insect groups are more adversely affected than others. Use selectivity may result from the timing of the application, the formulation used, the dosage level, or numerous there techniques. From a practical standpoint this approach offers the most immediate hope of being included in precision integrated pest management programmes. The selective use of an insecticide to kill a particular pest while permitting other insects to escape, particularly beneficials, achieves the same selective effect as the use of physiologically selective insecticides. The existence of both the pest and the desirable species, and utilize some difference in their habits, distribution, or biology to devise a discriminatory method of application, all these must be taken into a count by the planner if they use selectivity.

Although some insecticides are intrinsically more toxic physiologically to certain insects, and some are equally toxic to different species, they can be used selectively to affect pests more than beneficial insects. Insecticide selectivity thus can be achieved not only by restricted toxicity but also by precision in timing, the calculation of effective rates of application, the use of materials with short residual lives, methods of application such as seed treatment, spot treatments restricted to areas in which pests vastly outnumber beneficials, or use of those formulations that tend to enhance the survival of beneficials.

In general, selectivity is obtained by one of the following paractices : (1) use of physiologically selective insecticides that are more toxic to pest species than to others; (2) timing of applications such that detrimental effects on the pest's natural enemies are minimized; and (3) reduction of dosages to levels that adequately control the target pest while sparing relatively large numbers of natural enemies of the target pest and other potential pests. The selective use of insecticides is not simple.

In most instances it requires an intimate knowledge of not only the target pest but also the secondary pests beneficial species. Researchers are directing their efforts towards integrating the use of insecticides and beneficial insects. Effective insect control will require both toxicological research in the development of physiologically selective insecticides and applied ecological research in the use of available insecticides to achieve selective action. The selective use of insecticides requires an intimate knowledge of not only the target pest but also the secondary pests beneficial species.

Application of Insecticides

Insecticides are the chemical substances used in controlling the pests. The usefulness of any insecticide depends in a very large measure upon its proper application, and this is determined by the properties of the insecticide, the nature of the pest or pest complex to be controlled, and the site to which the application is to be made. The three general methods of applying insecticides are as sprays, in which water or oil is used as the carrier for the toxicant, as dusts, in which a fine dry powder is the carrier, and as fumigants, in which the insecticide in applied as a gas. The equipment for the application of insecticides ranges from such simple devices as the puff duster and "flit gun" to complex machines such as the mist blower and spraying helicopter, and this development is entirely a product of the last 100 years.

Before that time, liquid insecticides were applied by a bundle of twigs, a feather, or a brush broom, and dusts by a bellows or blowing tube. The progress which has been made is a product of detailed knowledge of physics and engineering coupled with a vast amount of trial and error. Much remains to be learned, and relatively simple devices may yet be produced which will effect vertiable revolutions, as for example, the development of the aerosol "bomb," which from its humble begining in 1942 has become a standard household article, of which 65,900,000 units were sold in 1959 for insect control.

Precautions in Insecticide Applications

Usually insecticides are applied when populations of a pest species reach the economic threshold. In this way economic damage is prevented and needless treatments are avoided. Other factor must often be considered as well, such as the timing of applications so as to coincide with the stage of development of the insect pest which is most easily killed or with the stage in the seasonal development of the plant when it well best withstand the treatment. Timing to observe the safe interval before harvest which is required to attenuate insecticide residues to legal levels and to avoid periods of bloom for protection of bees and other pollinating insects is also extremely important. Since proper timing varies widely with different crops, different insecticides, and in various areas of the country, the grower should follow only approved spray programmes of state and federal agencies, observe all label precautions and directions, and whenever in doubt should secure the advice of a trained entomologist who knows local conditions regarding spray programmes for different crops and insects. In general, the earlier the insecticide is applied after the insects appear, the easier it is to destroy them, but in most cases applications should be made only when populations reach economic thresholds.

Most of insecticides are violent poisons to human beings and livestock, as well as to insects. Therefore, they should be plainly labeled and, together with mixing vessels, kept out of reach of children. Animals must be kept away from liquid sprays and not allowed to pasture in treated areas. All discarded insecticide containers should be burned or buried. The spray applicator should familiarize himself with the hazards of various insecticides before use and determine the proper protective measures necessary for safe application, as furnished by the manufacturer.

In general, these will include the use of protective clothing, rubber gloves, and goggles when handling spray concentrates and a protective mask if there is any possibility of inhaling toxic dust, mist, or vapour. Prolonged wetting by sprays or other contamination should be avoided, and clothes changed at least twice daily, followed by the liberal application of soap and water to remove any skin contamination. In rainy weather the spraying and dusting should be avoided although many sprays will adhere satisfactorily to plants if the spray has time to dry before rain falls. Winter sprays should not be applied when the temperature is below freezing or when the trees are wet with snow or rain. When possible, spraying should also be avoided in very hot weather, above 90°F, because of increase susceptibility of plants to insecticide damage and because of the increased hazard to the operator from skin and inhalation absorption of poisons.

Sprays and Spraying

Historically, dusts have been the simplest formulations of pesticides to manufacture and the easiest to apply. In recent years, because of the increased effectiveness of the new organic insecticides, concentrate sprays have largely replaced dusts on many field and vegetable crops. Sprays are made up of solutions, emulsions, or suspensions of the toxicant. The liquid phase is usually water, but light oils are also employed. Insecticidal sprays are con- veniently described as space sprays or aerosols, directed against flying insects, and residual sprays, applied to surfaces of plants, animals, or structures frequented by insects.

Importance of Droplet Size

The size property of spray droplet is very significant and conveniently measured in microns (μ a or 0.001 millimeter). In every spray cloud there is a considerable spectrum of droplet sizes and the cloud is most conveniently characterized by a mean value which expresses the normal frequency distrubution of the droplets. The most commonly used characteristic is the mass median diameter MMD or D_m, which is that throretical droplet diameter which divides the volume of the spray into equal parts. The surface median diameter SMD or D_o is the theoretical droplet diameter which has the same surface to volume ratio as that of the total spray. D_m is a useful measure of the effectiveness of the spray in terms of the amount of active material deposited per unit area, while D_o defines the amount of exposed liquid surface and thus the coverage which may be obtained. D_m range from 1.1 to 1.6 times D_o, depending upon the nature of the atomizing device. However, with the common fan, swirl, and twin-fluid atomizers used in the application of insecticides. The average ration D_m/D_o is relatively constant with a value of 1.28, as studied by Fraser.

Droplet-size Requirements

The droplet-size of the sprays as per requirements for various insecticidal spraying operations cover a 300-fold range of droplet diameters and a 27,000,000 range of droplet volumes and extend from the 1-to30- micron size of true aerosols to the 100- to 500-micron size of hydraulic sprays. With contact sprays, the zone of efficiency for air-blast sprays is about 30 to 80 microns and for hydraulically driven sprays, between 100 and 300 microns. Special problems are encountered in aircraft spraying, and in general the most efficient range is from 100 to 300 microns.

Space sprays or aerosols

The principle of space spraying is to suspend a cloud of droplets in the space through which the insect pests are flying and thus to force the insect to accumulate a lethal deposit by colliding with the droplets. The successful application of this method is dependent upon physical laws governing the behaviour of aerosols, which are defined as colloidal suspensions of matter in air and involе an understanding of the insect flight behaviour. It is clear that the insecticide will be effective only as long as the droplets remain suspended in the free-air space.

Spraying Equipment

There are Narious types of spraying equipments. The size and type of sprayer to be used will depend upon the amount and kind of work to be done. However, spraying is at best a disagreeable task, and every effort should be made to have equipment of sufficient capacity so that the work will be done thoroughly and efficiently. The best spraying equipment can easily be ruined by failure to give it proper care. Clean water should always be pumped through the sprayer after use, the flush out all corrosive spray materials, and the metal parts should be oiled to prevent rust.

It is not advisable to apply insecticides with spraying equipment which has previously been used for the application of herbicides of the plant-growth-regulator type such as 2,4-dichloro-phenoxyacetic acid (2,4-D), because very small traces of such materials may seriously injure certain plants. There are many types of sprayers, but the most common sprays are hydraulic sprayers, which employ only pressure nozzles for the formation and distribution of the spray, and air-blast sprayers, which utilize an air stream for distribution and/or breakup of the spray.

Hydraulic Sprayers

The various types of hydraulic sprayers include (a) the compression sprayer, which is actually a special type in which compressed air is used to force the liquid through a pressure atomizing nozzle. The most familiar type is a 4-gallon cylindrical tank with a hand piston pump which is operated at 40 to 80 pounds per square inch and is very popular with the home gardener. The bucket or stirrup pump sprayer (b) has a single-or double-action plunger pump clamped into a 2-gallon bucket and operates at up to 150 pounds per square inch; (c) the napsack sprayer is a 2- to 5-gallon tank carried as a napsack and containing a diaphragm pump and agitator operations at 50 to 80 pounds per square inch by a lever carried under the operator's shoulder. These are especially suitable for small garden operations or treatments in inaccessible spots, as in mosquito larviciding or residual house spraying. The barrel pump (d) is a larger plunger-type spray pump which operates at a pressure of 200 to 300 pounds per square inch and uses a 15- to 45-gallon container as a reservior. The wheelbarrow sprayer (e) is essentially a portable barrel pump which is suitable for spraying small fields, orchards, or farm buildings. Most modern wheelbarrow sprayers have rubber tires and contain a two-cycle gasoline-engine-driven pump of 1.5 to 3.0 gallons per minute to 200 to 250 pounds per square inch and have a capacity of 15 to 50 gallons. Hydraulic power sprayers (f) range from the power whellbarrow sprayer through trailer-type field and row-crop sprayers to truck-mounted orchard sprayers. These power sprayers have 50- to 500-gallon capacities, operate up to 400 to 800 pounds per square inch, and have discharge capacities ranging from 5 to 80 gallons per minute. Discharge may be carried through one or more spray hoses or through a horizontal or vertical spray boon equipped with multiple swirl spray nozzles.

Parts of a sprayer

The important components of a spray outfit are similar for the various types, some of which are as follows: The cylinder is the portion of the pump in which the pressure is developed. It must be of noncorrosife material, and in large power sprayers, which have two to four cylinders, the lining is generally of acid-resistant porcelain. The plunger or piston forces the spray liquid through the cylinder and is made to fit tightly inside the cylinder by means of packing. Plungers may be fitted with molded rubber and fabric plunger cups which act as inside packing, expanding against the cylinder wall on the pressure stroke, or outside packing may be used which is compressed against stainless-steel plungers. Valves, usually of the ball type, and valves seats made of hardened stainless steel, direct the flow of spray liquid. They must be readily accessible for cleaning in case of clogging by particles of dirt or spray materials.

An air chamber is used to equalize the pressure and remove excessive strain on the pump. The discharge opening leading to the nozzle is located at the bottom of the air chamber. The tank, in which the spray ingredients are mixed and held, is usually constructed of metal with an enamel lining or of stainless steel. Since many spray ingredients, such as wettable powders, are not soluble in water, an agitator in necessary to keep the finely divided particles in suspension or the emulsion evenly distributed. In power sprayers this is commonly effected by two or more flat paddles or propellers on a rotating shaft near the bottom of the tank. A pressure gage indicates the pressure being maintained, and the pressure regulator maintains uniform pressure on the spray nozzles and allows the pump to operate at a greatly reduced load when spray is not being discharged. This is accomplished by a spring-and-ball valve, which can be set to lift at the desired pressure and permit the excess liquid to bypass to the tank. Strainers, over the intake to the tank and over the suction or intake pipe leading to the cylinders, keep the spray liquid free of troublesome foreign matter. The discharge pipe should be free from abrupt angles that cut down the pressure between pump and nozzles. The hose for orchard spraying shoed be at least 25 feet long, of ½ -to ¾ -inch inside diameter and four-to seven-ply strength.

The friction of the spray liquid against the inside surfaces of the spray hose results in a considerable loss of pressure at the nozzle. For example, with a 50-foot hose ½ -inch in diameter and a flow rate of 10 gallons per minute, the friction loos in 70 pounds per square inch. The pump of most power sprayers is of the direct-displacement-plunger type. The recent development of high-capacity, high-pressure boom sprayers for orchard spraying has been possible through the utilization of centrifugal pumps with capacities of up to 125 gallons per minute at pressures of 800 to 1,000 pounds per square inch. Such pumps, however, are subject to severe wear by abrasie wettable powders. Spray booms. The boom is a light hollow tube or pipe which is used to carry the spray liquid under pressure from the pump to one or more nozzles. The simplest type is the spray rod of lance, which is a tube from 3 to 8 or even 12 feet in length, with a cut off valve at one end and a simple swirl spray or flat fan nozzle at the other.

With power sprayers, a broom or fog-drive gun in sometimes used for spraying fruit trees. This consists of a spray rod from 3 to 5 feet long, with a cutoff at the base and a group of three to eight nozzles at the tip.

Residual spraying indoors

The spraying of houses and barns with a heavy deposit of a long-lasting residual insecticide is widely practiced for the control of flies, mosquitoes, bed bugs, cookroaches, silverfish, and ectoparasites of animals and is one of the principal weapons for the eradication of malaria. This type of application is most efficiently performed with a small-capacity, portable spray pump such as the napsack, compression, or bucket sprayer or the wheelbarrow power sprayer. The World Health Organization has specified that residual-spraying operations be conducted with a spray pump having a regulator to produce a constant pressure of 40 pounds per square inch and that a flash fan nozzle discharging at 0.2 gallon perminute with a spray angle of 60 to 65 degrees be used.

During the spraying operation, the nozzle should be held approximately 18 inches away from the surface being streated, and about 150 to 250 square feet of wall surface should be covered per minute. Adherence to the suggested conditions will avoid wasteful and unsightly discharge and produce a spray of maximum impingement and deposition.

Air-Blast Sprayers

In the air-blast sprayers a relatively large volume of high-velocity air is used to break up the spray droplets and to carry them to the target, as in the twin-fluid atomizers. For convenience, however, we shall also discuss air-blast equipment in which the spray may be produced by other means but dispersed by means of an air blast. The common hand atomizer or "flit gun" is perhaps the simplest familiar device of this type. Here a piston-type air pump furnishes the air blast, which also forces the spray liquid from a reservior up a small-diameter delivery tube, and atomization results from the apposition of the air and liquid streams, according to equation. The compressd air-paint spray gun is a larger example of the type. In the use of air-blast equipment for the treatment of field or orchard crops the 30- to 80-micron droplet range seems most suitable because of the questions of coverage and dynamic catch, and thus the matter of air volume and velocity is of great importance. This is particularly true in orchard spraying, where it is necessary to have sufficient air-moving capacity to agitate the air in all parts of the tree and to displace most of it.

Table 9.1 Air velocities at various distances

Diameter of Round outlet, in.	*Volume of air, cu. ft./min.*	*Velocity at indicated distance from blower, m.p.h.*				
		Outlet	*10 ft.*	*25ft.*	*50ft.*	*100ft.*
4	1,300	170	40	8	2	0
8	3,800	125	40	10	3	0
12	7,500	125	60	30	7	2.5
24	28,000	120	75	40	17	7
54	27,000	50	35	30	8	3

It has already been stated that volumes of air required with various-sized delivery tubes to produce adequate velocities at various distances from the blower. It is evidence that to secure adequate dynamic catch of 30-micron droplets at distances of 25 to 50 feet from the bolwer, very large volumes of air, ranging, from 7,500 to 28,000 cubic feet per minute are required. It has been shown that to spray a moderate-sized orchard tree 20 by 20 by 20 feet (8,000 cubic feet) with a mist blower travelling at 0.5 mile per hour, 17,600 cubic feet per minute of spray are required, and of course this requirement is directly proportional to the volume of the tree and the speed of travel of the sprayer. Two types of air-blast sprayers have been constructed the spray blower and the mist blower. In the spray blower, the liquid is broken up by hydraulic nozzles and carried to the target by the air blast, while in the mist blower the spray liquid is finely dispersed by means of air-atomizing nozzles or spinning disks and then carried by the air blast. Both types of equipment employ axial-flow turbines or centrifugal fans to generate from 4,000 to 60,000 cubic feet of air per minute at velocities ranging from 60 to 250 miles per hour.

The successful application of air-blast sprayers to field and row crops is difficult because of problems in equalizing spray distribution at varying distances from the point of discharge. Various devices, including manifolds with a number of outlets, fishtails, elongated slots, and rotating or oscillating outlets to sweep the area, have improved the evenness of coverage. Air blasts of 3,500 to 4,000 cubic feet per minute at 60 miles per hour have given effective results over swaths of 15 to 35 feet. The commercial units available, however, are generally modifications of orchard sprayers and use air blasts of 20,000 to 40,000 cubic feet per minute at moderate velocities of 70 to 90 miles per hour to avoid crop damage, damage, directed at right angles to the path of travel from either one or both sides. The rows over which the machine passes are simultaneously sprayed with a low-volume boom. Such equipment will treat a swath 40 to 60 feet wide suing 5o to 100 gallons per acre. A number of large spray blowers are in extensive commercial use for orchard spraying, and the Speed Sprayer is perhaps the best known. This equipment will apply from 50 to 500 or more gallons per acre and delivers up to 45,000 cubic feet per minute of air, with nozzle pressures of 50 to 70 pounds per square inch and pump capacities of 55 to 140 gallons per minute. The spray is delivered into the air blast from as many as 58 to 264 swirl spray nozzles.

The spray duster employs 40,000 cubic feet per minute of air passing through two vertical fishtail outlets to distribute high-pressure spray at 100 to 1,000 gallons per acre from vertical banks of nozzles. The air flow is periodically changed in direction by the oscillation of a hinged section of the outlet to aid the spray penetration by moving leaves and twigs. The most difficult problem with air-blast sprayers in orchard spraying is equalizing the coverage between the top and bottom of the tree.

Aerosol Generators

In the aerosol culture, we have raised being bombs, hair sprays, under arm deodorants, adhesives, car wax, fabnc, finishers etc. Aerosols for insecticidal use have been produced by a number of methods: (a) by burning the insecticide; (b) by spraying a solution of insecticide onto a heated surface; (c) by forcing the insecticide, dissloved in a liquefied, low-boiling gas through a capillary tube; (d) by spraying at high pressure through a swirl spray nozzle of low capacity; (e) by forcing a solution of insecticide between two closely apposed and rapidly spinning disks; and (f) by the use of a high-velocity stream of gas, air, or steam acting upon a stream of liquid in some type of twinfluid atomizer. The exhaust aerosol generator uses the hot exhaust gases from an internal-combusion engine to produce atomization.

The droplet size is directly proportional to the ratio of flow of the liquid to the flow of air and can be predicated. The use of the hot exhaust gases favours the production of droplets of aerosol dimensions because of the very high velocitics, 600 to 1,400 feet per second, obtainable with the hot gases and aerosols are not, however, produced at pressures below 25 pounds per square inch gage. These low-or moderate-pressure aerosols require a more critical nozzle design than the simple capillary tube of the high-pressure aerosoal. In order to introduce turbulence and the formation of small bubbles of gas which serve as nuclei for boiling, a mixing chamber is provided into which the liquefied gas flows through a 0.015-inch-diameter constriction and from which the mixture issues through a 0.020-inch orifice. The dispenser value consists of a ball of metal or nylon held by a spring against a valves seat of synthetic rubber. The use of low-pressure aerosols requires higher boiling propellants or mixtures such as 54 per cent dichlorodifluoromethane and 46 per cent trichlorofiuoro-methane, which develops 37 pounds per square inch gage to 70°F.

Other useful combinations include the addition of up to 25 per cent by volume of isobutane of n-pentane to the above mixture, II and III, I and IV, and I and V from Table. Mextures such as these form bubbles in the nozzle more readily than single- component propellants and produce better aerosol dispersion at low pressures. The insecticids most commonly dispersed by the liquefied-gas method are DDT, methoxychlor synergized pyrethrins or allethrin, lindane, chlordane and dieldrin. High purification is necessary to prevent clogging of the capillary nozzle and decomposition of the insecticide and consequent corrosion of the container, although this can be inhibited by including

0.1 per cent propylene oxide. Auxiliary solvents such as petroleum distillates, polymethylnaphthalenes, or cyclohexanone have been used to solubilize the insecticides in the propellant gas. A small percentage of nonvolatile material such as 10-W lubrication oil has been found effective in preventing too rapid volatilization of the aeroso droplets. The ultimate droplet size is determined by the concentration of nonvolatile materials, and the optimum for flying insects of 5 to 20 microns (D_m 10 to 5 microns) is obtained with about 20 per cent nonvolatile material in a high pressure formulation and about 15 per cent in a low-pressure formulation. Typical formulations of both types are:

High-pressure Aerosal	Per Cent
DDT	3
Pyrethrins, 20% extract	1
Alkylated naphthalenes	16
Dichlorodifluoromethane	80

Low-pressure Aerosol	Per Cent
Methoxychlor	2.0
Pyrethrins	0.25
Piperonyl butoxide	1.0
Alkylated naphthalenes	11.75
Dichlorodifluoromethane	30.0
Trichloromonofluoromethane	55.0

Such aerosols should be used at the rate of 2 to 4 grams per 1,000 cubic feet of space and the ischarge rate is approximately 1 gram per second. The most popular containers are the 1-pound size are available for larger enclosures and for greenhouse work. For the latter purpose a formulation of 5 to 10 per cent parathion or TEPP in methyl chloride, 90 to 95 per cent, is widely used in 10-pound bomb with a 2-foot rod and nozzle. This should be used only when wearing a protective mask. The liquefied-gas method has been used also to produce bombs delivering residual- type sprays for the control of cockroaches, ants and clothes moths and carpet beetles, at a low pressure of 20 pounds per square inch gage. For this purpose, the amount of propellant is reduced to 30 to 35 per cent of the total, and this produces a relatively coarse discharge of about 30 to 35 microns, which settles rapidly and has good wetting properties. Typical formulations are:

Roach and Ant Spray	Per Cent
Dieldrin	0.5
Pyrethrins	0.04
Piperonyl butoxide	0.1
Petroleum distillate	69.36
Propellant	30.0

Mothproofing Spray	Per Cent
DDT	3.0
Perthane	3.0
Petroleum distillate	59.0
Propellant	35.0

Dusts and Dusting

Dusting is the oldest and the simplest method of poisoning plants for insect control it has been much more extensively practices during the past 50 years. This is largely due to the following undesirable features of dusting: (a) decreased efficiency due to less efficient deposition on the plant, (b) increased drift problems with poisonous materials, (c) increased cost of dust diluents compared with water, (d) tendency of carrier and toxicant to separate in the air unless the diluent particles are coated with the toxicant, (e) difficulties in incorporating several insecticides or other agricultural chemicals into a single dust, and (f) incrased hazard to operator from inhalation of toxic dusts. Despite the disadvantages, dusting is in many cases much easier, lighter, and several times faster than thorough coverage spraying. The equipment is considerably simpler and lighter than the spray rig and can be better used in hilly territory and under muddy conditions.

Dusting is obviously the favoured means of insect control in regions with a limited water supply and is a useful means of achieving insect control just prior to harvest without exceeding insecticide-residue tolerances. Most insecticidal dusts are very finely divided, and typical surface mean diameters are talc 1.8 microns, pyrophyllite 2.2 microns, lead arsenate 6 to 10 microns, calcium arsenate 1 to microns, and ground rotenone roots 6 microns. Such dusts pass nearly completely through a 325-mesh screen of 44 micron aperture. Thus when failling free in air these dust particles should behave similarly to aerosol droplets.

However, dust particles not only very greatly in size and shape but also tend to agglomerate during the dusting operation, so that many particles adhere tightly together, and it is very difficult to predict their settling and drifting pattern. Since the settling velocity of these small particles is directly proportional to particle density, dusts of such materials as lead and calcium arsenates settle considerable faster than dusts of botanicals such as pyrethrum or rotenone. The density of thedust diluent and the presence of dush conditioners such as stabilizers and fluffing agents also affect the dusting behaviour. The electrostatic charge on dust particles has often been suggested as a means of importance in causing the dust to adhere to plant surfaces which are negatively charged. A number of devices have been designed to increase the charge on the insecticidal dusts, either by friction or by passing the dust particles through a flow of positive ions from a highpotential electrical discharge. Inorganic dusts such as arsenicals, fluorides, copper, lead, and sulfur assume positive charges, while botanicals assume negative charges. Diluents such as pyrophyllite and gypsum are also positively charged, and diatomite, clays, and tacls are negatively charged. The use of electrostatically charged dusts has been claimed to prevent agglomeration, to increase adherence and even distribution on both sides of leaves, and to increase the dust deposition as much as 4 to 10 times.

Insecticidal dusts are commonly applied to crop plants at rates of 10 to 50 pounds per acre. However, organic insecticides such as DDT give somewhat similar insect control as long as the amount of active ingredient and plant coverage are constant, regardless of whether applied as 3 to 10 pre cent dust. The use of dusts impregnated with oily materials such as mineral oil or polymethyl naphtalenes has sometimes been of advantage in increasing deposits and decreasing drift. For some insecticidal applications, such as in the control of the European corn borer in the whorls of the corn plant, ants and other soil-inhabiting insects, and mosquito larvae in water under dense vegetation, the use of relatively coars granular dusts of 30- to 60-mesh (250 to 590 microns) serves to prevent the insecticide from adhering to plant foliage and enables it to reach the soil or water surface. The use of these granular materials also greatly decreases drift problems.

Dusting Equipment

A good duster, like a good sprayer, is one that will spread the insecticide—in this case in dry form—in such a manner as to give the most uniform coating possible to the plants being treated. The simplest dusters are the small hand dusters adapted to the home garden in which a plunger discharges an air blast through a chamber containing from 0.5 to 2.0 pounds of dust and the dust cloud is emitted through a small flared nozzle or tip. Next in size are the blower dusters, which consist of an enclosed fan rotated by a hand crank which sends a continuous blast of air through a small chamber into which the dust is fed by an adjustable gate. An agitator stirs the dusts in the hopper and provides for an even feed of the dust into the discharge chamber.

The dust cloud is forced out through a delivery tube extending nearly to the ground, which may end in Y or fishtailed nozzle. Such a duster will contain 5 to 10 pounds of dust and can be used to protect 1 to 2 acres of truck crops. The most recent innovation in blower dusters is a back pack unit containing a small gasoline engine which drives a fan providing a steady stream of dust.

The bellous or knapsack duster of similar capacity uses the extension and compression of a bellows attached to the back of the duster to force air through the discharge chamber. This gives a discharge of the dust in puffs instead of the continuous stream as in the blower type. This design is best adapted to somewhat isolated plants such as small trees and shrubbery.

Traction dusters have been widely used for vegetable-and field-crop dusting, especially for small acreages of cotton. In this type the fan is driven by the wheel or wheels on which the duster runs. Such machines, containing 50 to 100 pounds of dust, may develop enough power to turn a 12- to 16-inch fan at 2,500 to 3,000 revolutions per minute and may operate with a boom containing up to eight nozzles for a four-row operation. Because the speed of the machines will vary with the rate at which they are drawn through the field and the wheels may slip in soft ground, it is always difficult to maintain an even discharge from traction-type dusters.

Power dusters operated by small gasoline engines are the most practical for orchard and field-crop work. The fan is operated at about 3,000 revolutions per minute. For field and row crops, booms up to 30 feet in length with from 8 to 18 delivery nozzles are used. These may be individually connected by flexible tubing to a peripheral manifold surrounding the fan, or they may be attached to a tapering hollow-boom manifold to equalize the discharge rates at each nozzle.

The *orchard dusters* commonly have a single discharge outlet or a double fishtail arrangement for discharging dust from both sides and a centrifugal or squirrel-cage fan to discharge large volumes of air, up to 20,000 to 40,000 cubic feet pet minute. This equipment will project dusts for considerable distances even to the tops of forest trees. In these machines the dust is mechanically fed by an agitator or metered by a worm or helix arrangement directly into the fan or blower, assuring a fairly even rate of discharge. However, in practice, the uniformity of discharge of power dusters may vary over a severalfold range because of the changes in the head of dust above the hopper opening and the tendency of dusts to cake and is greatly inferior in the respect to comparable spray equipment. The most recent developments to remedy this deficiency include equipping the hopper with an elevator which lifts the dust to the top, where it is force-fed into the fan, or the use of a rotating hopper which scoops up the same measure of dust at each revolution and feeds it into the fan, and the employment of a conveyer-belt system at the bottom of the hopper which transports an even layer of dust through a variable shutter into the fan. Such desings have reduced the variability of discharge to a few per cent.

The *spray duster* was developed to apply dusts simultane-ously with a mist spray from nozzles along the edge of the fishtail orifice, which served to wet the foliage and increase the initial deposit of dust. Certain types of mist blowers are also equipped for simile- taneous spray-dusting, and results have been secured by the simultaneous emission of 1 gallon of aqueous oil emulsion to each pound of dust, which were equal to or better than results of high- pressure spraying. However, this type of application has been superseded by the use of concentrate sprays. Many other innovations have been developed for dust applications.

For use with volatile materials such as nicotine an hopper with a mixing attachment in which the dust is mixed immediately before application has been used. The performance of nicotine dusts has also been improved by partially confining them under a gasp of canvas trailer 10 to 30 feet wide dragged behind the duster for a distance of 10 to as much as 100 feet. The trailler serves to confine the vapour of the insecticide given off by the dus and to thus expose the insects to it for a longer period.

Spraying of Insecticides by Aircraft

Practically the fust use of the airplane for insect control was in the application of lead arsenate dusts to control the catalpa sphinx, Ceratomia catalpae, in Ohio in 1921. The advantages of aircraft applications in rapidity, cheapness, and ease of treatment were readily apparent, and the uses in creased rapidly. Calcium arsenate dust was applied by air in 1923 to control the cotton boll weevil, Anthonomus grandis, and the cotton leafworm, Alabama argillacea, and paris green dust was applied by air to control anopheline mosquito larvae in the same year. By 1925, forest insects were being dusted by air in Germany, the United States and Canade.

During the nest 20 years, most aircraft operations employed dusts because of the difficulties in formulating suitable sprays with the water-insoluble arsenicals. During the Second World War, malaria control became an essential adjunct to military operations in many areas of the globe and DDT was found to be exceptionally effective for mosquito control. However, its poor dusting qualities resulted in the development of aircraft spraying and literally hundreds of devices were employed using a range of aircraft such as the Piper Cub L-4 and Stearman PT-17 equipped with booms and nozzles, fighter planes with wing tanks, and heavy multiengined ships such as the B-25 bomber and C-47 transport with large tanks emptying through vertical discharge pipes.

Both larviciding and adulticiding operations were proved practicable on the very largest scale and have since been extended to pest mosquito control in Alaska and Canada and to the control of the tsetse flies, Glossina spp., in Central Africa and to black flies, Simulium spp., in the Adirondack mountains. Aircraft have been employed extensively for the control of locusts and grasshoppers by a veriety of means.

Oil solutions of dinitro-o-cresol and BHC have been sprayed on swarms of *Locusta migratoria migratoriodes* both in flight and at rest on the ground. Aldrin, chlordane, and toxaphene have also been used both as sprays and baits for various species of Melanoplus, while arsenical-molasses baits have been applied on a very wide scale. Aircraft applications have been used very successfully for the control of forest insects where other means of application are completely impractical.

Notable campaigns have involved the use of calcium arsenate dust and, more recently. of DDT dusts and sprays for the control of the gypsy moth, Porthetria dispar, and of DDT sprays of the spruce budworm, Choristoneura fumiferana. It has been estimated that in the gypsy moth operations, the payload of a single C-47 airplane treats an area as large as that which could be covered by a truck-mounted spray rig in 4 years. Today, in the United States, the airplane plays a major role in the appication of pesticides or example, in 1955, 45,316,000 acres were treated by air for insect control and 265,808,000 pounds of dust and 51,274,000 gallons of spray were applied. On California farms in the same year, of a total of 5,756,941 acres treated by commercial pest-control operators, 4,853,462 were treated by air craft. The aircraft has brought an astonishing increase in efficiency to many insect-control operations. An ordinary dusting plane flying at 100 miles per hour can treat an area at the rate of about 10 to 40 acres per minute, depending upon the effective swath width, varying from 50 to 200 feet; allowing for an operating period of only about 3 hours per day and the time required for loading, from 500 to 2,000 acres are commonly treated. Even greater retes of treatment are obtained in specific operations with larger aircraft. Thus in mosquito-larviciding operations, several square miles may be treated in an hour, and in grasshopper-bait spreading operations with a C-47, 10,000 acres have been treated per day. In tussock moth, Hemerocampa leucosigma, control, 450,000 acres of forest were sprayed within a few weeks.

The aircraft operations have also made insecticidal applications possible in many areas which were previously inaccessible. These included swamps and marshes where mosquitoes and other biting insects breed, jungles and forests, vast plains areas where grasshoppers and locusts are found, and certain cultivated crops such as rice, sugarcane, mature corn, and cotton, where ground treatment is impractical or destructive. In assessing the place of aircraft in any insect-control operation, it is well to keep in mind that the advantages of aircraft are (a) speed and timeliness of application, (b) freedom from crop injury and soil compaction, (c) no need for farmer preparation of area to be treated, and (d) accessibility of all types of areas. Against these must be balanced the disadvantages: (a) dependency upon optimum weather conditions, (b) lack of uniform coverage, especially under leaves, (c) difficulty in confining application to treatment area which results in severe drift problems, (d) very hazardous operation, with equipment and sometimes human lives lost through accidents, (e) expense of operation and inefficency on small acreages, and (f) inflexibility after application has begun due to difficulty in communication with pilot. The use of the airplane or helicopter for the application of insecticides introduces and additional dimension of complexity into the operation, that of the turbulences imparted to the air by the rotation of the propeller and the down draft resulting from the flight of the aircraft.

The influence of these factors on sprays and dusts has been given a great deal of study, and the principles involved are well understood although they are often ignored in practice. A thorough appreciation of these factors is very important in designing and operating aircraft application equipment which will have the maximum effectiveness on the insects to be controlled are reduce hazardous or objectionable drift to a minimum.

Airflow

The lift of an aircraft in flight is obtained by imparting a downward motion to the air. This downwash or downdraft may amount to about 600 feet per minute for the ordinary biplane flying at 80 to 100 miles per hour and as much as 1,100 feet per minute for the helicopter. The downdrafts helps to carry the insecticidal discharge toward the ground and also moves foliage, which aids in penetration and distribution of the insecticide. The lift in secured by the airfoil of the wing and results in a decreased air pressure above the wing and an increased pressure below it. As a result, the high-pressure air under the wing flows out and around the wing tips to the low-pressure area above the wing, producing a rotary movement at each wing tip, the trailing wing-tip vortex, which may persist for several seconds after the passage of the aircraft. Superimposed on these forces in the rotary slipstresm of the propeller, which displaces the airflow in a counterclockwise direction. The velocities of the downdraft and the wing-tip vortices increase as the speed of the aircraft decreases, since the downward displacement of the air must equal the lift required to support the aircraft. With the helicopter, the forces are very similar. Pronounced vortices with outward and upward components occur at the ends of the rotor, while the downdraft is most pronounced under the central section of the rotor. The strength of these forces is greatest when the helicopter is hovering, and they are materially decreased under conditions of normal forward flight.

FUNGICIDES

Strictly speaking chemicals used to kill or halt the development of fungi. However, for our purposes we shall consider them as chemicals used to control bacterial as well as fungal plant pathogens, the causal agents of most plant diseases. Other organsim like viruses, rickettisias, algae, nematodes, mycoplasma like organism and parasitic seed plants causes plant diseases. In this chapter we discuss only those chemicals used for control of fungi and bacteria. Of plant diseases here are hundreds of examples which include storage rots, seedling diseases, root rots, gall diseases, vascular wilts, leaf blights, rusts, smuts, mildews, and viral diseases. These can, in many instances, be controlled by the early and continued application of selected fungicides that either kill the pathogens or inhibit their development. With today's fungicides most plant diseases can be controlled to some extent.

Phytophthora and phizoctonia rots Fusarium, verticillium and bacterial wilts and the viruses, are among those that are only beginning to be controlled with chemicals. The difficulties with these diseases are that they either occur below ground, and thus beyond the reach of fungicides, or they are systemic within the plant. Fungus is a plant living in close quarers with its host so fungal diseases are basically more difficult to control with chemicals than are insects. This explains the difficulty of finding selective chemicals that kill the fungus without harming the plant. Also, fungi that can be controlled by fungicides may undergo secondary cycles rapidly and produce from 12 to 25 "generations" during a 3-month growing season. Consequently, repeated applications of protective fungicides may be necessary, due to plant growth dilution and removal by rain and other weathering. Before there is any evidence of disease Fungicides must be applied to plants during stages when they are vulnerable to inoculation by pathogens, before there is any evidence of disease. Those fungicides referred to as chemotherapeutants can help to control certain disease after the symptoms appear. Also, protective fungicides are commonly used, even after symptoms of diseaese have appeared.

Eradicant fungicides are usually applied directly to the pathogen during its overwintering stage, long before disease has begun and symptoms have appeared. However, the fungicide must be applied as protect spray, in advance of the pathogens, to prevent the disease are in the case of crops, whose sale depends on appearance, such as lettuce and celery. In our present arsenal there are about 225 fungicidal materials, most of which are recently discovered. Most of these act as protectants, preventing spore germination and subsequent fungal penetration of plant tissues. Protectants are applied repeatedly to cover new plant growth and to replenish the fungicide that has deteriorated or has been washed off by rain.

The application principle for fungicides differs from that of herbicides and insecticides. Only that portion of the plant that has a coating of dust or spray film of fungicide is protected from disease. Thus a good uniform coverage is essential. Fungicides, with several new exceptions, are not systemic in their action. They are applied as sprays or dusts, but sprays are preferable, since the films stick more readily, remain longer, can be applied during any time of the day, and result in less off-target drift. Thanks to modern chemistry, many of the serious diseases of grain crops are controlled by treating the seeds with selective materials. Other are controlled with resistant varieties. Diseases of fruit and vegetable are often controlled by sprays or dusts of fungicides. Historically, fungicides have relied on sulfur, copper, and mercury compounds, and even today most of our plant diseases could controlled by these groups.

The development of organic fungicides vests in the reason, that sulfur and copper compounds can retard growth in sensitive plants. These sometimes have greater fungicidal activity and usually have less phytotoxicity. In organic forms of copper, sulfur, land mercury, until recently and metallic complexes of cadmium, chromium and zinc along with a variety of organic compounds are inclusions of the general purpose fungicides for agriculture whereas the general purpose lawn and garden fungicides are few in number and are usually organic compounds. The general-purpose lawn and garden fungicides are few in number and are usually organic compounds.

Inorganic Fungicides

Sulfur

Sulfur in many forms is probably the oldest effective fungicide known and is still a very useful garden fungicide. There are three physical forms of formulations of sulfur used as fungicides. The first is finely grounds sulfur dust that contains 1 to 5 per cent clay or talc to assist in dusting qualities. The sulfur in this form may be used as a carrier for another fungicides or an insecticide. The second is flotation or colloidal sulfur, which is so very find that it must be formulated as a wet paste in order to be mixed with water. It would be impossible to mix this with water when it is in its original dry, microparticle size, as it would merely flow but would merely float. Wettable sulfur is the third form; it is finely ground with a wetting agent so that it will mix readily with water for spraying. The easiest to use, of course, is dusting sulfure, applied when plants are slightly moist with the morning dew. The particle size of wettable sulfur should be no longer than 77 μm for controlling the most effective disease. A good grade of dusting sulfur should pass through a 32.5-mesh or finer screen.

Copper

The majority of inorganic copper compounds are practically insoluble in water and are pretty blue, green, red, or yellow powders sold as fungicides. The various forms include Bordeaux mixture, name after the Bordeaux region in France, where it originated. Bordeaux is a chemically undefined mixture of copper sulfate and hydrated lime, which was accidentally discovered when sprayed on grapes in Bordeaux to scare off "freeloaders." It was soon observed that downy mildew, a disease of grapes, disappeared from the treated plants. From this unique origin began the commercialization of fungicides. The fungicidal as well as phytotoxic and properties are provided by the copper, ion, which becomes available from both the highly soluble and relatively insoluble copper salts. A few of the many inorganic copper compounds used over the years are presented.

The EPA has determined that no tolerance level need be set for a large number of copper compounds : Bordeaux mixture, copper acetate, copper carbonate-basic, copper-lime mixtures, copper oxychloride, copper silicate, copper sulfate, copper sulfate-basic copper-zinc-chromate, cuprous oxide, cupric oxide, and copper hydroxide.

A comment on solubility is appropriate at this point. In general, protective fungicides have low ionization constant, but in water some toxicant does go into solution. That small quantity absorbed by the fungal spore is then replaced in solution from the residue. The spore continues to accumulate the toxic ion in sublethal doses, whose cumulative effect is lethal. Except for powdery mildews,, water in the penetration court—the place on plant tissue softened by fungal mycelium, which permits it to penetrate the plant cutlicle—thus permits spore germination and makes soluble the toxic portion of the fungicidal residue. To discourage killing all or portions as the host plant, the copper ion which being toxic to all plant cells must be used in dicrete dosages or in relatively insoluble forms. This is the basis for the use of relatively insoluble or "fixed" copper fungicides, which release only very low levels of copper, adequate for fungicidal activity, but not enough to affect the host plant. Copper compounds are not easily washed from leaves by rain, since they are relatively insoluble in water, and thus give longer protection against disease than do most of the organic meterials.

During spraying no special precaution is required as they are relatively safe to use. Although copper is an essential element for plants, there is some danger in an accumulation of copper in agricultural soils resulting from frequent and prolonged use. After using fixed copper for disease control a serious problem, in fact, of copper toxicity is being experienced by certain, citrus growers in Florida. The currently accepted theory for the mode of action of copper's fungistatic action is its nonspecific denaturation of protein. The Cu^{++} ion reacts with enzymes having reactive sulfhydryl group which would explain its toxicity to all forms of plant life.

Mercury

The inorganic mercurial fungicides are probably the most toxic of the fungicides. Mercury's fungicidal properties and toxicity to animals are due in part to the degree of association of divalent mercury ions, which are toxic to all forms of life. As a result, no mercury residues, are permitted in foods or feed. Over the past 30 years, many organic mercury compounds were developed, but they have been replaced by other organic fungicides.

Phenyl-mercury acetate (PMA) was useful for turf diseases, as a seed treatment, and as a dormant spray for fruit trees whereas ceresan is typical of those used as seed treatment. The mode of action for the mercurials is the nonselective or nonspecific inhibition of enzymes, especially those containing iron and sulfhydryl sites. With one or two exceptions, both organic and inorganic mercurial Fungicides have been banned from home and agricultural use by the EPA. This decision hinged on their toxicity to warm-blooded animals and accumulation of mercury in the environment.

Organic Fungicides

To replace the more harsh, less selective in organic materials many synthetic sulfur and other organic fungicides have been developed over the past 35 years. Most of them had no measurable build up effect on the environment after many years of use. Thiram, the first of the organic sulfur fungicides, was discovered in 1931; it was followed by many others. On 1943 and 1949 respectively, other new classes, the dithiocarbamates and dicarboxmides (Zineb and captan) was introduced. Now more than 200 fungicides of all classes are in use or in various stages of development. Several outstanding qualities are passessed by newer organic fungicides, that is being extremely efficient, smaller quantities are required than of those used in the past, usually lasts longer and are safe for crops, animals and the environment. Most of the newer fungicides also have very low phytotoxicity, many being at least ten times safer than the copper meterials. And most of them are readily degraded by soil microrganisms, thus preventing their accumulation in soils.

Dithiocarbamates

Among the dithiocarbamates we find the "old reliables" —thiram, meneb, ferbam, ziram, Vapam (SMDC), and zineb all developed in the early 1930s and 1940s. Such fungicides probably have greater popularity and use than all other fungicides combined. Except for systemic action, they are employed collectively in every use known for fungicides. The dithiocarbamates probably act by being matabolized to the isothiocyanate radical (– N = C = S), which inactivates the —SH groups in amino acids contained within the individual pathogen cells.

Thiazoles

The thiazoles, a class of compounds that offers a surprising chemical disposition, contains among others, ehtazol. The five-membered ring of the thiazoles is claved rather quickly under soil conditions to form either the fungicidal —N = C = S or a dithiocarbamate, depending on the structure of the parent molecule. Ethazol is used only as a soil fungicide and, as such, is exposed to the ring cleavage just mentioned. The probable mode of action is similar to that of the dithiocarbamates.

Triazines

The triazines structure, seen frequently in herbicides, is found in only one fungicide. A wide use for control of potato and tomato leaf spots and turf-grass diseases has been received by Anilazine which was introduced in 1955.

Substitued Aromatics

The substituted aromatics belong in a somewhat arbitrary classification assigned to the simple benzene derivatives that possess long-recognized fungicidal porperties. For seed treatment and as soil treatment to contract stinking smut of wheat, Hexachloro-benzene, was introduced in 1945. Pentachlorophenol (PCP) has been used since 1936 as a wood preservative as a seed treatment, and as a herbicide. Pentachloro-nitrobenzene (PCNB) was introduced in the 1930s as a fungicide for seed treatment and selected foliage applications. It is also availed as a soil treatment to restrain the pathogens of certain damping-off diseases of seedlings.

Chlorothalonil is a very useful, broad-spectrum foliage-protectant fungicide made available in 1964. And chloroneb, developed in 1965, is heavily used for cotton seedling and turf diseases. Dicloran (DCNA) is a highly useful fungicide against *Botrytis*, *Monilinia*, *Rhizopus*, *Sclerotinia*, and *Sclerotium* species, on a wide range of fruits and vegetables. Substituted aromatics are diverse in their modes of action. Being generally fungistatic, they reduce growth rates and sporulation of fungi, probably by combining with $—NH_2$ or —SH groups of essential metabolic compounds.

Dicarboximides (Sulfenimides)

Dicarboximides are three extremely useful foliage protectant fungicides. Captan appeared in 1949 and is undoubtedly the most heavily used fungicide around the home of all classes; folpet appeared in 1962; and captafol (Difolatan®) appeared in 1961. They are used primarily as foliage dusts and sprays on fruits, vegetables, and ornamentals. Recommended for lawn and garden use, as seed treatment and as protectants formildews, late blight, and other diseases, the Dicarboximides are some of the safest of all pesticides available. Remember the garden adage, "When in doubt, use captan."

Many compounds containing the—SCCl3 moiety are fungitoxic, a fact that indicates that group as a toxophore (molecular unit that account for the molecule's toxicity). Fungitoxicity of the dicarboximides is aparently nonspecific and is not a result of a single mode of action.

The dicarboximides' lethal effect on disease organisms is probably due to the inhibition of the synthesis of amino compounds and enzymes containing the radical.

Systemic Fungicides

Only in recent years have successful systemic fungicides been marketed, and very few are available. Systemics are absorbed by the plant and carried by translocation through the cuticle and across leaves to the growing points. Most systemic fungicides have eradicant properties that stop the progress of existing infections. They are therapeutic in that they can be used to cure plant diseases. To give prolonged disease control a few of the systemics can be applied as soil treatment and are slowly absorbed assimilated through the roots. These systemics offer much better control of diseases than is possible with a protectant fungicide that require uniform application and remain essentially where it is sprayed onto the plant surfaces.

There is, however, some redistribution of protective fungicidal residues on the surfaces of sprayed or dusted plants, giving them longer residual activity than would be expected. Systemic constitute of perfect method of disease control by attacking the pathogen at its site of entry or activity, and they reduce the risk of contaminating the environment by frequent broadcast fungicidal treatments. Undoubtedly, as newer and more selective systemic molecules are synthesized, they will gradually replace the protectants that compose the bulk or our fungicidal arsenal. Those systemics currently in commercial use are mentioned in order of their appearence. None are available for home use.

Oxathiins

The first of the systemics to suceed in practice, introduced in 1966, were the Oxathiins, represented by carboxin and oxycarboxin. They are used as seed treatments for the cereal crop, particulary those affected by embryo-infecting smut fungi, and they have potential for other used. They are selectively toxic to the smuts, rusts, and to Rhizoctonis (Thanatephorus), organisms belonging to the Basidiomycetes. The apparent mode of action of the oxathiins beings with their selective concentration in the fungal cells, followed by the inhibition of succinic dehydrogenase, an important enzyme to respiration in the mitochondrial systems.

Benzimidazoles

The benzimidazoles, represented by benomyl and thiabendazole (TBZ), were introduced in 1968 and have received wide acceptance as systemic fungicides against a broad spectrum of diseases. Benomy has the widest spectrum of fungitoxic activity of all the newer systemics, including the Sclerotinia, Botrytis, and Rhizoctonia species, and the powdery mildews and apple scab. Thiabendazole has a similar spectrum of activity to that of benomyl. Introduced in 1969, thiophanate, although not a benzimidazole in its original structure, is converted to that group by the host plant land the fungus through their metabolism. Thiophanate has a fungitoxicity similar to that of benomyl. All three compounds have been used in foliar application, seed treatment, dipping of fruit or roots, and soil application. Their mode of action appears to be the induction of abnormalities in spore germination, cellular multiplication, and growth, as a result of their interference in the synthesis of that vital nucleic material, DNA.

A later addition to the benzimidazoles is carbendazium, introduced in 1973. The interesting quality of this systemic is its proved usefulness in the control of the formidable Dutch elm disease, in the formulation of Lignasan, hydrochloride salt, it is injected into the trunks of diseased trees, and results in a slow curative action.

Pyrimidines

The pyrimidine systemic fungicides appeared in the late 1960s, and include dimethirimol, ethirimol and bupirimate. They are very active against specific types of powdery mildews : for instance, dimethirimol work well on cucurbits. Ethirimol is for cereals and other field crops, and bupirimate controls powdery mildews on apples and greenhouse roses.

Organophosphates

Among the newer systemic fungicides are two organo-phosphorous meterials, IBP and Conen. Developed in Japan and introduced in 1965, both are effective against rice blast, stem rot, and rice sheath blight.

Acylalanines

Acylalanines are another new group of systemic fungicides which includes metalaxyl and funalacyl. They are effective against soil-borne diseases caused by *Pythium* and *Phytophthora* and foliar diseases caused by the phycomycetes (downy mildews). They show promise as foliar, soil, and seed treatments for agricultural crops. They offer systemic and corrective activity as well as residual-protectant activity. However, against the Ascomycetes, Basidio-mycetes and Fungi Imperfecti they have little or no activity.

Triazoles

The sole systemic fungicide of the triazole group is the Triadimefon which carries both protective and curative actions and is effective against mildews and rusts on vegetables, cereals, coffee, deciduous fruit, grapes and ornamentals.

Piperazines

The piperazine systemic fungicides include only one member at this time, triforine, which is amazingly effective against powdery mildew on any host. Additionally it is active against scab and other diseases of fruit and berries, rust and black spot on ornamentals, powdery mildew and other leaf diseases on cereals, rust on cereals, several diseases of vegetables, land storage diseases of fruit.

Imides

A second group of dicarboximides, the imides, are structurally significantly different from the originals. These systemic chemicals appeared during the 1970s. The group now includes procymidone, iprodione and vinclozolin. They are particularly effective against *Botrytis*, *Monilinia* and *Sclerotinia*. Iprodione is also active against Alternaris, Heminthosporium, Rhizoctonia, Corticium Typhula and Fusarium species.

Dinitrophenols

We have mentioned the dinitorphenols are insecticides and as herbicides. Their mode of action as fungicides is the same : the uncoupling of oxidative phosphorylation in cells with an attendant upset of the energy systems within the cells. Dinocap (karathane) has been since the late 1930s, both as an acaricide and for powdery mildew on a number of fruit and vegetable crops. Dinocap undoubtedly acts in the vapour phase, since it is quite effective against powdery mildews whose spores germinate in the absence of water. This is popular home fungicide.

Quinones

The quinones are a facinating chemical group that offers countless numbers of molecules that are potential fungicides. Chloranil is the first of these to appear (1937). Until the decorboxi-mides it was used heavily as a seed treatment and Foliar application. The most popular of this group, however, is dichlone. It is used on a number of fruit and vegetable crops and for treatment of ponds, to control blue-green algae. Dichlone affects cellular respiration in many fungi and acts by attaching to the —SH groups in enzymes, thus inhibiting their action and indirectly uncoupling oxidative phosphorylation.

Aliphatic Nitrogen Compounds

A fungicide that has proved effective in controlling certain diseases such as apple and pear and cherry leaf spot was introduced in the mid 1950s and is known as Dodine having disease specificity and right systemic qualities. It has disease speificity and slight systemic qualities. Its mode of action is not clear, but it is taken up rapidly by fungal cells, causing leakage in these cells, possible by alteration in membrane permeability. The guanidine nucleus of dodine is also known to inhibit the synthesis of RNA.

Fumigants

As with the insecticides, there are distinct hightly volatile, small-molecule fungicides that have fumigant action. They are unrelated chemically but are handled alike and are deal with as a single class in this book. Chlorophicrin was mentioned in as a waring agent, in grain fumigants, but it is also an ideal fumigant itself : It controls fungi, insects nematodes, and weed seeds in the soil. Equally effective against fungi, nematodes and weeds. Methyl-bromide, is also listed as a fumigant insecticide. Methyl-isothiocyanate (MIT) is closely related to the dithiocabamates and has a similar mode of action against fungi, nematodes and weeds. SMDC was listed, but not illustrated, in the dithiocarbamate fungicide, where it belongs chemically. SMDC decomposes in the soil to yield methylisothio-cyanate. None of these is available for home use.

Antibiotics

Such antibiotics are penicillin, tetracycline, and chloramphenicol are used medically against human bacterial diseases. These are not involved in our present study. But we note in passing that the oxytetracylines are therapeutic against some of those mysterious mycoplasmalike diseases. As in the case of the medically important antibiotics the antibiotic fungicides are substance produced by microorganisms, which in very dilute concentrates inhibit growth and even destroy other micro-organisms. To date, several hundred antibiotics have been reported to have fungicidal activity, and the chemical structures of about half of these are already known. The largest sources of antifungal anitbiotics is the actinomycetes, a group of lower plants. Two antibiotics streptomycin and cycloneximide is obtained from one amuzion species, *streptomyces griseus*, which is found within the action mycetes.

Streptomycin is used as dust, spray, and seed treatment for the prevention of mostly of bacterial diseases such as blight on apples and pears, soft rot on leafy vegetable, and some seedling diseases. It is also effective against a few fungal diseases. Streptomycin, the mode of action of which is not clearly understood, probably interferes into synthesis of proteins. Despite the evidence of antibiotic-resistant strains, streptomycin has a place in the control of some bacterial diseases, and the tetracyclines may well play an important part in controlling some mycoplasmalike diseases of plant. Because streptomycin appears to have a single and specific site of action, resistance to it by both bacteria and mycoplasmas in inveitable.

Cycloheximide is a smaller, less complicated antibiotic, about which more is understood. First, cycloheximide is toxic to a wide range of organisms, including yeasts, filament-forming fungi, algae, protozoa higher plant and especially mammals. Surprisingly, it is inactive against bacteria, perhaps because the bacteria fail to absorb it. Cyclohximide cause growth inhibition in yeasts and filament forming fungi by inhibiting protein and RNA synthesis. Thus, with a fair amount of confidence, we can say that both streptomycin and cycloheximide act by inhibiting the synthesis of nucleic acids. In the control of powdery mildew, rusts, turf diseases and certain blights of Fungicide known as *Cycloheximide* was intorduced in 1948 which at present has become very popular. It is best known commercially under the name Acti-Dione. Because of its high acute toxicity, it cannot be purchased for home garden use. Blasticidin-S, discovered in 1955, is produced by the fermentation of Streptomyces griseo-chromogenses. Kasugamycin, introduced in 1963, is formed by Streptomyces kasugaenis, and Polyoxins are extracted from Streptomyces cacaoi. All three are Japanese contributions to the systemic antibiotics. The first two are effective against rice, blast, and the latter against rice sheath blight.

Actions of Fungicides

All agro chemical show their effects on different organic of the insect pests in different ways. Fungicidal action is usually expressed in one of two physically visible ways: The inhibition of spore germination or the inhibition of fungus growth. Most fungicides prevent spore germination or kill the spore immediately following germination. Some of these chemical inhibitors or toxicants also retard or halt fungus growth when applied after the infectious stage has developed. The newer systemics fungicides have eradicant properties and stop the progress of existing infections. What happens at the cellular level to cause these readily visible results? As currently viewed, all fungicides are metabolic inhibitors; that is, they block some vital metabolic process. For the sake of organization and simplicity, we can classify the modes of action into three broad groups: inhibitors of the electron transport chain, inhibitors of enzymes, and inhibitors of nucleic acid metabolism and protein synthesis.

Inhibitors of the Electron Transport Chain

Sulfur

Whenever certain pathogenic organisms and most mites come contact with sulfur and in also by its fumigant action at temperatures above 22°C they are killed. The fumigant effect is, however, somewhat secondary at marginal temperatures and under windy conditions. It is quite effective in controlling powdery mildews of plants that are not unduly sensitive to sulfur. Unlike those of any other fungus, spores of powdery mildews will germinate in the absence of a film of water in the penetration court (a spot on the tissue that is "softened" prior to penetration by spore).

Sulfur is absorbed by fungi in the vapour state, land its fumigant effect—acting at a distance—is undoubtedly important in killing spores of powdery mildews. Sulfur interferes in electron transport along the cytochromes and is then reduced to hydrogen sulfide (H_2S), a toxic entity to most cellular proteins. The way fungi respond to sulfur suggests that sulfur can stimulate enzyme activity.

The reduction of sulfur to hydrogen sulfide (H_2S) by fungi is a metabolic process that diverts protons (H^+) from steps of normal hydrogenation reactions. This reduction is aerobic (requires oxygen) and involves a cytochrome system, possibly cytochrome, a respiratory enzyme involved in the oxidation of nutritional components and foreign compounds. Sulfur presumably serves as an acceptor of hydrogen atoms from dehydrogenases, and then activates to full capacity the enzymic steps preceding sulfur reduction.

Oxathiins

Carboxin and oxycarboxin are the only two oxthiins currently in use. Carboxin inhibits glucose and acetate oxidation by intact fungi, and noncompetitively inhibits succinate, but not $NADH_2$, oxidation by mitochondria. This effect occurs in the electron transport chain of respiration. Succinate accumulates in carboxin-treated cells of fungi, suggesting that inhibition of succinate oxidation is the primary site of action. Oxycarboxin, though less toxic, has the same mode of action.

Enzyme Inhibitors

Copper

There are several inorganic copper compounds like protectants meterials applied before the pathogen appears, to protect plants from inoculation. They prevent the spread of an infection but cannot eradicate an existing one. Thus, the site of copper fungicidal action must be the fungal spore. Copper ions are concentrated by fungal spores from the surrounding medium, some up to 100-fold over that in the immediate environment. This high concentration of copper inside the spore supports the generally held view that the fungitoxic activity of copper ions is due to nonspecific attraction for various groups in the cell, such as imidazole, carboxyl, phosphate, or thiol, resulting in nonspecific denaturation of protein and enzymes. The currently accepted theory for the mode of action of copper's fungistatic action is its nonspecific denaturation of protein. The cupric ion (Cu^{++}) reacts with enzyme having reactive sulfhydryl (-SH) groups, which explains copper's toxicity to all forms of plant life, and especially its toxicity to the vulnerable copper-concentrating spores and cells.

Mercury

The fungicides of mercury compounds fungicides are purely historical and classical toxicologic interest, for they are no longer registered except for occassional use. Considerable concern over the widespread use of mercury pesticides focused on the toxic nature of mercury itself and the discovery in 1970 that methyl mercury (CH_3Hg^+)—which is more toxic to life than aryl (aromatic-ring derivatives) mercurials or inorganic salts of mercury—is synthesized in the biosphere. As a result, most registrations for mercury fungicides were canceled.

The biochemicla basis for mercury's toxicity is generally though to consist to interactions with the thiol groups of proteins, where mercury ions (HG^{++}) form compounds attaching to one or two sulfurs. Since there are so many functional -SH groups in a living system, it is virtually impossible to select one reaction of mercury is being the most important to its toxic effect. Organic mercury is different only in that its chemical form is R-Hg^+ rather than Hg^{++}. The organic mercurials are more soluble in lipid than the inorganic Hg^{++}, enabling them to penetrate more readily into cells and tissues. In general, inorganic mercury is a more effective enzyme inhibitor than organic mercury, though the organic forms remains as potent toxicants.

Dithiocarbamates

This category includes in the dithiocarbamates, the oldest group of the organic fungicides, are maneb, ferbam, zineb, manzate, dithana, polyram and ziram. The dithiocarbamates may act by being metabolized to the isothio-cyanate radical (–N = C = S), which inactivates the –SH groups in amino acids, proteins, and enzymes contained within the individual pathogen. It is difficult to establish exactly how they act since they are unstable, and the chemical nature of the fungitoxic agent(s) is not known for certain.

A more recent theory tends to discount the isothiocyanates as toxic products, suggesting instead that ethylene thiuram disulfide. An important function in both the lower and higher plants that may help explain the mode of action for not only the dithiocarbamate fungicides but also others, including those derived from the heavy metals, is *chelation*. A chelate is an organic ring structure composed of a metal atom linked to a ring by nitrogen, oxygen, or sulfur. Particularly as they involve enzymes, chelates are powerful and essential entities in the metabolic procesed of plant. Some of the metals required by the higher plants and fungi in trace amount assist enzymes in conducting their routine duties of metabolism. The metal may be active in this role as a chelate with the biological component.

One of the generally accepted theories to explain the fungicidal activity of copper, mercury, cadmium, and other heavy metals is the formation of chelates within the fungal cells. The chelates, in turn, disrupt protein synthesis and metabolism. And since the most critical protein of cells are enzymes the metals required in trace amounts appearing in abundance or excessive quantities are equivalent to the introduction of potent' poisons in the cells. If the chelation correct, it would account for the mode of action for the organic and inorganic heavy metal fungicides, for the formation of isothiocyanates from the dithiocarbamate molecules, and for the potency of the heavy-metal dithiocarbamates.

Certain fungicides are in themselves chelating agents. They attain to the scarce metal components, such as Fe, Mg, and Zn, within the cells, literally robbing cells of essential materials. In summary, chelation of heavy metals plays an important role in both the life and the death of cells.

Thiazoles

Ethazol, tricyclazole, and Busan-72 belong to the thiazole group. The unstable five-member ring of the thiazoles is broken rapidly under soil conditions to form either the fungicidal –N = C = S or a dithiocarbamate, depending on the structure of the parent chemical. The isothiocyanates inactivates –SH or –SR groups in amino acids, proteins, and enzymes contained within the individual pathogen. Thus the site of action is nonspecific.

Substituted Aromatics

The substituted aromatics are diverse in their modes of action. Being generally fungistatic, they reduce growth rates and sporulation of fungi, probably by combining with amino ($-NH_2$), –SH, or SR groups in essential amino acids, proteins, or enzymes. Hexachloro-benzene possesses some fumigant qualities, especially with wheat bunt, but is mode of action is unknown. Chlorthalonil is believed to work by inactivation of thiol groups in the fungal cell. Chloroneb may act by inhibiting DNA synthesis. PCNB is effective against many species of fungi having chitin in their cell walls, but is virtually inactive against fungi in which chitin is either absent or present in small quantities, for example, *Pythium* and Phytophthora. Thus, there is indirect evidence that PCNB may act by interferring with chitin synthesis. Pentachlorophenol (PCP) block the formation of ATP by uncoupling oxidative phosphoryl-ation.

Dicarboximides (Sulfenimides)

Captan, folpet, and captafol are the early members of the dicarboximide class of fungicides. They probably act non-specifically, because the nature of the reactions between the fungicides and components of the fungal cells are not definitely established. It is generally believed that –SH or –SR groups are likely reaction sites. However, folpet inhibits isolated chymo-trypsin, which does not contain thiol groups, so that reactions not involving thiol groups could be involved in fungicidal action.

Quinones

Belonging to the quinones are dichlone and chloranil. Fungal spores take up a large amount of dichlone before expiring, which suggests that dichlone exerts its toxicity by a simultaneous action on a variety of loci in major metabolic pathways. It is not clear what chemical reactions quinones undergo in the fungus. They do react readily with thiols, and they probably enter into a variety of other reactions in the cell, including combination with amino groups. The primary toxic effect of quinones therefore be due to reaction with –SH or –SR and $-NH_2$ groups of vital enzymes.

Inhibitors of Nucleic Acid Metabolism and Protein Synthesis

Benzimidazoles

The benzimidazoles include benomyl, thiabendazole and thiophanate. They are not toxic to fungi in their original state, but must be converted to their ester metabolites, which are known to be the toxic entitles. These metabolites cause morphological distortion of germinating spores and probably act by inhibiting DNA synthesis, or by interfering with some closely related aspect of cell or nuclear division.

Antibiotics

Cycloheximide is a protein synthesis inhibitor—it inhibit the incorporation of amino acids into protein. As a consequence of its interference with protein synthesis, it may also inhibit DNA synthesis. Cycloheximide is toxic not only to fungi but also to plants, and it has an oral LD_{50} in the rate of 2.5 mg/kg, by far the most toxic of the fungicides.

Streptomycin probably inhibits protein synthesis by binding to the ribosome, one molecule of streptomycin per ribosome. It also cause misreading of the genetic code, though this is not likely the primary effect.

Aliphatic Nitrogen Compounds

Dodine has a nonspecific mode of action, but it is generally agreed that it acts interfering with membrane structure. There is indirect evidence that its primary side of action is the mito-chondrial membrane. The quanidine nucleus of dodine is also known to inhibit the synthesis of RNA.

Triazines

Anilazine is the only fungicide belonging to the triazine class of fungicides. The ultimate site and mechanism of action are unknown. However, the fungicide does react with $-NH_2$ groups and less readily with -SH or -SR. Consequently, it appears likely that anilazine cause inhibition of a variety of cell processes by nonspecific combination with vital cell components.

10

POLYMORPHIC CHANGES

Population ecology is concerned with interrelations or coactions between individuals within and between species. Coactions may either be beneficial to the participants, *cooperation*, or harmful, *disoperation*. Interspecific cooperation includes mutualism, commensalism,'and many of the interrelations that establish the community as a dynamic unit. Coactions that are harmful, to at least one of the participants, include parasitism, predation, and competition. Many of these coactions act to regulate density or, along with climate, cause fluctuations or even more drastic changes in number of individuals from time to time. Established behavioral interrelations between individuals, characteristic of a species, are its *ethics* and as such are found in animals as well as in man.

INTRASPECIFIC COOPERATION

An early manifestation of cooperation in the evolution of animals is the grouping of freeliving protozoans to form colonies, and the further development of such colonies into multicellular metazoans that thereafter behave and respond as unit organisms. Whether the first gathering of protozoan cells to form colonies developed for better protection from some predator or environmental condition, improved utilization of food supplies, or more efficient reproduction, it is impossible to say. The colonial form, however, must have had survival value to persist.

Colonization quickly led to *division of labor* between somatic and reproductive cells, as occurs in *Volvox*, and later to division of labor between somatic cells themselves, so that different cells or organs became specialized to serve the particular functions of digestion, respiration, circulation, and so on. Cooperation among cells, tissues, and organs gave greater metabolic efficiency to the whole individual and resulted in evolution to the highest types of animals. Similarly, the aggregation of individuals must have survival value, because it persists. Hundreds, sometimes thousands, of spotted ladybeetles hibernate under leaves at the forestedge. Mayflies, midges, and mosquitoes swarm for mating purposes. Millions of bats roost together in large caves, notably in the Carlsbad Caverns in New Mexico. The migratory locust moves from one locality to another in immense hordes, and birds usually migrate in flocks. Highly organized societies are found in such insect groups as termites, ants, bees and wasps, as well as in some breeding colonies of birds and mammals. Even the biotic community exhibits division of labor in that dominant species control the microclimate and conditions in which other species live; producer organisms capture energy and manufacture food, which is then divided between all members of the community; predators help to control population densities of their prey; and so on.

Aggregations

All organisms react on their habitat in one way or another, and when they occur in numbers these reactions produce a conspicuous effect. *Water conditioning* occurs when physical or chemical changes occur as the result of organisms living in it. Compared with unconditioned water, these changes may have either a harmful or a beneficial effect on organisms introduced into the water after the original organisms have been removed. Water is said to be *homotypically* conditioned when the changes were previously produced by individuals of the same species as being studied and *heterotypically* conditioned when the changes were produced by a different species.

Experimental studies have demonstrated that goldfish grow faster in water that has been homotypically conditioned for 24 hours than in unconditioned water. Both fish and amphibian larvae also do better in water conditioned by the presence of mollusks than in unconditioned water. The marine flatworm *Procerodes wheatlandi* will survive much longer when transferred to fresh water conditioned by the presence of either live or dead individuals of the same species or by freshwater species of flatworms than they do in unconditioned fresh water. The longer survival in toxic solutions, faster growth, and greater reproduction of protozoans, snails, flatworms, cladocerans, amphibian larvae, and fish occurring in aggregations rather than as isolated individuals is attributable to water conditioning.

Colloidal silver is toxic to fish. Ten goldfish were simultaneously exposed to 1 liter of water dosed with colloidal silver. They lived an average of 507 minutes each. Fish individually exposed to a similar concentration of silver in the same volume of water lived an average of 182 minutes. The slime from the grouped fish was sufficient to precipitate much of the colloidal silver and render the solution less toxic. Photosensitive animals survive longer when exposed to excessive illumination in groups than singly because of partial shading of one by another, but freshwater planaria exposed to ultraviolet live longer in groups, even when no shading is involved. Marine flatworms *Procerodes* survive longer in fresh water in groups than singly because the first worms that die from the group release calcium into the water, conditioning it and giving protection to the animals that remain.

Various factors are involved in producing favorable conditioning: minute organic particles in suspension resulting from excreta, regurgitated food, or disintegration of dead animals previously present may become concentrated in the alimentary tract of the animals and serve as an unsuspected food resource; mucus or slime secreted by organisms may coagulate, precipitate, or reduce the potency of toxic substances; salts liberated from the body may change the osmotic properties of the culture medium; or there may be liberation of growthpromoting substances from one animal that affects other animals. Many of these effects, not all of which may be favorable, are doubtlessly at work in natural habitats and should be carefully studied as part of the internal dynamics of the biotic community. It may well be, for instance, that during the course of evolution organisms have become adapted to tolerate or take advantage of these external metabolites given off by their neighbors, with the result that the metabolites have become an important part of their environment.

Benefits derived from aggregating are shown in other ways by terrestrial animals. In honeybees, when hive temperatures drop below 14°C (57°F) during the winter, they form clusters and maintain a mass temperature several degrees above outside temperatures. This is brought about by increased metabolic oxidation of honey in their bodies and by increased muscular activity. Furthermore, the compact cluster presents a surface area for heat loss that is less than the total surface area of the individuals separately. When there is danger of overheating, the bees in the hive spread out on the combs and fan with their wings to create a circulation of air. They will also carry water into the hive and place small quantities both outside and inside the comb cells. The forced air circulation evaporates the water and cools the hive. Bees also cool themselves by constantly moving their tongues in and out of their mouths, exposing to evaporation the moisture that is present on them as a thin film. Temperature regulation is less well developed in other social Hymenoptera.

Coveys of bobwhite quail roost in close circles, at night. Perhaps this enables detection of predators approaching from any direction, but by that behavior the birds can tolerate lower air temperatures and for a longer time than isolated birds can. Similarly, mice huddle in low air temperatures, a behavior that reduces heat radiation and consequent need for frequent feeding.

A single muskox or bison may succumb to a pack of wolves. When in a group, the males form a circle facing outward with the females and young inside, whereby they are usually able to ward off the attack. By the same token, a single wolf has difficulty killing a deer; a single coyote, killing a pronghorn antelope. But in packs the wolves can overpower a deer, and by individually taking turns in relay fashion, a pack of coyotes can chase a pronghorn to exhaustion.

Whether an animal occurs singly or in groups may affect its learning rate and behavior. The common cockroach and the shell parakeet learn simple mazes less rapidly when other individuals are around than when alone, but goldfishes, minnows, and green sunfishes learn mazes faster in groups, phenomena spoken of respectively as negative and positive *social facilitation*. Many animals are more active and alert in groups than alone; in groups, individual imitations of others' behavior are common. Cormorants and pelicans fish more proficiently in groups than alone because group behavior is organized and each individual plays a certain role.

The beneficial effects of aggregation are lost if the aggregation is either too small or too large. For instance, the longevity of *Drosophila* is greatest with a population density of 35 to 55 flies per 1ounce (28g) culture bottle. Smaller densities are unable to control the growth of the yeasts on which they feed; greater densities exhaust the food supply and excessive amounts of excreta accumulate. Likewise an initial population of 4 *Tribolium* beetles per 32 g of flour reproduces more rapidly during the 25 days following than smaller or larger initial populations.

For all kinds of animals, competition for food and other resources of the habitat becomes more and more intense as populations increase in size above an optimum. The benefits resulting from an increase in the size of aggregations up to the optimum represents cooperation; the harmful effects resulting from aggregations that are too large is disoperation.

Social Organization

The simplest animal aggregations exhibit little social organization, for the individual organisms are brought together more or less ephemerally by chance, by sexual attraction, for reproduction, or because of a similar response to environmental factors. An evolution of organization may, however, be traced through intermediate stages to the complex division of labor found in some insect societies. Specialization occurs both in morphology and behavior. The three primary castes of termites and ants are the winged reproductive males and females, the wingless sterile soldiers that possess large mandibles and irritating glandular secretions, and the smaller, wingless, often sterile workers. The soldiers defend the colony against predaceous enemies; this function is assumed by workers in bees and wasps, among which a distinct soldier caste is lacking. In termites, the soldiers may be either males or females; in ants, they are females. The worker caste in ants is usually female, but in higher termites it may consist of either males or females. In primitive termites the nymphs of other castes substitute for the workers. The workers collect food, cultivate gardens of fungi, take care of domesticated aphids or coccids, feed the other castes, and build shelters. The earliest organized social life of primitive man was perhaps neither so highly organized nor so far advanced in an evolutionary sense as these complex societies of insects, even though it was from the greater psychological potentialities of primitive man that modern civilization arose.

In these social relations, indeed in all sorts of symbiotic relations between individuals, one or both partners must have specialized behavior to effect and maintain the relationship. Chemical stimuli are important in this respect and have received much study to date, but physical stimuli, such as color, shape, texture, temperature, and so on, may also have primary integrative importance as releasers for specific behavior responses, the products of long evolution.

MUTUALISM

Mutualism is an association between two or more species in which all derive benefit in feeding or in some other way. The term symbiosis has often been applied to this relationship, but *symbiosis* properly refers to the intimate association of two or more dissimilar organisms, regardless of benefits or the lack of them, and hence includes mutualism, commensalism, and parasitism.

Mutualism, as is true also with commensalism and parasitism, may be *facultative*, when the species involved are capable of existence independent of one another, or *obligate*, when the relationship is imperative to the existence of one or both species, Considerable study and experimentation is sometimes required to decide whether a particular relationship is facultative or obligate, or even whether it is truly mutualistic. Many examples of ecological interest of mutualism, commensalism, and parasitism are cited by Pearse and Allee; only a few will be given here.

Mutualism in plants is demonstrated in the associations of fungi and algae to form lichens, of nitrogenfixing bacteria with the roots of legumes, and of fungal mycorrhizae with the roots of many flowering plants.

There are many intimate relations between plants and animals. Mutualism is suspected in the presence of photosynthetic algal cells in the protective ectoderm of green hydra, and those associated with turbellarians, mollusks, annelids, bryozoans, rotifers, protozoans, and the egg capsules of salamanders. The algae give off oxygen, benefiting the animals, which in turn supply carbon dioxide and nitrogen to the plants. The thick growth of algae often found on the carapace of the aquatic turtles is important mostly as camouflage for the turtles. Certain beetles, ants, and termites cultivate fungi for food. Bacteria in the caeca and intestine of herbivorous birds and mammals aid in the digestion of cellulose. The crosspollination of flowers by the agency of insects and birds seeking nectar and pollen is of such great importance that many structural adaptations in both plants and animals fit the one to the other to ensure the success of the function.

Animals, especially birds and mammals, are of great importance as agents of plant distribution. Seeds, fruits, even entire plants become attached to feathers or fur, or ingested seeds are eaten and eliminated unharmed with the feces. When bare seeds are eaten they are usually macerated, digested, and entirely destroyed unless they have very hard coats. But fruits are fed upon primarily for pulp, and most of the seeds pass through the alimentary tract unharmed. Animal transportation of ingested seeds is perhaps the most important means by which fruit species are dispersed. Furthermore, germination of the seeds is frequently improved by mechanical abrasion in the stomach and thinning of the seed coat by digestive juices, making them more permeable to water and oxygen. Germination of acorns and nuts is improved if they are buried in the ground rather than left lying on the ground surface. Squirrels, chipmunks, wood rats, and some birds, particularly jays and woodpeckers, cache acorns and nuts as a winter food supply, hiding them in cavities and nooks or burying them in the soil. Perhaps most are recovered and eaten; Cahalane found that 99 per cent of the acorns buried by the fox squirrel in a locality where the animals were numerous were recovered by the animals, largely through the sense of smell. One per cent of the thousand of nuts produced by it during the lifetime of a tree that are buried but not recovered would be adequate to ensure the continuance of the forest. Invasion of oak and hickory trees into sandy areas is greatly accelerated by, and is sometimes dependent on, this coaction of squirrels, and the dispersal of forests up the slopes of mountains against gravity may also depend in large part on transportation of the heavy seed by animals. The interesting concept involved here is that plants have evolved fruits and nuts that are highly attractive to animals as food substances. However, the production of prodigious numbers of fruits and nuts during their lifetimes ensures that at least some will escape consumption and will be more widely and effectively dispersed.

Large populations of such herbivores as rabbits and deer sometimes do considerable damage to new propagation of herbs and trees, but the effects of overbrowsing cannot be dismissed as all bad. Removal of the lower branches of established trees by deer may not seriously affect the vigor of the trees; indeed, such pruning may actually increase their value as lumber. Deer pawing the leaflitter may thereby plant, so to speak, some seeds that would not otherwise become established. Thinning dense stands of young trees may allow residuals to grow more rapidly; much of new growth is doomed anyway because of root competition, and shading cast by established trees. Some species of shrubs and trees actually produce more annual growth under heavy than light browsing. Other species, however, may be killed when small by heavy browsing, although they tolerate considerable browsing when mature. Detrimental effects of both browsing and grazing become evident in an area in the form of excessive invasion of new species which are little used as food, and disappearance or stunting of the food species that are desirable.

Some tropical acacias have evolved foliar nectaries or other food bodies as well as enlarged hollow stipules, spines, or other structures to attract stinging ants. In return, the plants obtain protection from herbivorous mammalian and insect enemies.

Interspecific mutualism is nicely demonstrated by the flagellate Trichonympha, an obligate anaerobe in the gut of several species of woodeating termites (but not in the family Termitidae) where it digests cellulose. Trichonympha and related species also occur in the alimentary tract of the woodeating roach Cryptocercus. The termite and roach reduce the wood to small fragments, passing them through the alimentary canal to the hindgut, where the protozoans digest the cellulose, changing it into sugar. The host benefits the protozoa by removing harmful metabolic waste products and maintaining anaerobic conditions in the intestine.

The ruminant stomach and the horse caecum contain enormous numbers of ciliates and bacteria, some of which digest cellulose. The microorganisms reproduce the equivalent of their biomass each day. Digestion of these organisms provides the host with about 20 per cent of its nitrogen requirement.

Commensalism

Commensalism defines the coaction in which two or more species are mutually associated in activities centering on food and one species, at least, derives benefit from the association while the other associates are neither benefited nor harmed. It is often difficult to establish definitely the nature of the relations between species; and phenomena considered at one time to be commensalism have been later found to be parasitism or mutualism. The concept of commensalism has been broadened, in recent years, to apply to coactions other than those centering on food; cover, support, protection, and locomotion are now frequently included.

The remora fish are remarkable for having the spinous dorsal fin modified to form a sucking disk on top of the head by means of which they become attached to the body of the shark, swordfish, tunny, barracuda, or sea turtle. They are of small size and are not burdensome to the host. The host benefits the remora, however, for when the host feeds, the scraps of food floating back are swept up by the remora. Many small animals become attached to the outside of larger ones, such as the protozoans Trichodina and Kerona on Hydra, vorticellids on various other aquatic organisms, branchiobdellid annelids on crayfish, and so on. Commensals may also be internal; consider, for instance, the harmless protozoans that occur in the intestinal tract of mammals, including man.

The yellowbellied sapsucker drills holes into the phloem of trees from which oozes a flow of sap. The woodpecker uses the sap for soaking or mixing with the insects that it captures for food or may drink the sap separately. The sap and the insects that it attracts are also fed on extensively by other birds, especially hummingbirds, and by other insects and a few mammals.

The pitcher of the pitcher plant found in bogs furnishes a breeding site or home for certain species of midge flies, mosquitoes, and tree toads. Many kinds of microorganisms, both plant and animal, live in the canal system of sponges. The nest of one species often furnishes shelter and protection for other species as well. Ant nests may contain guest species of various other insects. Large hawk nests sometimes have nests of smaller species tucked in their sides; some birds place their nests close to wasps, bees, or ants for the protection offered by these insects. Wood chuck burrows are used also by rabbits, skunks, and raccoons, especially in the winter. During dry periods the water in crayfish burrows, a meter below the ground surface, often teem with entomostraca.

Community Organisation

The final stage in the evolution of cooperation is the biotic community. Analogous to a multicellular individual, the community is composed of organic units, in this case organisms and species rather than cells and tissues. It has a definite anatomy in its stratification, niches, and food chains. The community, too, is a thing born, and it exhibits the same characteristics of growth and maturity as do individuals. There is succession of stages to the climax community like the series of instars in the life cycle of an insect. If the community is injured, it heals the wounds in its structure through secondary succession. The community is selfsustaining in that it absorbs energy from the sun and metabolizes it at various trophic levels in order to do work. There is division of labor, analogous to the functions of the various organs in the body of a single individual; plant species manufacture the food that animals need, and dominant species create environment conditions within the community suitable for the existence of other species. There is transmission of stimuli, intercommunication, between individuals and species by voice, odor, sight, and contact. There is control over the numbers of individuals of each species in the balance of nature. The result is that the biotic community is a highly integrated recognizable unit in which species exhibit various degrees of interdependency. The existence of each component depends to a certain extent on cooperation between them all, so that the community functions as if it were an organic entity. These complicated interrelationships have come about through evolution because of their survival value for the component species involved.

Parasitism

Parasitism is the relation between two individuals wherein the *parasite* receives benefit at the expense of the host; parasitism is therefore a form of disoperation. Parasitism is mainly a food coaction, but the parasite derives shelter and protection from the host, as well.

A parasite does not ordinarily kill its host, at least not until the parasite has completed its reproductive cycle. Were the parasite to kill its host immediately on infecting it, the parasite would be unable to reproduce and would quickly become extinct. The balance between parasite and host is upset if the host produces antibodies or other substances which hamper normal development of the parasite. In general the parasite derives benefit from the relation while the host suffers harm, but tolerable harm.

Classification

Parasites are commonly classified as *ectoparasites*, those which live on the outside of the host, and *endoparasites*, those which live in the alimentary tract, body cavities, various organs, or blood or other tissues of the host. Ectoparasites may be parasitic only in the immature stagesthe hairworm larvae, parasitic in aquatic insects; only the adults parasiticfleas, on birds and mammals; or both larvae and adults may be parasiticthe bloodsucking lice and flies, biting lice, mites, and ticks that occur on birds, mammals, and sometimes reptiles, and the monogenetic trematodes on fish. Similar relations obtain among endoparasites, although it is more common to have all stages parasitic: entozoic amoebae, trichomonad flagellates, opalinid ciliates, sporozoans, pentastomids, nematodes, digenetic trematodes, acanthocephalons, cestodes, and some copepods.

Animals may also be parasitic on plants. Nematodes infest the roots of plants. Galls are formed by wasps or gnats especially on oaks, hickories, willows, roses, goldenrods, and asters. Mites stimulate formation of witches' brooms in hackberry. A variety of insects the larvae of which are leaf miners, wood borers, cambium feeders, and fruit eaters, should be included here. Plants themselves may be parasites either on other plants or on animals. Bacteria and fungi are among the most important diseaseproducing organisms in animals.

Social parasitism describes the exploitation ofone species by another, for various advantages. Old World cuckoos and the brownheaded cowbird of North America do not build nests of their own; rather, they deposit their eggs in the nests of other species, abandoning eggs and young to the care of foster parents. The bald eagle sometimes robs the osprey of fish that it has just caught. One species of ant waylays foraging workers of another species and snatches away the food they are transporting, or the robber species may deliberately rob another nest of food. Some species of ants make slaves of the workers of other species. Various other types of dependency of one species on another have evolved, not only between ants, but also in other social insects, such as termites, wasps, and bees. Social insects are apparently the only animals other than man to have succeeded in domesticating other species, and of cultivating plants, particularly fungi, for food.

Evolution and Adaptations

The ancestors of ectoparasites'were clearly freeliving forms. It is not difficult to imagine how a small organism living freely in water or vegetation could accidentally have settled on the outside of a larger species and found conditions favorable for survival. There would even be selective advantage in such a niche if the organism found a rich source of food. The biting lice probably evolved from psocid insects that live beneath the bark of trees. They may have transferred from this niche to bird nests and then to the birds themselves. Most ectoparasitic insects probably are derivatives of carnivores, saprovores, or suckers of plant juices.

Endoparasites may in some cases have evolved from ectoparasites; more likely, they came directly from freeliving ancestors or from commensals. For example, freeliving nematodes and scavenger beetles both feed upon decaying organic material, and it is easy to visualize how the beetles could have accidentally consumed one or more nematodes. Many kinds which have since become parasites, such as protozoans and flatworms, could have had their first entrance into the alimentary tracts of prospective hosts via drinking water, and subsequently invaded other organs in the body. The invaders would have found their hosts abundant

Fig. 10.1. Life cycle of a snake tapeworm Ophiotaenia perspicua.

food sources, but would have needed some preadaptation to live at the low oxygen concentrations character-istic of digestive tract, to resist being consumed by the digestive juices of the host, and to keep from being carried out with the feces. As succeeding generations of parasites became increasingly adapted to live either on or in their hosts, many kinds lost the capacity for a freeliving existence. Specialization to internal parasitism has cost the loss of locomotor, sensory, and digestive organs, none of which is needed, and led to the development of organs of attachment, increased reproductive capacity, and, in several forms, to polyembryony, interm-ediate hosts, and a complicated lifecycle.

Some parasitic species are more highly evolved than others. Many parasites, for instance, pass their entire existence in a single host; others require one, two, even three intermediate hosts. It is of ecological significance that both primary and intermediate hosts of a parasite occur in the same habitat or community. Even then the hazards to successful passage from one host to another are so great and mortality so high that prodigious quantities of offspring are produced to ensure that at least a few individuals will complete the cycle.

Parasites are transferred from one host to another by active locomotion of the parasite itself; by ingestion, as one animal sucks the blood of or eats another; by ingestion, as an animal takes in eggs, spores, or encysted stages of the parasite along with its food or drinking water; as a result of bodily contact between hosts; or by transportation from host to host by way of vectors. As an illustration of vectors, the bacteria that cause tularemia in man are carried from rabbit to rabbit by ticks. Man contracts the disease when he handles infected rabbits, but the incidence of infection is greatly reduced in the autumn when cold weather forces the ticks to leave the rabbits and go into hibernation.

Host Specificity

Copepods are of all animal parasites the most ubiquitous in their host relationships, being reported from various invertebrate groups and from fish. Most parasitic genera, however, are adapted to hosts of one phylum only. The acanthocephalans *Gracilisentis* and *Tanarhamphus* are yet more specific, normally found only in the gizzards of shad fish; *Octospinifer* is found only in catostomids; *Eocollis*, only in centrarchids. Each order of birds possesses its own particular species of tapeworms; this is true even when several orders of birds live in the same habitats, as do, for instance, grebes, loons, herons, ducks, waders, flamingoes, and cormorants. Species of flagellate protozoans that occur in termite alimentary tracts are largely hostspecific. However, considerable caution needs to be exercised in assigning host specificity to protozoans. Many species have invaded more than one taxonomic host group; and often several species of a single genus of Protozoa frequent the same host species. Some species of gall wasps attack only one species of oak. Where a single species parasitizes two or more host species, the shape and structure of the gall formed around the egg and larva on both hosts is essentially similar. When several insects are found on the same oak, each kind of parasite produces its own characteristic gall form. Apparently the characteristics of the gall that develops depend more on the kind of enzyme secreted by the parasite than on differences of host tissues.

Segregation of parasites to special niches is demonstrated by species of biting lice restricted to the head or body regions of birds. Some nematode species are found throughout the body in connective tissue, but not in the gut; some occur only in the digestive tract and associated organs; certain species occur in the glandular crop of birds, but others only in the caecum; many species occur exclusively in the lungs or in the frontal sinuses. Such fine restriction of parasites to particular hosts or organs is a consequence of precise physiological and morphological adaptations that permit the parasite to survive and complete the life cycle only under very special conditions. Likewise, the life cycle of the parasite is often closely synchronized in time with that of its host.

Hostspecificity can make the taxonomies of many parasites useful for corroborating phylogenetic relationships of their hosts. The South American bird *Cariama cristata* has been shifted from one order to another, and was at one time even put into a special order. A study of its helminth parasites disclosed two species of nematodes and two genera of cestodes present which occur together elsewhere only in Eurasian bustards. The occurrence of these forms in groups so far removed geographically from one another could be coincidental or the result of parallel evolution, but for a number of reasons it seems more likely that *Cariama* and the bustards are derived from a common ancestor which became infected with these parasites, the parasites persisting in spite of evolutionary divergence and geographic separation of hosts. It is interesting to note that this relationship of the hosts is sustained by recent taxonomic study of them by ornithologists.

Effect on Host: Disease

By *disease* we mean a condition which so affects the body or a part of it as to impair normal functioning. Parasites may not cause immediate mortality, but they cause damage to body structures which, should it become excessive, may cause death. We may perhaps better visualize the role parasites play in producing disease by listing some of the more common agents of mortality in organisms, in addition to predators and parasitoids, which will be described beyond.

1. Worm parasites, such as tapeworms, nematodes, and acanthocephalans may wander through the host's body doing mechanical injury as well as destroying and consuming tissues. The host may respond by forming a fibrous capsule or cyst around an embedded parasite.
2. Protozoan parasites are especially important in the alimentary tract and in the blood. A sporozoan species of *Eimeria* damages the walls of the intestine in upland game birds, producing coccidiosis; *Toxoplasma* becomes encysted in the brain of rodents; *Leucocytozoon* is a blood parasite common among waterfowl and game birds.
3. Bacteria cause a variety of diseases, notably tularemia, paratyphoid, and tuberculosis among birds and mammals, as well as other diseases in lower types of organisms.
4. Viruses are so submicroscopic in size that many kinds pass through the finest filters. They are the potent agents of hoof and mouth disease in deer, spotted fever in rodents, encephalitis and distemper in foxes and dogs.
5. Fungus spores of *Aspergillus* that occur in moldy pine litter may be drawn into the lungs of groundfeeding birds, where they germinate and grow, causing aspergillosis. Fungus may also develop on, the external surface of animals.
6. External parasites such as ticks, fleas, lice, mites, and flies do not commonly produce serious mortality by themselves, but they are often vectors transmitting protozoa, bacteria, and viruses from one animal to another. Heavy infestations of external parasites may, however, lower the vitality or vigor of an animal and cause diseases of fur (mange) or feathers.
7. Nutritional deficiencies in vitamins or minerals, or improper balance among carbohydrates, proteins, and fats may produce malformations, lack of vigor, even death. Variations in amount, composition, and intensity of solar radiation may affect the vitamin content of the food an animal consumes. Long restriction to emergency foods of low energy content and outright starvation often cause considerable loss of life during periods of climatic stress.

8. Food poisoning, *botulism*, occurs when certain foods become contaminated with the toxins released by the bacterium *Clostridium botulinum*. Many waterfowl are stricken in some localities. Waterfowl also often pick up and swallow gunshot from marshes in which there has been much hunting, and get lead poisoning.
9. Physiological stress is a term that has come to be applied to changes produced in the body nonspecifically by many different agencies which may accompany any disease. Effects of stress include loss of appetite and vigor, aches and pains, and loss of weight. Internally, the stress syndrome is characterized by acute involution of the lymphatic organs, diminution of the blood eosinophiles, enlargement and increased secretory activity of the adrenal cortex, and a variety of changes in the chemical constitution of the blood and tissues.

   Stress gives rise to abnormal conditions, but it simultaneously elicits from the body defense mechanisms against those abnormal conditions. It is presently believed that the anterior pituitary gland and the adrenal cortex are chiefly responsible for integrating the defense mechanisms. Three stages are involved: the *alarm reaction*, in which adaptation has not yet been acquired; the *stage of resistance*, in which the body's adaptation is optimum; and the *stage of exhaustion*, in which the acquired adaptation is lost. Characteristic of the exhaustion phase are, among others, hypoglycemia, adrenal cortical hypertrophy, decreased liver glycogen, and negative nitrogen balance.
10. Accidents, ageing, starvation, and so on, must also be included as important causes of mortality.

Organisms that produce disease generally fall into one or two categories. They are either present in the body at all times but not normally virulent, or they are normally absent but are virulent from the moment the host is infected by them. Even the healthiest animals chronically entrain many parasites and noxious organisms in the body, but these organisms wreak overt harm only when they become unusually abundant, when virulent mutant strains develop, or if, for one or another reason, the host's vitality and resistance decline to the point where the host is no longer able to withstand the effects of their presence. Any animal suffering an unusually heavy infestation of parasites will show the tax thus put upon its vitality as a loss of vigor and weight, decreased growth rate, and low resistance to vicissitudes of its natural environment. Normally, a more or less mutual tolerance exists between host and parasite such that the demands of the parasite are in equilibrium with the host's capacity to meet them. Hosttolerant parasites have been naturally selected for; mutant strains that are exceptionally virulent quickly die out because they kill the host, without which they cannot survive.

A single attack, even a mild one, of some diseases often confers a partial of complete *immunity* from further attacks of the same disease, even though the agent of the disease may still be carried in the body of the recovered victim. Immunity is an acquired physiological adaptation by which the immune is able to withstand the presence of an otherwise noxious organism, suffering little or no deleterious consequence of that presence. The fact of immunity is demonstrated when parasites not conspicuously harmful to their normal hosts are introduced into a species to which they are normally exotic. The novel host has had no prior occasion or opportunity to adapt immunitively to the alien parasite, and may sicken, even die, of the effects of the parasite's presence, the same effects in kind and intensity which the normal host easily takes in stride. For instance, a trypanosome that is a natural parasite in many of the larger wild mammals of Africa evokes no spectacular effects in its usual hosts. But when the parasite

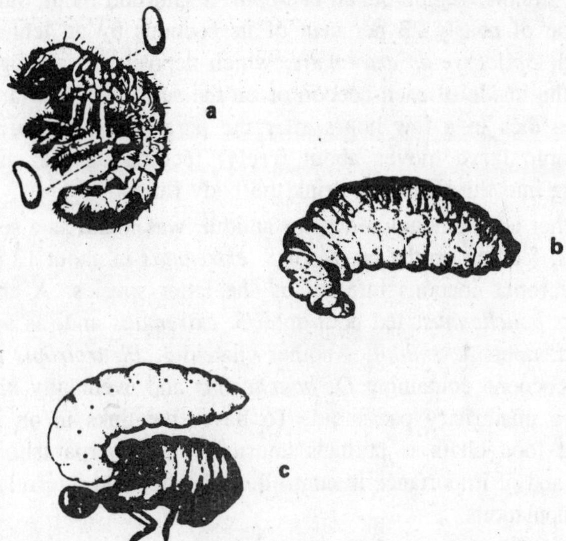

Fig. 10.2. Development of a parasitoid black digger wasp: (a) eggs in position on the host larva; (b) the developing larva; (c) the fully grown larva devouring the remainder of the host.

is vectored by the tsetse fly to man, it causes sleeping sickness; to cattle, nagana. *Zoonosis* is the study of diseases common to animals and man. Animals often serve as reservoirs for disease organisms that may under proper circumstances be conveyed to man. The threat of plague is ever present in some parts of the world, where the organism prevails among the resident rodents. A good discussion of diseases in wild animals is a symposium edited by McDiarmid.

Parasitoidism and Predation

Some Diptera and Hymenoptera deposit their eggs in the immature stages of other insects; the larvae on hatching feed on the host until they are fully grown. The relation of the larva to its host is frequently described as one of parasitism. But it is fundamentally different from parasitism in that the host generally dies of the larval depredations before the larva emerges, but the larva generally lives in spite of the host's death. The relationship resembles that of predator to prey, except that, unlike the true predator, the larva lives within the body of its prey and kills it slowly as it feeds, not suddenly before it feeds. Such larvae are, for these reasons, best thought of as *parasitoids*.

Parasitoids may in turn be infested with *hyperparasitoids*. In the Chicago, Illinois, region *Samia cecropia*, a saturniid moth, suffers the destruction of nearly 23 per cent of its cocoons by an ichneumonid parasitoid, *Spilocryptus extrematis*, which deposits an average of 33 eggs on the inside of each cocoon or on the surface of the larva. The host larva dies in a few hours after the parasitoid hatches, and the ichneumonid larva moves about freely, feeding on the cuticle or burrowing into the tissues to drink the body fluids.

Another ichneumonid, *Aenoplex* smithii, was found as a secondary parasitoid, feeding on the larvae of *S. extrematis* in about 13 per cent of the cecropia cocoons infested by the latter species. A chalcidid, *Dibrachys boucheanus*, fed both upon *S. extrematis* and, as a tertiary parasitoid, upon *A. smithii*. Another chalcidid, *Pleurotropis tarsalis*, infected cocoons containing *D. boucheanus* and eventually killed the larva as a quaternary parasitoid. To have five links in an inverted parasitoid food chain is perhaps unusual, but hyperparasitoidism is common and of importance in controlling the size and interrelations of animal populations.

Predation is a form of disoperation, at least in point of immediate effects, since one animal kills another for food. Predation is important in community dynamics in so many ways that we will divide discussion of it to food coactions, productivity, and regulation of population size.

Competition

Competition is the more or less active demand in excess of the immediate supply of material or condition exerted by two or more organisms. The materials and conditions sought by animals include food, space, cover, and mates. When these materials are in more than adequate supply for the demands of those organisms seeking them, competition does not occur; when they are inadequate to satisfy the needs of all the organisms seeking them, the weakest, least adapted, or least aggressive individuals are forced to do without, or go elsewhere. Competition may result in death for some competitors, but this is from fighting or being deprived of food or space rather than being killed for food as in predation, or by disease as in extreme parasitism.

Competition may be either direct or indirect. It is *direct* where there is active antagonism, struggle, or combat between individuals; *indirect*, when one individual or species monopolizes a resource or renders a habitat unfavorable to the establishment of other organisms having similar requirements. Direct competition, or *interference*, is evident in the fighting of bull seals for larger harems and of grouse for a better position in the social hierarchy; in chasing and color displays (a sort of saberrattling) by fish and birds for defense of territories; in the singing and calling of birds, some mammals, and frogs as bids for mates; and in the excretion of chemicals that affect the behavior or health of other organisms (allelochemistry).

Indirect competition, or *exploitation*, is common among plants when certain species monopolize the water and nutrient resources of the soil or available light so that competing species cannot maintain themselves. Once' an area is well saturated with established individuals, it is often more economical of energy for new individuals to seek homes elsewhere, even in less favorable situations, than to intrude. To be successful by indirect competition, a species needs to get established in an area first, or if the invasion of various species is nearly simultaneous, then to have a more rapid rate of reproduction and growth, or a greater longevity, so as to utilize the resources of the habitat to the fullest possible extent. It is desirable, but not always possible, to distinguish between and quantify the relative roles of interference and exploitation when species compete.

Competition is usua!ly keenest between individuals of the same species, *intraspecfic competition*, because they have identical requirements for food, mates, and so on, and because they are more nearly equal in their structural, functional, and behavioral adaptations.

Interspecific competition occurs where different species require in common at least some materials or conditions. The severity of competition depends on the extent of similarity or overlap in the requirements of different individuals and the shortage of the supply in the habitat. It is generally the case that the more unlike the kinds of competing organisms, the less intense the competition. Yet birds compete with squirrels for acorns, nuts, and seeds; insects and ungulates compete for food in grassland; the bladderwort plant competes with small fish for entomostraca and other plankton.

The immediate test of success in competition is survival; the ultimate test is leaving the largest number of established offspring. Aside from allelochemistry, which is only beginning to be understood, competition has five important effects in the animal community:

1. Establishment of social hierarchies
2. Establishment of territories
3. Regulation of population size
4. Segregation of species into different niches
5. Speciation

The first two effects are chiefly intraspecific and, along with allelochemistry, will be considered in this chapter. Regulation of population size involves both intra and interspecific competition, and many other types of coaction as well. The last two effects are interspecific. It is important to realize that, when these effects are fully manifested, there is a decrease in tension and intensity of competition as each individual takes its place in the structural and functional organization of the community.

Allelochemistry

Included here are the coactions whereby chemicals secreted by one organism affects the growth, health, or behavior of other organisms: *Allelopathy* is produced in plants when toxins are liberated that inhibit seedling growth in the vicinity. This may affect succession of plant species, especially important in the early stages. Some pioneer species in the abandoned field sere produce substances inhibitory to nitrogenfixing and nitrifying bacteria. This retards invasion of other species that require higher nitrogen concentration in the soil. Volatile inhibitors are generally more prevalent than watersoluble ones, and relatively more prevalent in arid than humid climates. Antibiotics produced by bacteria, fungi, actinomycetes, and lichens are widespread in nature and may be one of the reasons why bacteria pathogenic to

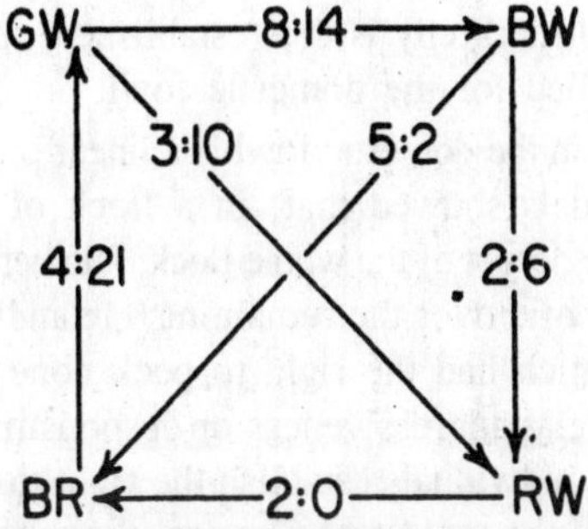

Fig. 10.3. Peck-dominance between the lowest four birds in a flock of seven common pigeons. All four birds were dominated by the three birds of the flock.

man cannot multiply well in soils. A number of antibiotics, such as penicillin, have been used extensively in human medicine. The use of allelochemic coactions in agriculture has possibilities but has yet to be exploited.

Among animals, overcrowding of tadpoles in culture dishes is associated with the occurrence of peculiar round vacuolated cells in the intestinal tract and feces that appear responsible for curtailment of further growth. Under laboratory conditions, killer stocks of *Paramecium aurelia* produce a toxin, paramecin, at the rate of one unitparticle per animal per 5 hours. One unitparticle is enough to kill one individual of sensitive stock of the same species as well as being lethal to other species of *Paramecium*. Conditioning that becomes unfavorable homotypically may sometimes be favorable, or at least tolerable, heterotypically. Thus in protozoan infusions there is a microsere of one species succeeding another.

Allelochemic effects are of great variety in both plants and animals: repellants, escape substances, suppressants, venoms, inductants, counteractants, attractants, signals, stimulants, autotoxins, antoinhibitors, and so on. *Pheromones* are chemical messages between members of a species especially important in reproductive behavior, social regulation and recognition, alarm and defense, territory and trail marking, food location, and so on. Many of these effects are beneficial to the individual; others serve for competitive purposes.

Social Hierarchies

When groups of individuals of certain animal species are confined to limited areas, frequent fights or pecking of one another occur. By way of these encounters, the more aggressive and successful individuals establish a hegemony to which the more submissive individuals acquiesce. A social hierarchy is thus established; the phenomenon was first clearly described for the domestic fowl.

The peckorder in the domestic fowl is a linear one. Close observation of marked individuals showed that, in a flock of 13 birds, one bird became the supreme despot of the whole flock; another bird was submissive to the first but despotic over the remaining 11; and so it went on down to the last bird, which had the right to peck none but was pecked by all. This *type* of social aggressiveness or despotism is called *peckright*. In practice, certain individuals establish the right to peck others and not get pecked back. In the middle of a series, the order is sometimes less fixed, and reversals or triangles occasionally occur. Although most easily demonstrated in the crowded conditions of captivity, peckright has also been observed to obtain under free natural conditions. The peckright type of social hierarchy has been found to occur in various degrees of perfection in several other species of birds, in several species of mammals and fish, and in a few lizards, frogs and toads, crayfish, and insects.

Possession of the following characteristics usually gives an individual at least some advantage in gaining a high position in the despotic order: strength, good health, maturity, relatively large size, hegemony over own territory, responsibility of acting to protect young, accompaniment by members of his own group when meeting a stranger, male over female during the nonbreeding season, female mated with a strong male, the hormone testosterone, and innate aggressiveness. A high position in the social order is advantageous to the individual, as it gives him priority over food, mates, territory, and other resources of the habitat and is sometimes, but not always, correlated with leadership in the group.

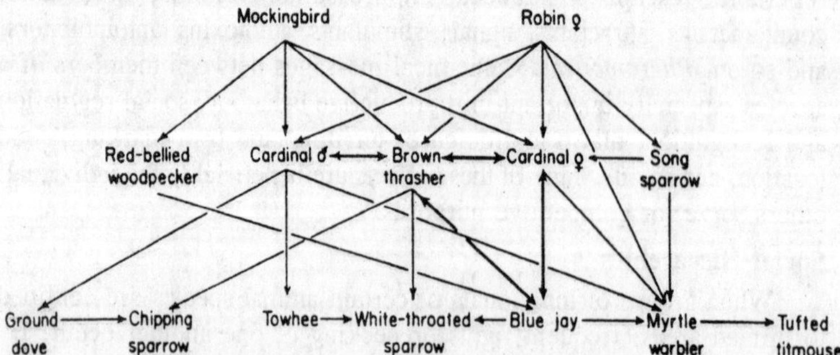

Fig. 10.4. Social hierarchy among species of birds visiting a feeding station during the winter.

In some species the social hierarchy is not as overt as that we have described. In *peck dominance* the individual that is usually subordinate is successful in a certain number of conflicts. Position in the despotic order is a function of ratios of success in continuing conflicts rather than of the results of the initial contact an individual has with each member of the group. Bird BR successfully subdued GW in 21 encounters but was subdued by GW in 4 encounters. Thus, an individual can occupy any position in the hierarchy as long as he is able to maintain that position against all challengers. The individual whr' has successfully challenged a higher position from a lower moves up to the higher. Plainly the positions in the hierarchy are fixed in order, but occupancy of those positions is fluid. A still more fluid form of social aggressiveness is supersedence, in which a successfully challenging individual usurps the position of another individual, momentarily possessing special advantages in the presence of food or some other thing. This type of relation has been described for the goldencrowned sparrow and may likely be found in many other species. There have been few studies of social despotism as an interspecific phenomenon. The range of aggressiveness between individuals within any species is so wide that strong individuals of one species may be despotic over weak individuals of another even though the majority of individuals in the first species are submissive. However, the sharptailed grouse is usually dominant over the ringnecked pheasant, and the latter is usually dominant over the prairie chicken. The manner in which different species fit into a social hierarchy may, be a key to structure and organization of communities.

Territory and Home Range

The establishment of territories, especially during the breeding season, is another expression of despotism, but a special one in that, it determines the spatial relations between motile animals. A territory is any area defended against intruders. It may be the entire home range over which the animal is active, or only a small portion around the nest. Although many animals tend to be gregarious during the nonbreeding seasons, they frequently take up isolated positions and become intolerant of the close presence of others when undertaking reproduction. A home range is that area regularly traversed by an individual in search of food and mates, and caring for young but is not defended.

The establishment of territories is best developed in birds, but also occurs in some other vertebrates, including man and certain inverte-

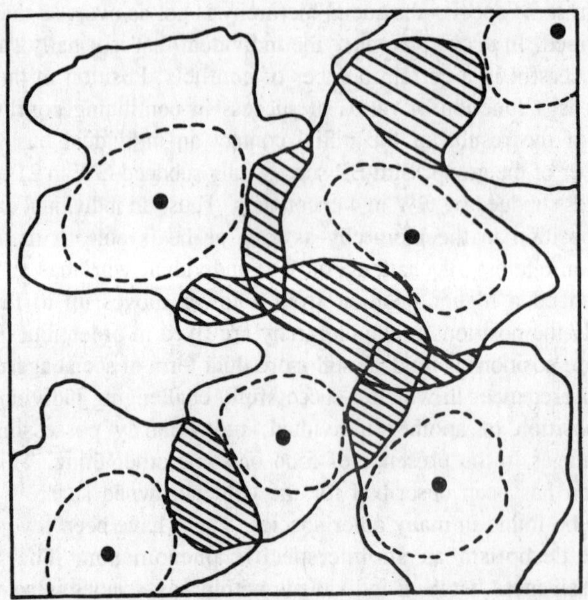

Fig. 10.5. Theoretical relation between home ranges (area enclosed within solid lines) and territories (area enclosed within broken lines). The black dots represent nesting sites.

brates. There is increasing evidence that most adult animals, except for small aquatic species, establish home ranges, if not territories, at least during the breeding season. Immature animals, species in migration, or shifting populations during the nonbreeding seasons commonly do not have definite areas to which they confine their activities. An area should not be called a territory unless one can ascertain that it is defended against intruders of the same species or at least exihibits exclusive possession in the presence of others. A territory is usually defended by individuals, but it may be defended by a closely integrated group of individuals, even over a period of years, as in lemurs. The size of home ranges and territories varies proportionally to weight (W) in lizards ($W^{0.95}$), birds ($W^{1.16}$), and mammals ($W^{0.09}$). Larger animals have higher metabolic requirements and therefore need larger areas over which to hunt for food.

Territoriality has become so ingrained in the behavior of some types of animal that simple advertisement of possession constitutes adequate defense. Such advertisement takes the form of song or other vocal expression in birds, some mammals, and some frogs, or the deposition of scent or chemical cues (pheromones), as in many mammals and insects. If an intruder persists in invading a territory, however, the owner will variously display bright threatening coloration, scold or growl, give chase, or actually engage in physical combat.

Maintenance of a definite territory has several benefits : a definite breeding location in which the nest can be confidently established and protected is afforded; it aids the acquisition of mates; it ensures an area of sufficient size to provide food both for the adults and, later, for the young; and it frees the possessor of the onus of despotic interference by other individuals. The extent to which these advantages are attained varies with the species. Although competition for territory is most keen between individuals of the same species, it also occurs between different species with similar requirements for food and reproduction. A home range, on the other hand, only provides a breeding location. Possession of territory lessens the pressure of competition during the reproductive period, particularly for the female, when the entire energy and attention of animals needs to be devoted to the production of offspring.

Polymorphism in Cepaea nemoralis

Cepae nemoralis is a common European land snail. The distribution extend from Scandinavia in he north to northern Spain and Italy in the south, from Ireland in the west to Germany in the east. Like Drosohila pseudoobscua and Man almos all populations which have been examined exhibit polymorphism. In the case of C. nemoralis the polymorphism studied is visual, affecting the colour and pattern of the shell.

There are six main loci which control the polymorphism. The most important determine the ground colour of the shell and the presence or absence of bands. There are three main shell colours, brown, pink and yellow, and more than one shade of each colour. These are controlled by a series of multiple alleles at one locus, the darker shades being dominant to the paler ones throughout the series. The banding locus has two alleles, for absence of bands or presence of five brown bands. These two loci appear to be about 0.2 map units apart. Closely linked to them are two loci which modify the appearance of the bands. One interupts the pigment in the bands so that it appears as a series of brown spots. The other has an allele, hyalozonate, which removes the pigment altogether so that the band appears as a colourless stripe, and two other alleles which modify band and lip colour. Two genes, unlinked to this group or to each other, alter the number of bands present. One of these, trifasciata, removes the two upper bands completely, so that the shell has three bands lying on the equator and lower half of the shell (the phenotype 00345), while the other, midbanded, leaves only band three present, running round the widest part of the shell (phenotype 00300). Variation at one of the last four loci is undetectable if the shell is unbanded. There is some evidence of interaction between loci. Dark browns are never banded, whereas all the other colours may be banded or unbanded, apparently with equal facility. In addition, the presence or absence of individual bands and the fusion of one band to its neighbour are under multifactorial control, producing a bewildering variety of different phenotypes.

The question raised by this remarkable polymorphism are how it is maintained and why it should be universally present. The genus Cepaea is now one of the most intensively studied groups of organisms from a population genetic point of view. The story provides a fascinating account, not only of a biological problem but also of the development of the ideas and concepts of population genetics. In this chapter several of the explanations and approaches will be introduced, more or less in their historical order.

Effect of Sampling Drift

One of the early investigators of C. nemoralis from a population genetic viewpoint was C. Driver.He pointed out that the polymorphism is very ancient. Banded and unbanded shells may be found in samples from deposits going back in time as far as the Pleistocene. He was also one of the first people to survey the distribution of morphs in a limited area, rather than simply collecting the unusual forms, and made detailed studied at several sites in England and ireland, including one of a stretch of sand dune at Aerrow in Somerset. In these surveys he noted a rapid change in morph frequency over short distances, sometimes no more than a few feet. There was apparently haphazard variant which bore no relation to any measurable changes in the environment. As a result of much careful work of this nature, which failed to reveal an environmental correlation or direct causative agent, Driver came to the conclusion that the explanation of the patterns probably lay in random drift in frequency of alleles which had almost equal fitness. Working as he was in the 1930'' this came as no surprise, for the general climate of opinion was that most selective differentials were very small. Haldane, who had already demonstrated massive selection in Biston betularia, drawson Diver's work in his 1932 book when he says that the difference in fitness between banded and unbanded must be of the order of 10^{-5}.

Stationary Gene Frequence Distributions

Shortly after the war M. Lamotte was studying the distribution of morphs in several regions of France, including the valley of the Ariege in the Pyrenees, and Aquitaine in the south west. He worked on a larger scale than Diver, but like him uncovered patterns of distribution of morph frequency which could not obviously be accounted for by reference to any of the observed features of the environment. Lamotte then made on innovation where such a distribution is concerned, by drawing on the theoretical studies of Wright on stationary gene frequency distributions.

If a series of populations is started at some common frequency and there is no force operating to maintain a polymorphism the gene frequencies will fluctuate until eventually all the populations are fixed at one extreme or the other. After a long period of time the graph of number of populations against gene frequency would have an extreme Ushaped form, indicating that most of the populations were monomorphic. The smaller the population were monomorphic. The smaller the population the faster this position is reached. Although the haphazard fluctuation in frequency form population to population claimed by Diver favours drift, theaction of random drift alone implies a greater degree of monomorphism than is actually found unless some event such as hybridization between different groups has occurred in the recent past. Such an occurrence cannot be common, and so cannot provide a general explanation of the polymorphism.

If there was some factors operating to maintain polymorphism, however, a series of populations would be subject to random drift tending to disperse gene frequencies and the systematic force tending to move the populations to a single equilibrium value. The dispersive force depends on population size, while the centripetal force depends on the type and effectiveness of the systematic factros involved. Whatever the starting frequencies, the series must eventually arrive at a distribution dictated by these contrary forces, known as a stationary gene frequency distribution. There would be a mode at or near the equilibrium, and the population would be more or less closely grouped round the mode, depending on the relative importance of sampling drift. Sewall Wright has investigated the mathematics of stationary gene frequency distributions in detail. The theory is discussed in Crow and Kimura's book. Now if an apparently stationary distribution is found in practice, as in the case of the sample distributions in Lamotte's suveys of cepaea, and some of the parameters controlling it are known, it is possible to estimate the unknown ones. This Lamotte proceeded to do.

The first area he examined was an environmentally homogeneous region in Aquitaine. There was little direct evidence of selection, but the populations were almost always polymorphic, suggesting that a systematic factor operated to maintain an equilibrium. The distribution of the gene unbanded is assymetrical with a mean of about 0.18 (the frequency q), and a range from 0 to 0.5. When the mean is near one extreme, a restricted population size is likely to lead to random fixation. Once it has occurred fixation cannot be reversed by selection, so that mutation must generate sufficient variability to keep the populations polymorphic. This focusses attention on the mutation rates. Suppose the mutation rate from unbanded to banded is u, the reverse rate is v and the effective population size is *Ne*. Migration has an effect similar in kind to mutation, and another factor v' may be defined which combines the effect of both forces, such that v' = v + mq, the q in this expression being the mean for all the populations. From Wright's equation it is then possible to generate a very similar distribution to the observed one if v' is approximately equal to 1/2Ne. If v' or Ne are smaller there will be more monomorphic populations, if they are larger there will be fewer. Lamote estimated Ne to be about 1000 (later work has given a range from 300 to 10000), and the migration rate to be about 0.003. Using the figure of 1000 we get as estimate of v' of about 0.0006, and adjustment for the effect of migration provides v = 0.0001. The estimate of u comes out at the much larger figure of 0.001 or more, depending on the value of other parameters.

Actually, Lamotte's calculations are both more extensive and more precise than outlined here. They include estimates of selection which suggest that the unbanded homozygote has about an 8 per cent selective advantage over banded and that the hetrozygote is intermediate, and they lead to an even higher value for u. The conslusion is that selection does not keep the populations polymorphic, but some polymorphism maintaining factor must operate. The only force able to do so in mutatuion and that must occur at very high rates.

There are two substantial objections to this viewpoint. The genes are considered one by one in isolation, whereas in reality they are closely linked and affect the same aspects of the phenotype. The picture may be different if morphs, rather than genes, are considered. But more importantly, the mutation rates calculated are far higher than those normally observed. Some loci with high mutation rates are known - for example, the defect chondrodystrophy in Man has a rate of about 4×10^{-5} but the usual figures are 10^{-5} or less. Nevertheless, Lamotee's analysis was a comprehensive one which sought to identify the reason for the polymorphism as well as the factors modifying gene frequency.

SELECTIVE PREDATION

Working at the same time as Lamotte, A.J. Cain and P.M. Sheppard examined populations in the rich agricultural land and deciduous woodland around Oxford. They observed there the powerful effect of selective predation by the song thrush, Trudus ericetorum. Thrushes take Cepae during the winter, when the snails are hibernating at ground level or just below it. In Spring, when the birds are nesting, snails form an important part of their diet. During this period heaps of broken shells ar left by the birds on stones, called anvils, at which they extract the animals inside. Such thrushpredated samples may be compared with those collected by the human investigator.

Cain and Sheppard found that there was ample evidence that a large fraction (sometimes 20 percent or more) was taken by thrushes in many populations in the Oxford area.They also observed that some morphs were highly cryptic against particular elements of the environment; yellow unbanded is camouflaged on short grassland, yellow banded in long grass and herbage and pink and brons unbanded on the uniform floor of beech woods. When the samples from the thrush anvils were compared with the populations from which they were derived they were seen to contain a deficiency of the most cryptic morphs and an excess of the conspicuous ones. Evidently thrush predation is selective.

Other lines of investigation filled out a picture indicating the great importance of selective predation. There is a rapid change in composition of the samples as one moves from one habitat to another, for example, from the one habitat to another, for example, from the edge of woodland to short grass. In each type of habitat the morphs which are most cryptic in that habitat predominate. This was demonstrated very clearly by Cain and Sheppard in a graph showing the frequency of yellow in the samples plotted against the frequency of yellow in the samples plotted against he frequency of effectively unbanded individuals – those in which bands 1 and 2 are missing.

This method of display emphasises the role of the phenotype rather than of genes at particular loci: there are many ways in which an effectively unbanded phenotype can be formed. When plotted in this way the samples from herbage and grassland cluster in the upper left portion while those from woodland fall in the lower right. Populations from light and disruptive backgrounds tend to be yellow and banded while those from dark and uniform backgrounds tend to be pink or brown and effectively unbanded. At the colour and banding loci the favoured combinations are likely to be pink unbanded and yellow banded. The browns in this region are always unbanded.

Sheppard made direct observations of predation by thrushes. Early in spring when the ground was brown in colour the birds discovered yellows more easily than the other colours. As the year proceeded and new vegetation developed the background became paler and greener and the selective trend reversed. There was no doubt that relative crypsis affected the composition of the samples removed.

These lives of investigation present us with a new picture. Selection through visual predation acts to favour the cryptic morphs and tends to match the composition of the population to its background. The selection acts on morphs according to their conspicuousness and not on individual loci. The selection pressures are large, commonly about 10 to 20 per cent. Against this kind of force mutation and sampling drift can have little effect, unless the mutation rate is uniquely high. Whatever maintains the polymorphism must be some selective agent. Similar patterns or distribution have been observed elsewhere in England. Lamotte had observed predation by birds in France, but concluded that it was not an important selective agent there.

Digestion: Visual Selection on Cepae Hortensis

C. hortensis is a sibling species of C. nemoralis. In the Oxford area it occurs intersperesed with C. nemoralis, often in mixed colonies, and B.C. Clarke investigated whether it behaved in a similar way in response to visual selection. The two species have the same range of morphs, but usually at different frequencies when they occur together at the same site. Clarke found that when he took sample data and plotted it on the Cain and Sheppard diagram the points tends to cluster in a small area at the top. Most samples were yellow and effectively banded. Nonyellow is always less frequent in C.hortensis than in C. nemoralis and it is possible that for reasons unconnected with visual appearance they are less fit in the one species than in other. In a similar way, dark brown with banding in C. nemoralis is rare, perhaps because of some intrinsic dysfunction, whereas brown and unbanded or pink and banded are not. How, then, can C. hortensis achieve a dark phenotype without being genetically pink or brown?

Clarke concluded that the best combination would be yellow banded with extreme fusion of bands. Accordingly, he plotted frequency of yellow unbanded shells against frequency of fusions in bandeds and got a separation like that obtained by Cain and Sheppard.The grassy, pale habitats clustered in the top left while the closed dark ones fall in the lower right portion. This species, like C. nemoralis, responds to visual selection but the response is based on different elements of the genetic repertoire. We shall see further apparent examples of different responses to the same pressures by similar groups oor organisms in subsequent chapters. A nice corollary to the present instance is that where C. nemoralis has been introduce into Virginia, USA, yellow is the only colour present and the species responds to background in the same way as C. hortensis in Britain.

Further Evidence of Selection

In 1959 and 1960 B.C. Clarke and J.J. Murray reexamined the populations of the Berrow sand dunes, first surveyed by Diver in 1926. Between 10 and 15 generations had elapsed in the intervening period. Driver's study was so detailed that it was possible to return to the same sampling sites and make a series of repeat collections from them. The dune is divided into patches of high density (populations) with low densities between them. If the morph frequency pattern really was the result of random drift the changes occurring in different populations should be very poorly correlated and the second pattern should bear little resemblance to the first.

If fact, Clarke and Murray found essentially the same pattern as before, implying that some stabilizing forces were at work. There were also secular changes taking place. The frequency of brown decreased throughout the area, and the frequency of midbanded increased. The selective differentials estimated from the changes indicated that over the period concerned brown suffered a disadvantage of about 6 per cent compared to nonbrown, while midbanded was at an advantage of about 5 per cent over its allele. Here again there is evidence of quite strong selection, although the selective agent is unknonw.

Direct Selection of Coadaption?

To the west of the Oxford area is a region of high and covered with open grassland, copses and areas of herbage including nettle. Part of this region, the Marlborouh Downs, was examined by A. J. Cain and J.D. Currey after the completeion of the work around Oxford. What they found was a completely different morph frequency pattern. There were areas of distinct homogeneous morph composition separated from each other by narrow boundary areas of intermediate frequency.

These patches were too large to be the effect of random fluctuation since they comprised numbers of individuals many times the effective population size, and they did not coincide with any identifiable changes in the environment. The authors called them *area effects*, to indicate that a previously unknown phenomenon had been identified, and favoured the possibility that the snails were reacting to changes in the habitat from place to place which have not yet been identified by the observers.

Both Sewall Wrigth and C.B. Goodhard have raised an alternative possibility to explain the distributions on the Marlborough Downs. According to this theory, in the past different areas have been colonized independently by small numbers of individuals of different constitutions. As their numbers expanded there was first an adjustment of genes at numerous loci so as to produce combiantions which functioned successfully together. Because each colony started with a different set of alleles available their responses to the same environmental pressures have taken a different form. In a similar manner, C. nemoralis and C.hortensis respond in distinctive ways to theneed to become phenotypicaly dark, presumably because of the interactions of the colour and banding genes with genes elsewhere in the genome which are present at different frequencies.

On the Marlborough Downs the environmental pressures may be uniform but the several microraces respond differently. As their numbers and range increased the groups came into contact with eachother and formed narrow bands of inteargradation because their different modes of adjustment resulted in hybrid inviability. On this theory the process is potentially one of incipient speciation, but in a case like the one discussed the divergence had not proceeded sufficiently far before contact was again made. Emphasis is laid on the epistatic interaction between loci, which makes it impossible to consider a gene on its own; its fitness may be altered radically by the presence or absence of other alleles at another locus.

There is plenty of evidence which may be used to support this view where closely related species are concerned. An analogy to the situation proposed is the zone of hybridization between the carrion crow, Corvus corone and the hooded crow, Corvus cornix. These two birds are widespread in Europe and occupy a range of similar habitats. In each the pattern of plumage is quite constant. They do not overlap in range, but have a narrow zone of hybridization where they meet, in which variation of pattern occurs. It must be said, however, that there is no evidence that the area effect pattern in C. nemoralis arises from coadaptation of this kind, rather the contrary because the zones of gene frequency change at the colour and banding loci do not coincide.

CLIMATIC SELECTION

A number of studies have been carried out which support the belief that average morph frequency is related to climatic conditions. For example, R.W. Arnold surveyed the populations in several valley systems in the Pyrenees. In the extreme conditions prevailing there frequencies tend to be associated with aspect, sunshine and exposure. In general, the colder and more exposed the site, the greater the frequency of unbanded and yellow, and while lip, a character unrepresetned in lowland areas. As before, the patterns ar constant over quite large areas, so that sampling drift is not an important determining factor, and the results suggest that the gene frequencies ar affected by selection imposed by climatic conditions. Cain has assembled much evidence from other regions and habitats for selection by microclimate.

Experimental investigations of the animals and their shells have been carried out by a number of authors. They show that there are physical differences between shells of different morphs. Browns and pinks absorb radiant heat faster than yellows, banded more than unbandeds. In many experiments animals of different morph types behave in different ways, for example, in their response to humidity gradients or the height to which they will climb. These results are not consistent from one experiment to another, however, probably because physiological adaptation has a marked effect on response.

Nevertheless, experiments on survival under extreme conditions agree in suggesting that of the four principal morphs yellow unbanded is the morph best able to survive extreme heat and extreme cold, while pink banded is the least able, pink unbanded and yellow banded being intermediate.

So far in this discussion all the investigations support the view that there is a relatively small effective population size and large selective pressures, of the order of 10 per cent or more. Several studies indicate that selection affects the morph frequency distribution through the action of predation, climate and probably other, unidentified, factors. If we reject mutation as the factor maintaining polymorphism, because its rate of occurrence is too small to counteract the selective effects, then polymorphism must result from some kind of selective balance.

One way in which it could occasionally operate is through opposing selection in different seasons. On the basis of known evidence this could occur, for example, in a dark woodland habitat where the snails were subject to selective predation favouring pinks during the summer,, and to climatic selection tending to eliminate them during the winter. Since the animals live for several years it would be possible for the two forces to cancel one another. This system is not likelyto persist, however, and even if it did at one site it is inconceivable that it could function successfully to maintain polymorphisms under the wide range of habitats in which they occur. There must be heterozygote advantage. It could operate because the heterozygotes have an intrinsic advantage whatever the environment, the most likely situation to explain the universal polymorphism, or because the dominance differed between different pleiotropic effects. For example, in the example suggested above, pink is visually dominant. If the disadvantage in cold weather were confined to pink homozygotes then the net result of both types of selection would be heterozygote advantage.

Another Polymorphism Maintaining Factor– Apostatic Selection

Since Lamotte proposed a balance arising from mutation only one other system for maintaining polymorphism has been discussed. It was connsidered by Cain and Sheppard and by Haldane, but the argument was developed fully by B.C. Clarke. Clarke studied selective predation and observed that there were several lines of evidence suggesting that predators tend to take relatively more of the common types and fewer of the rare types of prey out of a mixed prey population. The reason for this is concerned with the strategy of predation. It may be because the rare items are not recognised by the predators as food, or because their hunting procedure is such that they concentrate on the types of prey which have already yielded the highest return. Whatever behaviour pattern underlies the response there is no doubt that many predators, especially birds, do act in this way.

If the song thrush may be included among the birds which exhibit frequencydependent feeding behaviour and if the different morphs of Cepaea nemoralis are sufficiently distinct to them, then a new generation of bird arises every year which tend to remove the commonest morphs and to neglect the rare ones. Whatever morphs are present this behaviour tends to maintain a polymorphism. It also acts to favour any gene combination which makes the morphs more distinct, an interesting property because although some morphs are more cryptic than others on a given background none of them mimics any particular environmental element very closely and the morphs are strikingly distinct. The selection would cause the morphs to diverge in appearance, and for this reason Clarke called it *apostatic selection*.

The idea that polymorphism in *Cepae* is maintained by apostatic selection is an attractive one. So far, however, the experimental evidence that birds behave in the correct manner does not come from thrushes, and the evidence that thrushes behave in the appropriate way in nature is not conclusive. A more serious objection to this force as the universal factor retaining polymorphism is that the distribution of suitable predators with colour vision does not always coincide with the distribution of Cepae. Many populations are not attacked to any noticeable extend by birds. At the present time we are therefore left with heterozygote advantage, undemonstrated and of unknown origin, as the most likely general cause of polymorphism.

11

BIOGAS

Various methods of methane generation from biomass are described including: complete mix and stratified regimes, anaerobic contact process, upflow anaerobic sludge blants, rotating biological contractors, packed columns and supported growth fluidized bed. The advantages and limitations of each are discussed and specific applications are suggested. In general, suspended growth systems are preferable for wastes with high solids content and supported growth systems are indicated for soluble organics and, in some case, smaller volumes. The character, volume and location of the waste will usually determine the choice of system. Pretreatment methods for enhancing gas production and digestion temperatures are also discussed.

Natural gas, which is mostly methane, is the cleanest, most nonpolluting fuel currently in use. It provides about 25% of the energy in the United States today, heating about one-half of the homes and running clothes dryers, air conditioners, dishwashers and a whole range of appliances for which the only substitute is electricity, that can be five or even ten times as expensive. The methane delivery system consists of 350,000 miles of underground high-pressure pipelines, 650,000 miles of distribution mains, and connections to 45 million users. This system, which is also connected to a storage system of exhausted underground wells, is capable of carrying a volume of gas considerably larger than present usage. Methane from biomass can be fed into this system, as it is being done from feed lots in Oklahoma and a landfill in California.

A choice of the most economical method for the destruction of an organic waste material depends on many factors, including the capital investment in equipment required as well as operating costs and energy requirements for aeration, pumping, heating, sludge dewatering and credits for usable or saleable byproducts.

In aerobic treatment, the organics are generally oxidized to carbon dioxide via the citric acid cycle, producing a high energy yield. Usually this results in rapid cell growth and substrate degradation, with a concurrent large production of sludge. In anaerobic digestion, much of the chemical energy in the substrate is retained in the methane produced. Only limited amounts of energy are available for cell growth, resulting in low growth rates and a small sludge production. Due to the slower growth rates of the methane forming bacteria the hydraulic residence time as well as the mean cell residence time is greater than that for aerobic digestion. With cell recycle, anaerobic treatment is possible even with highly dilute wastes; however, the economics of building larger reactors must be considered.

There seems to be general agreement that wastes with BOD (*biochemical oxygen demand*) > 3000 and COD (*chemical oxygen demand*) > 4000 are more economically treated by anaerobic than aerobic methods, and at higher concentrations (> 20,000 mg/l COD) anaerobic methods cost about 25% of equivalent aerobic methods. An upper economic limit of COD strength for anaerobic digestion has not been established, but it is probable that at concentrations > 50,000 mg/l evaporation and solids recovery may well compete with anaerobic digestion.

In summary, advantages of the anaerobic process as compared with the aerobic process are:

1. Energy recovery is possible through the use of methane generated.
2. Higher organic loadings are possible.
3. A smaller volume of sludge is produced.
4. Energy consuming aeration is unnecessary.
5. It can be cost effective at loadings between 4000 and 50,000 mg/l COD.

Disadvantages are:

1. Digesters (or influent) must be heated.
2. The process is somewhat more susceptible to upset.
3. The slower growth rate of methane bacteria requires longer detention times.
4. The process may not be economical for dilute wastes (BOD < 3000 mg/l).

In individual cases, whether or not anaerobic treatment and energy recovery is feasible depends on many secondary factors including:

1. Quantity of waste to be treated,
2. Continued availability of waste (e.g., fruit and vegetable cannery wastes are often only seasonably available),
3. Toxic materials in waste stream that may require specific pretreatment procedures,
4. Pretreatment that may be required to enhance biodegradability,
5. Physical characteristics of waste that may pose special problems such as foaming, flotation, etc., and
6. Transportation costs to the treatment facility.

Description of Systems

Anaerobic digestion systems may be divided into two general categories: suspended growth, that is where the microorganisms are suspended in the liquid of the reactor, and supported growth systems where the biomass grows as a film attached to some sort of support media. The character of the waste will probably suggest the most appropriate system(s) design.

Suspended Growth Systems

Suspended growth systems may be operated in unmixed (stratified) or mixed regimes, or a combination of both in series. When solids are recycled from a secondary stratified reactor to a primary complete mix reactor the system is known as the "contact" process. Upflow anaerobic sludge blanket reactors are also suspended growth systems.

Unstirred (conventional rate) digesters are stratified, and may be heated or unheated. Solids loading rates are usually 0.03-0.10 lbVSS/

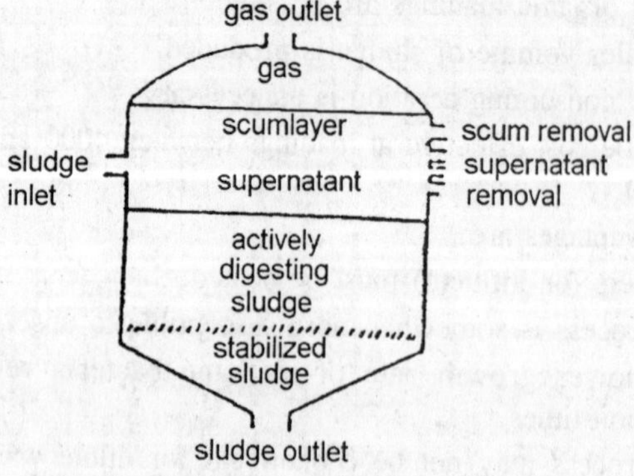

Fig. 11.1. Standard rate (stratified) digester.

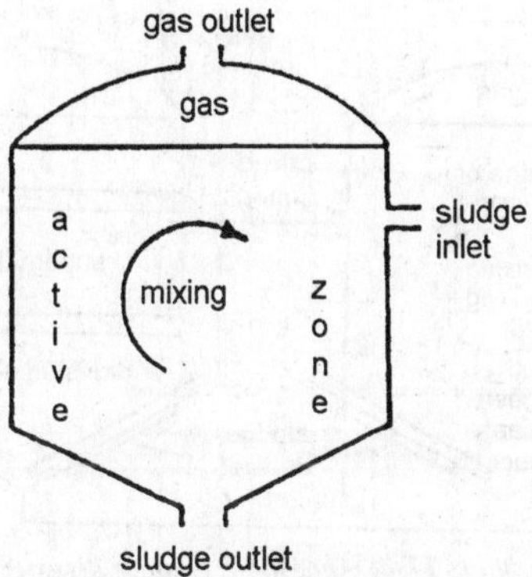

Fig. 11.2. High rate (complete mix) digester.

ft^3/day and they are operated as batch reactors with intermittent feeding and withdrawal. Detention times are 30-60 days. Variations of this system have been designed for rural, agricultural installations.

Complete mix (high rate) digesters may be operated at higher loading rates, 0.1-0.2 Ib VSS/ft^3 /day, and shorter detention times, 15-20 days. They are usually heated (30-35°C) and operated in a continuous mode. First order rate constants based on carbon concentrations have been found to be 0.086/day as compared to 0.054/day in unstirred reactors. Multiple stirred reactors (up to 40) placed in series have demonstrated high methane yields at short detention times. Complete mix and unstirred reactors may also be operated in series, with the stratified reactor effecting solids separation. When a portion of the settled sludge is recirculated to the first tanks (as in the activated sludge process) the system is called an anaerobic contact process. This practice increases the rate of waste stabilization and gas

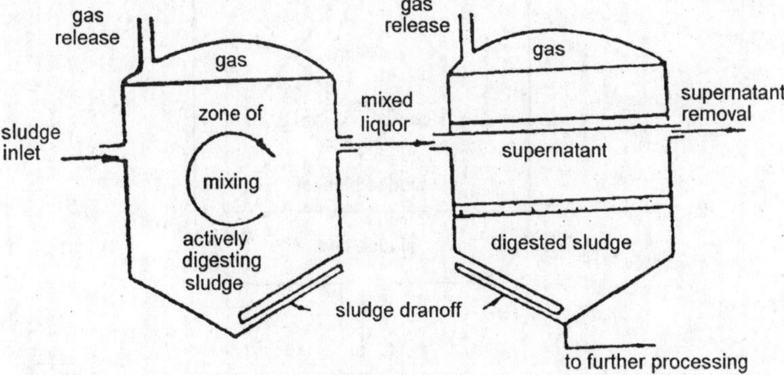

Fig. 11.3. Two-stage anaerobic digester.

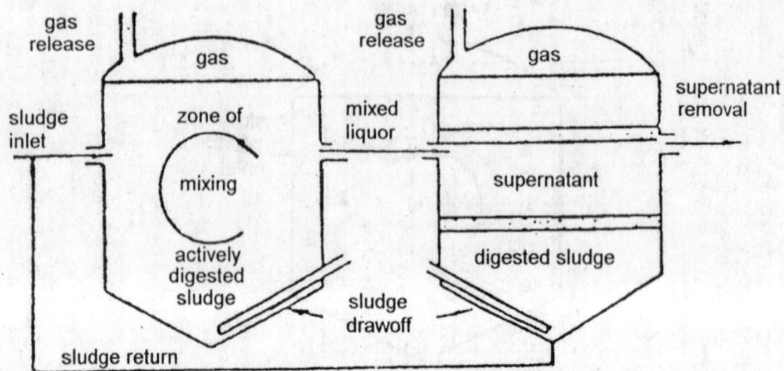

Fig. 11.4. Anaerobic contact digester.

production. A critical solids retention time of 10 days has been suggested.

The *upflow anaerobic sludge blanket* (USAB) was developed based on the concept that anaerobic sludge inherently has superior settling characteristics. This is probably a result of the shape of the particles (oval), and their relatively high density probably caused by the presence of ferrous sulfide. This is important to prevent "wash-out" of the reactor. Mixing is effected by the circulation of flow and by rising gas bubbles. Settlers are mounted inside the reactor. Flow in the settlers is laminar and retention time is low to prevent gas formation;

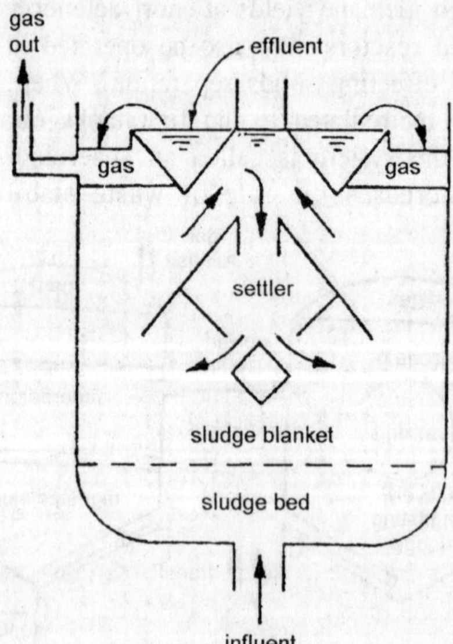

Fig. 11.5. Upflow sludge blanket reactor.

the thickened sludge is constantly being fed back into the reactor. USAB reactors are able to tolerate high organic and hydraulic loadings without process failure. Overloading takes place when the linear fluid velocity in the settler exceeds the settling velocity of the sludge, or if the recirculation of the sludge is exceeded by the sludge input into the settler. Removal efficiencies of 90% at loading rates of 10 to 25 kg/m^3-day and hydraulic retention times of 5-6 hrs. at 30°C have been suggested.

Supported Growth Systems

Supported growth systems are those in which the micro-organisms grow as a biomass attached in some manner within the reactor. This provides for longer solids (microbial) retention times, a well acclimated biomass and a resistance to wash-out at low hydraulic retention times. The attached film of biomass gradually increases in thickness as the organics are metabolized until nutrients can no longer diffuse to the organisms in contact with the support. That portion of the biofilm will then be sloughed off and carried into the effluent. For this reason a final settling chamber is usually required. This chamber need not be large, for the settling characteristics of this sludge are excellent. The thickness to which the film may grow before sloughing is, to some extent, controlled by the hydraulic regime (i.e., shear stresses) in the reactor. Types of attached growth systems include submerged rotating biological contactors, upflow anaerobic filters and fluidized beds.

Rotating biological contactors consist of many rotating corrugated polyethylene discs. The discs are usually relatively small (50 cm) to

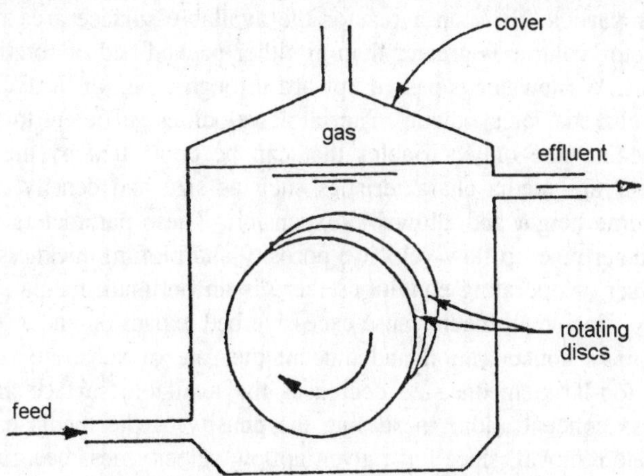

Fig. 11.6. Submerged rotating contractor.

minimize the velocity differential along the face; however, several units may be operated in series. A typical rotational velocities is 13 rpm for a 47 cm diameter disc. Active biomass is attached to the reactor walls as well as to the discs, and substrate removal is proportional to biomass area. These units have been shown to be effective in the treatment of high-strength soluble substrates and hydraulic retention times of 4.4 to 8.8 hours have been suggested.

Upflow anaerobic filters provide a support media for the biofilm. Gravel, Raschid rings, limestone chips, rough glass or polystyrene beads, and even oyster shells have been used. Oyster shell and limestone packing contribute buffering capacity to the system. Good removals have been reported for high-strength (COD 10,000 mg/l) wastes. Substrate removal increases with increased *hydraulic retention times* (HRT). HRT's of 10 to 24 hrs. usually effected COD removals between 70 and 80% while HRT's of 10 days or more 97% COD removals have been reported. Recirculation is often used to increase HRT and protect the system from shock loads as well. Addition of surfactants has been reported to increase gas production by as much as 40%, possibly by enhancing the separation of gas bubbles from the biomass and/or by activating the cell membrane. Up flow filters have been suggested for secondary treatment of municipal waste waters as well as for high strength commercial soluble waste streams. These systems may be operated between 20°-26°C as well as in the mesophillic range.

The *fluidized bed biofilm reactor* (FBBR) is a supported growth system in which the micro-organisms grow as a film on a free moving carrier particle. In such a reactor the available surface area per unit of reactor volume is greater than in either packed bed or rotating disc systems. Wastewater is passed upward through a bed of media such as sand, charcoal or synthetic material at velocities sufficient to fluidize the media. The only variables that can be controlled by the design engineer are media characteristics such as size and density, as well as column height and allowable expansion. These parameters will, in turn, determine up flow velocity, porosity and biofilm thickness. For a given set of operating conditions there is an optimum media size and density. Too-small media cause excessive bed expansion and a decrease in biomass concentration and thus in the rate of substrate removal, while too-large media size decreases the available surface area and biomass concentration. Increasing the density of the media enhances substrate removal, since for a given upflow velocity, less bed expansion and therefore greater biomass concentration will result. This advantage must be weighed against the higher energy requirements for fluidizing heavier media.

These systems have been shown to be capable of achieving high organic removal efficiencies at low temperatures (10-20°C) and treating relatively lowstrength (COD-600 mg/I) at short HRT's (several hours) and high organic loading rates (up to 8 kg. COD/m^3-d). High substrate concentrations can sometimes result in bed flotation, suggesting that diffusion of gaseous products may be rate limiting, for if gas production rates exceed diffusion, gas bubbles will be formed, disrupting the system. Recirculation may be used to "dilute" high COD wastes. As with other supported growth systems FBBR's require that the waste streams be low in suspended solids.

APPLICATIONS

Treatment Plant Sludges

Wastewater sludge is being generated in enormous quantities at sewage treatment plants, particularly at activated sludge facilities. Recent regulations have mandated both the end of ocean dumping of sludges and provisions for full secondary treatment. These regulations will result in increased production of sludge and necessity to treat and dispose of it in an acceptable manner. Anaerobic digestion is one of the processes employed in the stabilization of these sludges, to remove from the raw sludge its odour, pathogens, putrescibility and other offensive characteristics.

Using the methane produced by the anaerobic digestion of sludge to supply power for the sewage treatment plant is not new. This practice was used in the 1940's and the 1950's by many municipalities, but was gradually abandoned in the 1960's when electricity became inexpensive. Aerobic digesters were chosen in place of anaerobic ones because of their relative ease of operation and resistance to upset. However, they were energy users. Increased energy costs have resulted in a renewed interest in anaerobic sludge digestion since the methane produced can often both heat the digester and supply the bulk of the power needed by the entire treatment plant.

Due to the high solids content of the sludges suspended growth systems are necessary, and sludge digesters currently in use in the United States are either stratified, complete mix, contact stabilization, alone or in combination. Although supernatant and thermal pretreatment liquors are generally returned to the head of the treatment plant, upflow filters have been suggested to ease the load on stressed facilities.

Since sludge handling may represent as much as 30-40% of the capital cost and 50% of the operating cost of a treatment plant, investigations to optimize the process continue. Thermal pretreatment (180°C for 30 min.) results in improved biodegradability (more gas, less sludge), improved dewaterability, odour control and sterilization. Thermal pretreatment prior to anaerobic digestion may actually result in an increase in net energy production, based on expected increase in biodegradability and hence in gas production.

The addition of activated carbon to the digesters (45 kg/day) improved the overall treatment efficiency and allowed less chlorine to be used. Savings in filter chemicals, labor and chlorine more than paid for the carbon used. Processes have also been investigated using a two-stage combination of anaerobic and aerobic steps using solar energy for heating requirements and under pressure of two to three atmospheres.

The principal use of digester gas is heating the digester. It has been estimated that approximately 30% of the gas production is used for direct heating in temperature climates. The amount of gas necessary to heat a digester depends on many factors, including insulation, siting, climate and exposure. A conventional hot water heat exchanger system using direct gas heating is about 70% efficient, although use of waste heat from a gas engine, turbine or other source would make the digester gas available for other uses. Two of the main London sewage works, Beckton and Mogden, were producing 2.22 and 2.03 million ft^3 gas/day in 1971-1972.

A small amount of this gas is purified and sold, but nearly all is used to run dual-fuel engines or gas turbines driving air compressors and generators which provide 80-90% of the total power needed in the sewage works.

An $83 million Yonkers Joint Treatment plant in Westchester County, New York, is demonstrating the cost effectiveness of using waste heat and sludge gas. Sludge gas is used to produce steam and heat and to fuel dual-fuel engines (92% methane, 8% diesel oil) which drive process blowers. Of the gas produced, 90% is used, of which 10% goes to produce steam and 90% to fuel engines. Other conservation efforts at this plant are related to heat recovery. Engine exhaust gas (1200°F) is used to generate steam, contributing 20% of the total building heating requirements. Energy recovered from engine jacket and lubricating oil (via heat exchangers) is used to heat digesters. Fuel savings at Yonkers are estimated to be about $600,000/yr.

Feed Lot Wastes

Modern intensive farming systems where animals are kept in feed lots are ideal candidates for methane from animal waste production systems. The amounts of wastes produced by farm animals varies, but typical figures are much higher per capita than for humans, so that a farm of 1000 pigs has a sewage disposal problem equivalent to a town of 4000 people. The amounts of gas produced per pound of dry-weight waste will vary with the type and age of the animal as well as with the diets. Animal wastes generally contain more lignocellulose material than domestic wastes and thus more resistant to attach by micro-organisms. The percent of organics digested anaerobically has been reported as 50-70% for poultry wastes, 50-60% for swine and 10-26% for dairy cows. The digestability of pig and cattle manure may be increased by 170% by sodium hydroxide pretreatment. This treatment consists of adding approximately 8% of a 50% NaOH solution and allowing it to reactor for 14 days before neutralization and digestion. Digestion proceeds satisfactorily, although more slowly at temperatures below the mesophillic range. The economics of larger reactors vs. heating costs in a particular climate will determine the best design.

A system for the conversion of animal wastes to methane called "Anox" is valued at £30,000 and can digest about 120 m^3 of pig manure with a digester volume of 1500 m^3. The gas is used to generate electricity at the rate of 1 kW/hr/150 pigs. The solids are dewatered and used for fertilizer and the supernatant is treated with ozone. Economic evaluation indicates an annual cost of £62,250 as compared to credits of £96,099.

The Calorific project in Guyman, Oklahoma, processes about 500 ton/day of cattle manure producing not only liquid fertilizer and methane (160,000 ft^3/day which is upgraded to pipeline quality) but also cattle feed from the dried solids. Studies conducted to evaluate the efficacy or recovered biomass as a high protein feed have found it to be comparable to soybean meal as a feed supplement. Experiments to determine optimum operating conditions suggest the following:

1. Digesters cannot be started up using "neat" feces and urine. Microbial seed from other digesters must first be acclimated using slurries of approximately 5% volatile solids. As the micro-organisms become acclimated, solids loading can be increased.
2. Detention times of stabilized digesters should be maintained at about 10 days. If the SRT falls below 7 days, performance can drop off rapidly.

3. 35°C is a reasonable digestion temperature (mesophilic). Digesters adapted to 30°C exhibited performances about 5% below those at 35°C. Sudden changes in temperature are upsetting to performance.
4. Scum formation can be a major problem. Scum in farm waste digesters consists of animal hairs, etc., and fibrous feed residues which, when not mixed in, dry on top of the digester contents to form an impermeable layer. In one case a scum of hen feathers blocked the top of the digester and pipes and caused an explosion. Scum formation can be prevented by stirring digester contents.
5. Scroll and flexible stator pumps have been found best for farm waste which is more abrasive on pumps and pipes than is domestic sewage sludge.
6. The methane produced is sufficient to heat digesters, stir digesters and operate pumps.
7. In-ground holding and digester tanks are more thermally efficient.
8. Other wastes, when available, can be added to the digester to increase gas production. These include silage effluent, rotten potatoes, crop residues and waste vegetable matter.
9. Of the number of possible designs for farm digester plants, the single-stage, stirred tanks are the most useful. Two stage or feed-back systems would not appear to justify the increased cost and energy requirements.
10. Thermal pretreatment of the animal wastes might be indicated if an economic use for the additional methane produced is available.
11. Shapes of digestion tanks greatly influenced circulation of tank contents and scum formations. Egg-or-pear-shaped European digesters have fewer scum problems and better circulation than cylindrical digesters, however, they are more expensive to construct.

Rural Installations

Small package plants have been designed and are in use in small villages in undeveloped countries. Large number of these plants are in use in China and more than 20,000 have been installed in India over the past 15 years. One design used in Taiwan, South Korea, India and elsewhere has two digester compartments. The first is about 3' 8" × 3' 8" × 6' and the second 8' 2" × 4' 10" × 6'. This unit was designed for a 20 pig farm and is said to provide fuel for cooking for a family of twelve "if operated properly." The efficiency of these units varies considerably, not only with the skill or the operators, but also with climatic conditions.

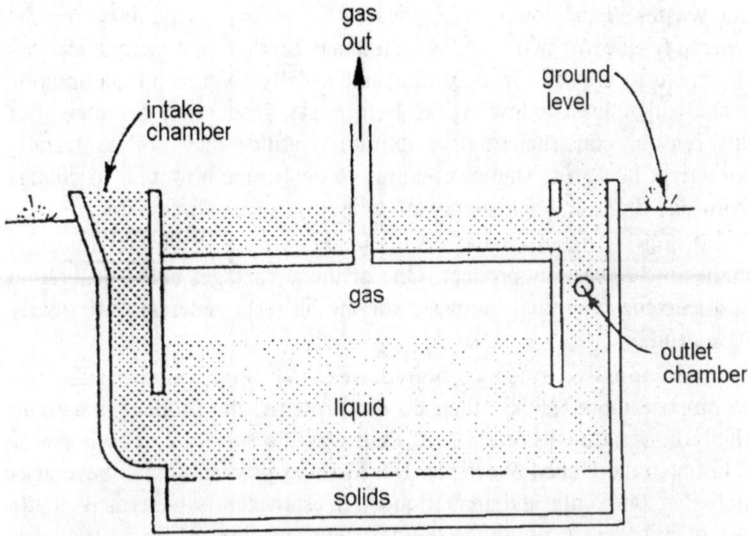

Fig. 11.7. Typical rural biogas reactor.

This type of reactor can be built and operated with unskilled labor. Generally, the input and proportions are human wastes 10%, animal feces 30%, and crop stalks 10%, mixed in 50% water. It has been found that if the moisture content drops below 45% gas production decreases substantially. Before plant matter is put into the digester, it should be composted for at least 10 days so that decomposition will have begun.

These reactors are installed in-ground and are not heated so that gas production rates vary seasonally. During summer and autumn digester temperatures average 23°C and considerable gas is produced. In winter, with ambient temperatures ranging from 0°C to 7°C, digester temperatures stay at about 10 °C and gas is produced more slowly. The unit should be desludged once a year. If the removed sludge is to be used as fertilizer it should be stabilized at pH 12 to destroy pathogens and parasitic eggs.

Food Processing Wastes

Disposal of processing wastes is a major problem for the food processing industry. Nearly all types of food processing wastes have been shown to be suitable substrates for methane generation. Suspended growth systems are more appropriate for wastes with high solids while supported growth systems may be preferable for soluble or settled wastes. A major drawback, in many cases, is the seasonal nature of the wastes. Duration of the "season" for any particular crop is, typically, one or two months. Methane producing reactors are not amenable to changes in substrate and usually require an acclimation period of at least a few weeks before gas production resumes. For this reason, construction of elaborate facilities may not be fiscally attractive; however, studies continue to determine how best to change from one digester feed to another without process disruption.

Promising results have been obtained from anaerobic digestion studies of brewery by-product. One of the advantages is the generation of a saleable microbial biomass, for the bacteria generated are nearly 70% protein.

Additionally, since carbohydrates and lipids in the substrate decompose more quickly than do the proteins, the product is a more highly concentrated protein feed than was the substrate. The brewery industry in the United States has 185 facilities producing malt beverages including beer, ale and malt liquor. Wastewater is generated at the rate of 8.4 liters (excluding cooling water) per liter of beer produced. Average wastewater loadings for a large facility are 1,400 mg BOD/l and 570 mg SS/l. It is further proposed that since the cooling water has a temperature of approximately 70°C, it be used to heat the digesters.

Food processing and crop wastes are often deficient in nitrogen and phosphorus, and addition of these nutrients is required. Studies indicate that for maximum methane production a COD/N/P ratio of 300/5/1 is usually adequate.

PRETREATMENT

Anaerobic digestion of primary and secondary sludges generally results in reductions of organic matter of about 50 and 35% respectively. If this efficiency could be improved, benefits would be accrued not only from the greater volume of methane produced, but also from the smaller quantity of sludge requiring disposal. Thermal pretreatment has been studied by several investigators and has been shown to have the following advantages:

1. Improved degradability and thus increased gas production,
2. Improved dewaterability,
3. Destruction of pathogens, and
4. Reduced sludge volumes remaining for disposal after digestion.

Temperatures between 150°C and 250°C have been suggested, but a typical pretreatment program is 175°C for 30 minutes. At higher temperatures there is danger of the formation of toxic substances such as furan compounds, which inhibit microbial action. Addition of sodium hydroxide prior to thermal pretreatment often further enhances the degradability of the wastes. The effect of the pretreatment is the solubilization of as much as 60 to 70% of the suspended solids and the breakdown of structural lignocellulose materials that may physically restrict the extracellular hydrolizing enzytnes excreted by the microorganisms.

12

ENERGY CYCLE OF BIOSPHERE

The energy that sustains all living systems is solar energy, fixed in photosynthesis and held briefly in the biosphere before it is reradiated into space as heat. It is solar energy that moves the rabbit, the deer, the whale, the boy on the bicycle outside my window, my pencil as I write these words. The total amount of solar energy fixed on the earth sets on limit on the total amount of life; the patterns of flow of this energy through the earth's ecosystems set additional limits on the kinds of life on the earth. Expanding human activities are requiring a larger fraction of the total and are paradoxically making large segments of it less useful in support of man. Solar energy has been fixed in one form or another on the earth throughout much of the earth's 4.5-billion-year history.

The modern biosphere probably had its beginning about two billion years ago with the evolution of marine organisms that not only could fix solar energy in organic compounds but also did it by splitting the water molecule and releasing free oxygen. The beginning was slow. Molecular oxygen released by marine plant cells accumulated for hundreds of millions of years, gradually building an atmosphere that screened out the most destructive of the sun's rays and opened the land to exploitation by living systems (see "The Oxygen Cycle," by Preston Cloud and Aharon Gibor; Scientific American Offprint 1192). The colonization of the land began perhaps 400 million years ago. New species evolved that derived more energy from a more efficient respiration in air, accelerating the trend.

Evolution fitted the new species together in ways that not only conserved energy and the mineral nutrients utilized in life processes but also conserved the nutrients by recycling them, releasing more oxygen and making possible the fixation of more energy and the support, of still more life. Gradually each landscape developed a flora and fauna particularly adapted to that place. These news arrays of plants and animals used solar energy, mineral nutrients, water and the resources of other living things to stabilize the environment, building the biosphere we know today. The actual amount of solar energy diverted into living systems is small in relation to the earth's total energy budget (see "The Energy Cycle of the Earth," by Abraham H. Oort; Scientific American Offprint 1189).

Only about a tenth of 1 percent of the energy received from the sun by the earth is fixed in photosynthesis. This fraction, small as it is, may be represented locally by the manufacture of several thousand grams of dry organic matter per square meter per year. Worldwide it is equivalent to the annual production of between 150 and 200 billion tons of dry organic matter and includes both food for man and the energy that runs the life-support systems of the biosphere, namely the earth's major ecosystems: the forests, grasslands, oceans, marshes, estuaries, lakes rivers, tundras and deserts. The complexity of ecosystems is so great as to preclude any simple, single-factor analysis that is both accurate and satisfying. Because of the central role of energy in life, however, an examination of the fixation of energy and its flow-through ecosystems yields understanding of the ecosystems themselves. It also reveals starkly some of the obscure but vital details of the crisis of environment.

More than half of the energy fixed in photosynthesis is used immediately in the plant's own respiration. Some of it is stored. In land plants it may be transferred from tissues where it is fixed, such as leaves; to other tissues where it is used immediately or stored. At any point it may enter consumer food chains. There are two kinds of chain: the grazing, or browsing, food chains and the food chains of decay. Energy may be stored for considerable periods in both kinds of chain, building animal populations in the one case and accumulations of undecomposed dead organic matter and populations of decay organisms in the other. The fraction of the total energy fixed that flows into each of these chains is of considerable importance to the biosphere and to man.

The worldwide increase in human number not only is shifting the distribution of energy within ecosystems but also requires that a growing fraction of the total energy fixed be diverted to the direct support of

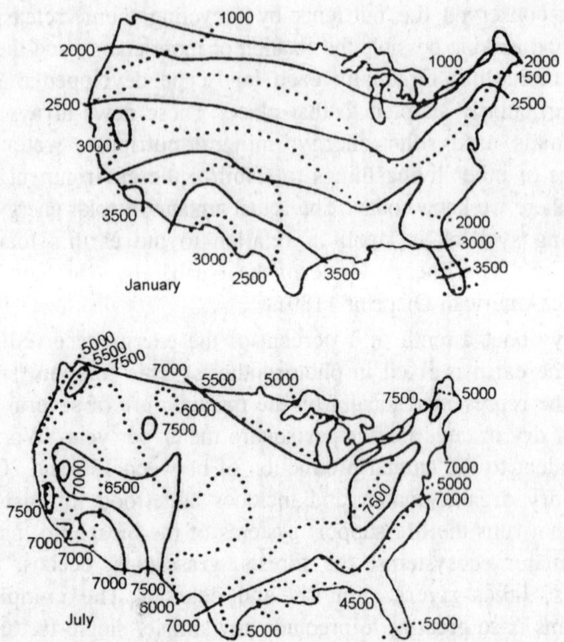

Fig. 12.1. Lines of equal daily solar energy at the ground on cloudless days (solid lines) and on days of average cloudiness (dash lines) in January and July. Units are kilocalories per square meter per day.

man. The implications of such diversions are still far from clear. Before examining the fixation and flow of energy in ecosystems it is important to consider the broad pattern of their development throughout evolution. If one were to ascribe a single objective to evolution, it would be the perpetuation of life. The entire strategy of evolution is focused on that single end. In realizing it evolution divides the resources of any location, including its input of energy, among an ever increasing number of different kinds of users, which we recognize as plant and animal species. The arrangement of these species in today's ecosystems is a comparatively recent event, and the ecosystems continue to be developed by migration and continuing evolution. Changes accrue slowly through a conjoint evolution that is not only biological but also chemical and physical.

The entire process appears to be open-ended, continuous, self-augmenting and endlessly versatile. It builds on itself, not merely preserving life but increasing the capacity of a site to support life. In so doing it stabilizes the site and the biota. Mineral nutrients are no longer leached rapidly into watercourses; they are conserved and recirculated, offering opportunities for more evolution. Interactions among ecosystems are exploited and stabilized, by living systems adapted to the purpose. The return of the salmon and other fishes from years at sea to the upper reaches of rivers is one example; impoverished upland streams are thus fertilized with nutrients harvested in the ocean, opening further possibilities for life. The time scale for most of these developments, particularly in the later stages when many of the species have large bodies and long life cycles, is very long.

Such systems are for all practical purposes stable. These are the living systems that have shaped the biosphere. They are self-regulating and remarkably resilient. Now human activities have become so pervasive as to affect these systems all over the world. What kinds of change can we expect? The answers depend on an understanding of the patterns of evolution and on a knowledge of the structure and function of ecosystems. And the fixation and flow of energy is at the core. Much of our current understanding of ecosystems has been based on a paper published in Ecology in 1942 by Raymond L. Lindeman, a young colleague of G. Evelyn Hutchison's at Yale University. (It was Lindeman's sixth and last paper; his death at the age of 26 deprived ecology of one of its most outstanding intellects.) Lindeman drew on work by earlier scholars, particularly Arthur G. Tansley and Charles S. Elton of England and Frederick E. Clements and Victor E. Shelford of the U.S., to examine what he called the "trophic-dynamic aspect" of ecology. He called attention to the fixation of energy by natural ecosystems and to the quantitative relations that must exist in nature between the different users of this energy as it is divided progressively among the various populations of an ecosystem.

Lindeman's suggestions were provocative. They stimulated a series of field and laboratory studies, all of which strengthened his synthesis. One of the most useful generalizations of his approach, sometimes called "the 10 percent law," simply states that in nature some fraction of the energy entering any population is available for transfer to the populations that feed on it without serious disruption of either. The actual amount of energy transferred probably varies widely. It seems fair to assume that in the grazing chain perhaps 10 to 20 percent of the energy fixed by the plant community can be transferred to herbivores, 10 to 20 percent of the energy entering the herbivore community can be transferred to the first level of carnivores and so on. In this way what is called a mature community may support three of four levels of animal populations, each related to its food supply quantitatively on the basis of energy fixation. No less important than the grazing food chains are the food chains of decay. On land these chains start with dead organic matter: leaves, bits of bark and branches. In water they originate in the remains of algae, fecal matter and other organic debris.

The organic debris may be totally consumed by the bacteria, fungi and small animals of decay, releasing carbon dioxide, water and heat. It may enter far more complex food webs, potentially involving larger animals such as mullet, carp, crabs and ultimately higher carnivores, so that although it is convenient to think to the grazing and decay routes as being distinct, they usually overlap. They decay food chain does not always function efficiently. Under certain circumstances it exhausts all the available oxygen. Decay is then incomplete; its products include methane, alcohols, amines, hydrogen sulphide and partially decomposed organic matter. Its connections to the grazing food chain are reduced or broken, with profound effects on living systems. Such shifts are occurring more frequently in an increasingly man-dominated world. How much energy is fixed by the major ecosystems of the biosphere? The question is more demanding than it may appear because measuring energy fixation in such diverse vegetations as forests, fields and the oceans is most difficult.

Rates of energy fixation vary from day to day-even from minute to minute-and from place to place. They are affected by many factors, including light and the concentration of carbon dioxide, water and nutrients. In spite of the difficulties in obtaining unequivocal answers several attempts have been made to appraise the total amounts of energy fixed by the earth's ecosystems. Most recently Robert H. Whittaker and Gene E. Likens of Cornell University have estimated that in all the earth's ecosystems, both terrestrial and marine, 164 billion metric tons of dry organic matter is produced annually, about a third of it in the oceans and two-thirds of it on land. This "net production" represents the excess of organic production over what is required to maintain the plants that fixed the energy; it is the energy potentially available for consumers.

Virtually all the net production of the earth is consumed annually in the respiration of organisms other than green plants, releasing carbon dioxide, water and the heat that is reradiated into space. The consumers are animals, including man, and the organisms of decay. The energy that is not consumed is either stored in the tissues of living organisms, or in humus and organic sediments. The relations between the producers and the consumers are clarified by two simple formulas. Consider the growth of a single green plant, an "autotroph" that is capable of fixing its own solar energy. Some of the energy it fixes is stored in organic matter that accumulates as new tissue. The amount of the new tissue, measured as dry weight, is the net production. This does not, however, represent all the energy fixed. Some energy is required just to support the living tissues of the plant. This is energy used in respiration.

The total energy fixed, then, is partitioned immediately within the plant according to the equation $GP—Rs_A = NP$. The total amount of energy fixed is gross production (GP); Rs_A is the energy used in the respiration of the autotrophic plant, and the amount of energy left over is net production (NP). The growth of a plant is measurable as net production, which can be expressed in any of several different ways, including energy stored and dry weight. The same relations hold for an entire plant community and for the biosphere as a whole.

If we consider not only the plants but also the consumers of plants and the entire food web, including the organisms of decay, we must add a new unit of respiration without adding any further producers. That is what happens as an ecosystem matures: consumer populations increase substantially, adding to the respiration of the plants the respiration (Rs_H) of the heterotrophs, the organisms that obtain their energy from the photosynthesizing plants. For an ecosystem (the total biota of any unit of the earth's surface) NEP equals $GP—(Rs_A + Rs_H)$. NEP is the net ecosystem production, the net increase in energy stored within the system. $Rs_A + Rs_H$ is the total respiration of the ecosystem. This last equation establishes the important distinction between a "successional," or development, ecosystem and a "climax," or mature, one. In the successional system the total respiration is less than the gross production, leaving energy (NEP) that is built into structure and adds to the resources of the site. (A forest of large trees obviously has more space in it, more organic matter and probably a wider variety of microhabitats than a forest of small trees).

In a climax system, on the other hand, all the energy fixed is used in the combined respiration of the plants and the heterotrophs. NEP goes to zero: there is no energy left over and no net annual storage. Climax ecosystems probably represent a most efficient way of using the resources of a site to sustain life with minimum impact on other ecosystem. It is of course such ecosystems that have dominated the biosphere throughout recent millenniums. These general relations are clarified if one asks, with regard to a specific ecosystem, how much energy is fixed and how it is used, and how efficient the ecosystem is in harvesting solar energy and supporting life. The answers are found by solving the simple production equations, but in order to solve them one must measure the metabolism of an entire unit of landscape.

Such studies are being attempted in many types of ecosystem under the aegis of the International Biological Programme, a major research effort designed to examine the productivity of the biosphere. The example I shall give is drawn from research in an oak-pine forest at the Brookhaven National Laboratory. The research has spanned most of a decade and has involved many contributors. A most important contribution was made by Whittaker, who collaborated with me in completing a detailed description of the structure of the forest, including the total amount of organic matter, the weight and area of leaves, the weight of roots and the amount of net production.

The techniques developed in that work are now being used in many similar studies. Such data are necessary to relate other measurements, including measurements of the gas exchange between leaves and the atmosphere, to the entire forest and so provide an additional measurement of net production and respiration. A major problem was measuring the forest's total respiration. We used two techniques. First, Winston R. Dykeman and I took advantage of the frequent inversions of temperature that occur in central Long Island and used the rate of accumulation of carbon dioxide during these inversions as a direct measurement of total respiration. The inversions are nocturnal; this eliminates the effect of photosynthesis, which of course proceeds only in daylight. During an inversion the temperature of the air near the ground is (contrary to the usual daytime situation) lower than that of the air at higher elevations.

Since the cooler air is denser, the air column remains vertically stable for as much as several hours; the carbon dioxide released by respiration accumulates, and its buildup at a given height is an index of the rate of respiration at that height. The calculation of the buildup during more than 40 inversions in the course of a year provided one measure of total respiration. A second measurement came from a detailed study of the rates of respiration of various segments of the forest (including the branches and stems of trees) and the soil. The estimates available from these studies and others are converging on the following solution of the production equations, all in terms of grams of dry organic matter per square meter per year: The gross production is 2,650 grams; the net production, 1,200 grams; the net ecosystem production, or net storage, 550 grams, and the total respiration, or energy loss, 2,100 grams, of which Rs_A is 1,450 and Rs_H is 650.

The forest is obviously immature in the sense that it is still storing energy (NEP) in an increased plant population. The ratio of total respiration to gross production (2,100/2,650) suggests that the forest is at about 80 percent of climax and confirms other studies that show that the forest is "late successional." The nest production of the Brookhaven forest of 1,200 grams per square metre per year is in the low middle range for forests and is typical of the productivity of small-statured forests. The efficiency of this forest in using the annual input of solar energy effective in photosynthesis is about .9 percent.

Large-statured forests (moist forests of the Temperate Zone, where nutrients are abundant, and certain tropical rain forests) have a net productivity ranging up to several thousand grams per square metre per year. They may have an efficiency approaching 3 percent of the usable energy available throughout the year at the surface of the ground, but usually not must more. Sugarcane productivity in the Tropics has been reported as exceeding 9,000 grams per square metre per year. The new strains of rice that are contributing to the "green revolution" have a maximum yield under intensive triple-cropping regimes that may approach 2,000 grams of rice per square metre per year. Over large areas the yield is much lower, seldom exceeding 350 to 400 grams of milled rice per square metre per year. These yields are to be compared with corn yields in the U.S., which approach 500 grams. (The rice and corn yields are expressed as grain, not as total net production as we have been discussing it.

Net production including the chaff, stems, leaves and roots is between three and five times the harvest of grain. Thus the net production of the most productive agriculture is 6,000 to 10,000 grams, probably the highest net production in the world. Most agriculture, however, has net production of 1,000 to 3,000 grams, the same range as most forests.) The high productivity of agriculture are somewhat misleading in that they are bought with a contribution of energy from fossil fuels: energy that is applied to cultivate and harvest the crop, to manufacture and transport pesticides and fertilizers and to provide and control irrigation. The cost accounting is incomplete; these systems "leak" pesticides, fertilizers and often soil itself, injuring other ecosystems. It is clear, however, that the high yields of agriculture are dependent on a subsidy of energy that was fixed as fossil fuels in previous ages and is available now (and for some decades to come) to support large human populations. Without this subsidy or some other source of power, yields would drop. They may suffer in any case as it becomes increasingly necessary to reduce the interactions between agriculture and other ecosystems. One sign is the progressive restriction in the use of insecticides because of hazards far from where they are applied. Similar restraint may soon be necessary in the use of herbicides and fertilizers. The oceans appear unproductive compared with terrestrial ecosystems.

In separate detailed analyses of the fish production of the world's oceans William E. Ricker of the Fisheries Research Board of Canada and John H. Ryther of the Woods Hole Oceanographic Institution recently emphasized that the oceans are far from an unlimited resource. The net production of the open ocean is about 50 grams of fixed carbon per square metre per year. Areas of very high productivity, including coastal areas and areas of upwelling where nutrients are abundant, do not average more than 300 grams of carbon. The mean productivity of the oceans according to this analysis, would be about 55 grams of carbon, equivalent to between 120 and 150 grams of dry organic matter. Inasmuch as the highest productivity of enriched areas of the ocean barely approaches that of diminutive forests such as Brookhaven's, the oceans do not appear to represent a vast potential resource.

On the contrary, Ryther suggests on the basis of an elaborate analysis of the complex trophic relations of the oceans that "it seems unlikely that the potential sustained yield of fish to man is appreciably greater than 100 million (metric) tons (wet weight). The total world fish landings for 1967 were just over 60 million tons, and this figure has been increasing at an average rate of about 8 percent per year for the past 25 years... . At the present rate, the industry can expand for no more than a decade." Ricker comes to a similar conclusion. Neither he nor Ryther appraised the effects on the productivity of the oceans of the accumulation of toxic substances such as pesticides, of industrial and municipal wastes, of oil production on the continental shelves, of the current attempts at mining the sea bottom and of other exploitation of the seas that is inconsistent with continued harvesting of fish. The available evidence suggests that, in spite of the much larger area of the oceans, by far the greater amount of energy is fixed on land.

The oceans, even if their productivity can be preserved, do not represent a vast unexploited source of energy for support of larger human populations. They are currently being exploited at close to the maximum sustainable rate, and their continued use as a dump for wastes of all kinds makes it questionable whether that rate will be sustained. A brief consideration of the utilization of the energy fixed in the Brookhaven forest will help to clarify this point.

The energy fixed by this late-sucessional forest is first divided between net production and immediate use in plant respiration, with about 55 percent being used immediately. (The ratio of 55 percent going directly into respiration appears consistent for the Temperate Zone forests examined so far; the ratio appears to rise in the Tropics and to decline in higher latitudes). The net production is divided among herbivores, decay and storage. In the Brookhaven forest herbivores populations have been reduced by the exclusion of deer, leaving as the principal herbivores insects and limited populations of small mammals. Our estimates indicate that only a few percent of the net production is consumed directly by herbivores (a low rate in comparison with other ecosystems).

Practically all this quantity is consumed immediately in animal respiration, so that the animal population shows virtually no annual increase, or contribution to the net ecosystem production. The principal contribution to the net ecosystem production is the growth of the plant populations, which accounts for more than 40 percent of the net production. The remainder of the net production enters the food chains of decay, which are obviously well developed. Clearly the elimination of deer, combined with poorly developed herbivore and carnivore populations, has resulted in a diversion of energy from the grazing chain into the food chains of decay. This is precisely what happens in aquatic systems as they are enriched with nutrients washed from the land; the shift to decay is also caused by the accumulation of any toxic substance, whether it affects plants or animals. Any reduction in populations of grazers shifts the flow of energy toward decay. Any effect on the plants shifts plant populations away from sensitive species toward resistant species that may not be food for the indigenous herbivores, thereby eliminating the normal food chains and also shifting the flow of energy into decay.

These observations simply show that the structure and function of major ecosystems are sensitive to many influences. Clearly the amount of living tissue that can be supported in any ecosystem depends on the amount of net production. Net production, however, is coupled to both photosynthesis and respiration, both of which can be affected by many factors. Photosynthesis is sensitive to light intensity and duration, to the availability of water and mineral nutrients and to temperature. It is also sensitive to the concentration of carbon dioxide; on a worldwide basis the amount of carbon dioxide in the atmosphere may exert a major control over rates of net production. Greenhouse men have recognized the sensitivity of photosynthesis to carbon dioxide concentration for many years and sometimes increase the concentration artificially to stimulate plant growth. Has the emission of carbon dioxide from the combustion of fossil fuels in the past 150 years caused a worldwide increase in net production, and if so, how much of an increase? With equipment specially designed at Brookhaven, Robert Wright and I supplied air with enhanced levels of carbon dioxide to trees and determined the effect on net photosynthesis by measuring the uptake of the gas by leaves.

The net amount of carbon dioxide that was fixed increased linearly with the increase in the carbon dioxide concentration in the air. Such small increases in carbon dioxide concentration have virtually no effect on rates of respiration. The data suggest that the increase of about 10 percent (30 parts per million) in the carbon dioxide concentration of the atmosphere since the middle of the 19th century caused by the industrial revolution may have increased net production by as much as 5 to 10 percent. This increase, if applicable worldwide and considered alone, would increase the total energy (and carbon) stored in natural ecosystems by an equivalent amount, and would result in an equivalent improvement in the yields of agriculture. The increase in net production also tends to stabilize the carbon dioxide content of the atmosphere by storing more carbon in living organic matter, particularly in forests, and in the nonliving organic matter of sediments and humus. Such changes have almost certainly occurred on a worldwide basis as an inadvertent result of human activities in the past 100 years or so.

Such simple single-factor analyses of environmental problems, however, are almost always misleading. As the carbon dioxide concentration in the atmosphere has been increasing, many other factors have changed. There was a period of rising temperature, possibly due to the increased carbon dioxide concentration. More recently, however, there has been a decline in world temperatures that continues. This can be expected to reduce net production worldwide by reducing the periods favourable for plant growth. Added to the effects of changing temperature-and indeed overriding it-is the accumulation of toxic wastes from human activities. The overall effect is to reduce the structure of ecosystems. This in turn shortens food chains and favours (1) populations of small hardy plants, (2) small-bodies herbivores that reproduce rapidly, and (3) the food chains of decay. The loss of structure also implies a loss of "regulation"; the simplified communities are subject to rapid changes in the density of these smaller, more rapidly reproducing organisms that have been released from their normal controls.

Local increases in water temperature also give rise to predictable effects. There is talk, for example, of warming the waters of the New York region with waste heat from reactors to produce a rich "tropical" biota, but such manipulation would produce a degraded local biota supplemented by a few hardy species of more southerly ecosystems. Such circumstances again favour productivity not by complex, highly integrated arrays of specialized organisms but by simple arrays of generalized ones. Energy then is funneled not into intricate food webs capped by tuna, mackerel, petrels, dolphins and other highly specialized carnivores but into simple food webs dominated by hardy scavengers such as gulls and crabs and into the food webs of decay.

As the annual contribution to decay increases, these webs in water become overloaded; the oxygen dissolved in the water is used up and metabolism shifts from the aerobic form where oxygen is freely available to the much less efficient anaerobic respiration; organic matter accumulates, releasing methane, hydrogen sulphide and other noxious gases that only reinforce the tendency. The broad pattern of these changes is clear enough. On the one hand, an increasing fraction of the total energy fixed is being diverted to the direct support of man, replacing the earth's major ecosystems with cities and land devoted to agriculture-the simplified ecosystems of civilization that require continuing contributions of energy under human control for their regulation. On the other hand, the leakage of toxic substances from the man-dominated provinces of the earth is reducing the structure and self-regulation of the remaining natural ecosystems. The trend is progressive.

The simplification of the earth's biota is breaking down the insulation of large units of the earth's surface, increasing the interactions between terrestrial and aquatic systems, between upland and lowland between river and estuary. The long-term trend of evolution toward building complex, integral, stable ecosystems is being reversed. Although the changes are rapid, accelerating and important, they do not mean that the earth will face an oxygen crisis; photosynthesis rate in certain places, stimulated by increased carbon dioxide concentrations in air and the availability of nutrients in water. A smaller fraction of the earth's fixed energy is easily available to man, however.

The energy flows increasingly through smaller organisms such as the hardy shrubs and herbs of the irradiated forest at Brookhaven, the scrub oaks that are replacing the smogkilled pines of the Los Angeles basin, the noxious algae of eutrophic lakes and estuaries, into short food chains, humus and anaerobic sediments. These are major man-caused changes in the biosphere. Many aspects of them are irreversible; their implications are poorly known. Together they constitute a major series of interlocking objectives for science and society in the next decade focused on the question: "How much of the energy that runs the biosphere can be diverted to the support of a single species: man?"

13

Population Ecology

A population is an interacting and interbreeding group of individuals of the same species, for example, all the perch in a, small lake. Although they may also occupy a particular area—in this example, the lake—certain populations, such as the elk, a migratory species, live in different locales at different times, of the year. Because the members of a population interbreed, they have a common gene pool; thus, populations are the units, upon which natural selection acts. Ecologists are interested primarily in *population dynamics*: the various factors contributing: to the increase, stability, decline, and extinction of populations including those of our own species. Understanding these factors is not only important in itself but is essential to an understanding of evolution. The term *population* is sometimes used to refer to larger or smaller groups than the fishes in a small lake. For instance, all the foxes in England or all the beech trees in North America can be regarded as populations. Even such small groups as all the bacteria living in some-one's mouth or all the ascarid worms parasitizing a dog can be considered populations. Although we often refer to the human population of a country, its members not only interact with but even interbreed with individuals of other countries. Moreover, within a single country, people may be divided into various subpopulations, some of which are quite isolated from each other. The term *population* thus has several connotations. The connotation intended is usually clear from the context in which the term is used.

Population Growth and Structure

The growth of a population in a given area is the result of two opposing forces:

1. the biotic potential or reproductive capacity, of the species and
2. the environment resistance. Biotic potential is the maximum rate of population increase under ideal conditions, which are seldom realized in nature for long periods of time.

Environmental-resistance includes all factors that limit population size, such as nutrient and habitat availability, disease, and predation.

Exponential Growth Versus Population size Stability

An important basis for the natural election theory of Darwin and Wallace was the fact that all organisms are capable of exponential, or logarithmic, growth in numbers. In a population growing exponentially, each successive total is a multiple of the one preceding it. Put another way, populations are capable of doubling at a constant rate, such as once every so many hours, weeks or years. (An example of logarithmic increase is provided by the series 2, 4, 8, 16, 32, 64, and so on. An arithmetic increase is an increase by a constant number, exemplified by the series 10, 20 30, 40, 50 and so on.) Many organisms have enormous reproductive potential. Some bacteria can double in Dumber every 15 minutes. A single mushroom produces hundreds of thousands of spores, each of which could, if provided the space, nutrients, and so on, give rise to a huge colony of mushrooms, each of which could produce many more mushrooms. In many insect and fish species, one individual can lay thousands of eggs. In some species of clams, one individual can produce 3 million eggs in a season. A single pair of mice, given the necessary food and space and an absence of disease

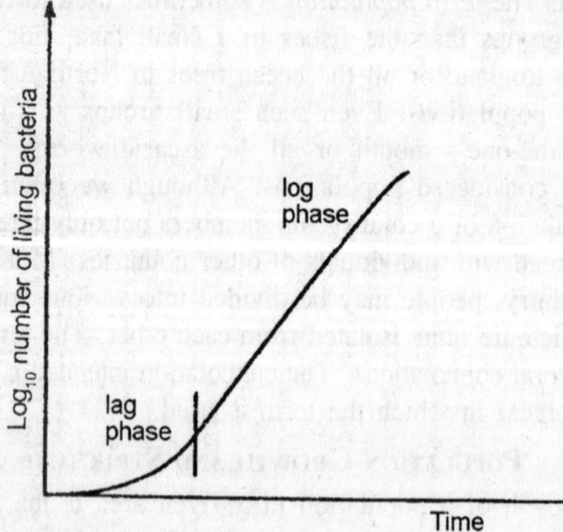

Fig. 13.1. Exponential growth of bacteria in laboratory culture.

and predators, could theoretically, have 3,000 descendants in one year—or 6.75 million in 3 years!

To emphasize the universality of the potential for logarithmic increase among organisms, Darwin analyzed the most slowly reproducing animal species, elephants, and found that even they were potentially capable, within a few years, of producing more offspring than the earth could accommodate. However, as we are well aware, this enormous potential is, except for brief periods, never realized. Darwin further observed that, with occasional fluctuations, populations of a given species tend to remain about the same size from one year to the next. He concluded that "this geometric tendency to increase must be checked by destruction at some period of life." In the more than a century since those words were written, we have learned much—but far from all—about the factors that regulate population growth.

Analysis of Exponential Growth

Some insight into factors that determine population size has been achieved by studying simplified ecosystems. For example, when a culture vessel containing a warm broth composed of a balanced mixture of all the nutrients needed for the growth of one particular species of bacterium is inoculated with a few hundred individuals derived from a thriving culture, the cells continue dividing at a steady rate. If the culture's growth is graphed as the logarithm of the number of cells against time, the result is a straight line. Such a culture is said to be in exponential growth, in a logarithmic growth phase, or, simply, in log growth. To predict the growth of a given population over a particular time period, we need to know the initial size of the population (symbolized by N) and the potential growth rate of that species, called its *intrinsic rate of increase* (symbolized r). Intrinsic rate is the

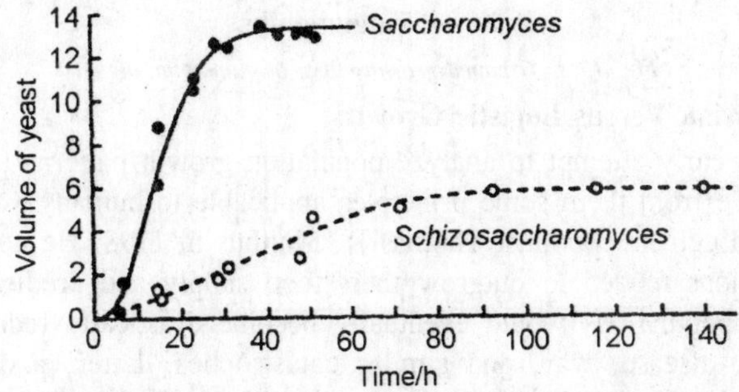

Fig. 13.2. Sigmoidal growth of pure culture of the yeasts Saccharomyces and Schizosaccharomyces.

difference between the population's *natality* the annual rate of births, hatchings, and the like—and its *mortality*—the annual rate of deaths—under what could be called ideal conditions, with adequate food and shelter, no predation, and so on. Both natality and mortality rates are largely genetically determined and can differ greatly from one species to another.

Expressed mathematically the increase in size of a population undergoing exponential growth is stated as $\Delta N \ \Delta t = rN$, in which ΔN represents the difference (Δ) in population size (N), Δt is the length of the time interval (difference in time) during which exponential growth occurred, and rN is the intrinsic rate of increase (r) for a starting population of size *N.* Stated in words, the equation reads as follows: the actual increase in numbers (ΔN) during a time interval of Δt (some fraction of a year) equals the potential maximum rate of increase per year (*r*) times the number of individuals (*N*) at the beginning of the time interval Δt. So it can be seen that, under ideal conditions, the actual amount of increase may equal the potential growth rate and, for brief periods, this may be case.

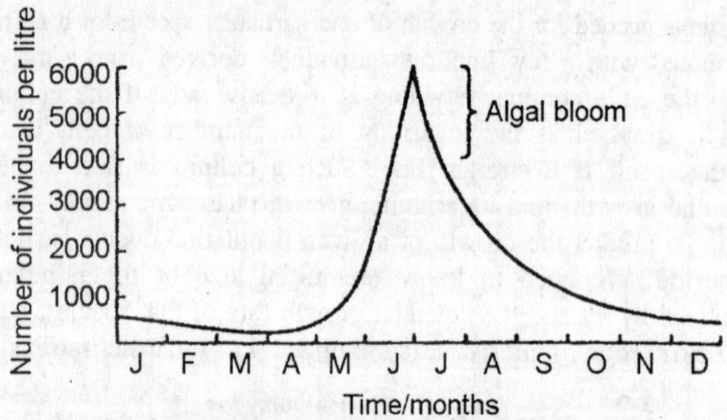

Fig. 13.3. J-shaped growth curve of planktonic algae.

Malthusiau Versus Logistic Growth

An early attempt to analyze population growth patterns and thus to derive from them some principles applicable to humans was made by the English economist Thomas R. Malthus in 1798. He noted that populations tended to outgrow their food supply and predicted that human populations would eventually become drastically reduced by famine, disease, war, and similar catastrophes. Later, a differing concept, termed *logistic growth*, emerged. According to it populations eventually tend to level off at a relatively high level.

Malthusian growth

The Malthusian doctrine can be illustrated by the hypothetical growth curve which a period of exponential growth is climaxed by a series of catastrophic reductions in population size due to massive famine, disease, and other factors: This pattern is referred to as *irruptive growth* or *Malthusian growth*. Malthus idea about humanity's future was poorly received, not only at the time it was published but also in subsequent years. The concept was decried by religious leaders as an irreverent attempt to curb human procreation and by scientists as a lack of faith in the capabilities of science. Malthus was viewed as an alarmist, and his idea was generally ignored. Only in the latter half of this, century, as his predictions have begun to come true, has his theory been accorded much respect. The strongest reason for ignoring Malthus idea during the nineteenth century was cogent: the experience of so many observers to the contrary.

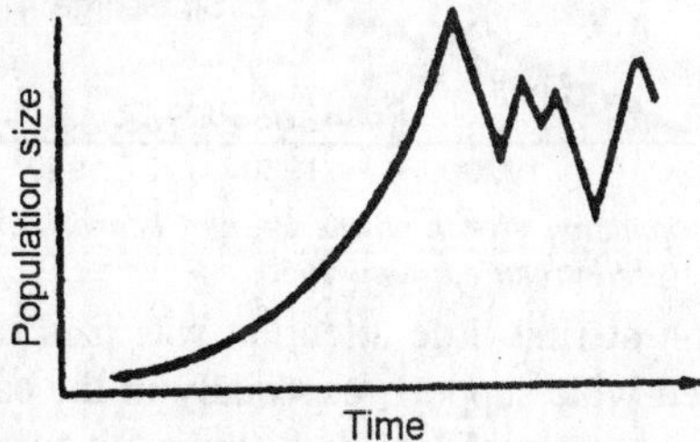

Fig. 13.4. Malthusian, or irruptive, growth curve. The upslope of the curve is steep and is followed by a series of partial population collapses.

By the mid-1800s, considerable data had accumulated on the growth of natural and laboratory populations of many species of organisms. In natural populations, in which predation and other limiting factors were operative, growth of the population often leveled off for an indefinite period, exhibiting a sigmoid (S-shaped) growth pattern. In laboratory cultures of microorganisms, a similar sigmoid growth curve was often obtained. Unfortunately, these bacterial experiments were terminated before what is called the death phase began; otherwise, the results would have been more typical of Malthusian predictions.

Logistic growth

The many examples of sigmoid growth curves in laboratory and natural populations led to the concept of logistic growth. This is a pattern typified by a slow beginning, a log growth phase, and a gradual slowing to a smooth, orderly, predictable pattern of population limitation at a level indefinitely supported by the ecosystem, said to be its *carrying*

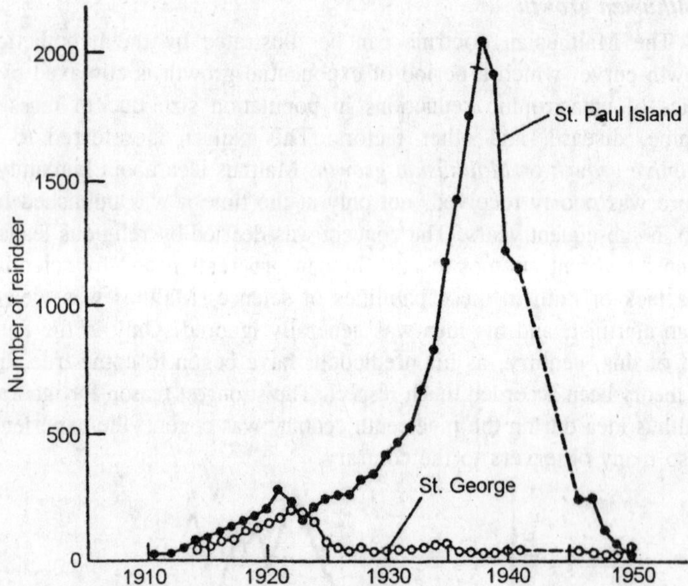

Fig. 13.5. Reindeer population growth on the Pribilof Island, Bering Sea, from 1911, when they were introduced, until 1950.

capacity. Although at first little attention was paid to this concept, it eventually received wide support, especially in the early 1900s. Some even hailed it as a universal law and applied it to the growth of all populations, including humans.

One enthusiast, invoking the principle of logistic growth, predicted that the world human population would naturally stabilize at 2.6 billion (2.6×10^9) people in AD 2100. As we know, due to changes in several factors since that time, this population size was exceeded before 1960, passed the 4 billion mark in 1975, and is still increasing rapidly. In a bacterial culture, after a period of log growth, the length of which depends on the size of the initial population and the total volume and nutrient concentration of the broth, growth of the culture slows. It then gradually enters a period of variable length in which the number of cells remains relatively constant.

In a bacterial culture, this is termed the maximum stationary phase. It represents a balance between the production of new cells and the death of older ones. A similar pattern is exhibited by populations of most species in most environments. The factors that control population size in such situations are examined later in this chapter. The final population size achievable by a species in a particular

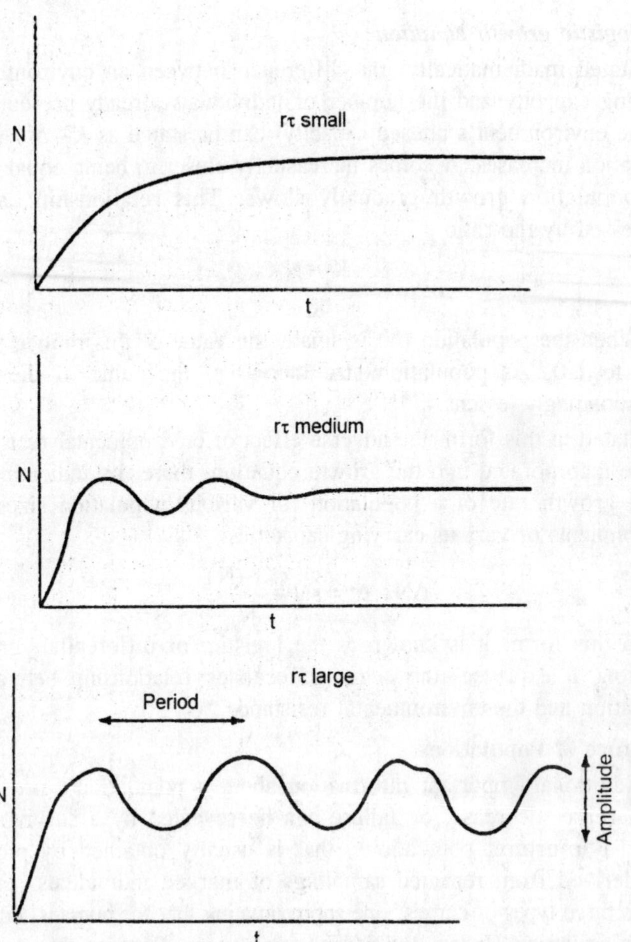

Fig. 13.6. Logistic growth curves with a time lag for species with overlapping generations. Growth depends on the value of rτ, the product of the intrinsic rate of increase and the time lag. In this figure, rτ increases from top to bottom. At a certain value of rτ, the population oscillation become stable and no longer coverage on the carrying capacity.

environment; that is, the carrying capacity of that environment, or the environmental resistance, is symbolized as K. Although the carrying capacity of a given environment is a constant, it is subject to change as the long-term climatic or certain other factors in an environment may change over the years. Moreover, by its evolutionary adaptation to a particular environment, a species can, over many years, increase the value of K.

The logistic growth equation

Stated mathematically, the difference between an environment's carrying capacity and the number of individuals already present in it (or the environment's unused capacity) can be stated as *K—N*. As the population increases, *N* comes increasingly closer to being equal to *K*, and population growth gradually slows. This relationship can be .expressed by the ratio.

$$\frac{K-N}{K}$$

When the population (*N*) is small, the value of this ratio is high, close to 1.0, As population size increases, the value of the ratio correspondingly lessens.

Stated in this form, the adverse effect of environmental resistance can be incorporated into the growth equation, there, by indicating the actual growth rate of a population for various population sizes and environments of various carrying capacities:

$$\Delta N / \Delta t = rN\frac{(K-N)}{K}$$

In this form, it is known as the logistic, or differential, growth equation. It expresses the negative feedback relationship between a population and the environmental resistance.

Structure of Populations

Additional important information about a population's prospects for increase, decrease, or failure can be revealed by a survivorship curve. For natural populations, that is usually obtained by plotting data derived from repeated samplings of marked individuals. Figure depicts three types of curves, one approximating that for humans, another for the cnidarian *Hydra*, and a third for oysters. Because they are not cared for, the mortality rate for oyster eggs and larvae is extremely high, although any that survive early development have a good chance of living to maturity. Humans, on the other hand, with extended parental care, have a relatively high life expectancy at birth, with a sharp increase in mortality in the last 15 of their expected 70 years. Hydras exhibit the same death rate at every age. Although survivorship curves provide some insights into the prospects of a population of a given species for survival, much more useful in this regard are the age structures of particular populations. In one way of representing age structure, the population is divided into three age groups, each a percentage of the entire population:

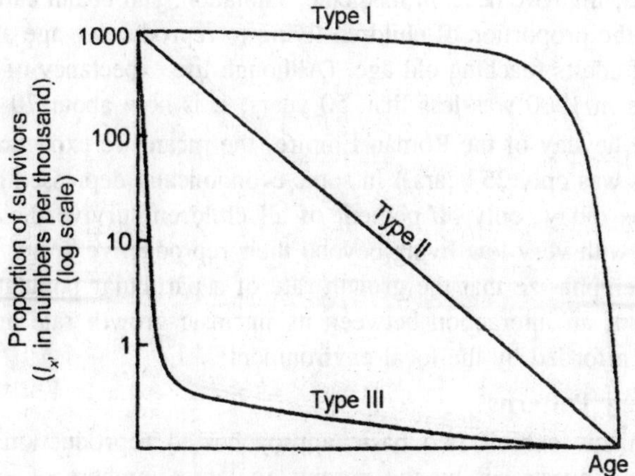

Fig. 13.7. Hypothetical survivorship curves.

1. pre-reproductives, or immature individuals;
2. reproductives, or fecund adults; and
3. post-reproductives, or nonfecund adults.

The proportions of a population in each class are key factors in determining whether that population can be expected to increase, decrease, or stay the same. The diagram for an expanding population, such as the present worldwide human population, has a pyramidal outline, its base being composed of pre-reproductives. The diagram for a stable population has a bell-shaped outline; that for a declining population has an urn-shape. An urn-shaped distribution, with its small proportion of pre-reproductives, usually indicates that a population is dying off. While greatly affected by environmental factors, age structures are also genetically influenced and can differ greatly from species to species.

Some insects, for example, spend months or even years as larvae and then, as adults, as little as one day in reproduction. In contrast, the reproductive years of humans in a favourable environment comprise about 42 percent of a mean life expectancy of 70 years. Some species reproduce throughout adult life, and many animals, including various worms, insects, and fishes, die almost immediately after reproducing. Populations of such species have no post-reproductive component in their age structure. The relative proportions of the component stages in the life history represented by each of the three classes can be greatly affected by both biotic and environmental factors and may therefore differ geographically for a given species. With many human populations, improvements in medicine, sanitation, and health care have increased the proportion of children living to reproductive age and the number of adults reaching old age. (Although life expectancy of North Americans in 1900 was less than 50 years, it is now about 70 years.

In the heyday of the Roman Empire, the mean life expectancy of its citizens was only 25 years.) In some economically depressed human populations today, only 50 percent of all children survive their fifth birth day, with very few living beyond their reproductive years. These examples emphasize that the growth rate of a particular population is the result of an interaction between its intrinsic growth rate and the resistance afforded by the local environment.

Life History Patterns

Organisms exhibit two basic approaches to reproduction. One involves an investment by the parents in large numbers of spores, seeds, eggs, or offspring that receive little or no care. In such species, sheer numbers tend to ensure that some individuals survive to reproductive age. The other approach involves the production of a few offspring that receive Substantial parental assistance. In the former case, much of the parental body is given over to reproductive structures; in the latter, most of the body is vegetative, that is, nonreproductive, biomass. This principle applies to all forms of life. Examples of plants that produce few individuals but invest much it each of them are those that send out either stolons (runners) or rhizomes (underground stems) from which new individuals bud off and reach maturity while still attached to the parent plant. An example of animals that produce few individuals are birds that lay one to three eggs and provide the hatchlings with food .and other parental care until they are almost fully grown.

Between the two extremes of one offspring that receives much care and many offspring that receive no care is a full spectrum of intermediate situations. For example, some organisms produce a moderately large number of offspring that receive some care and protection in the early stages of development. All of these behaviours, of causes, are genetically determined, although learned components are often incorporated into them. Like many other hereditary traits, they represent adaptations to particular environmental situations. What determines, for a particular species, which of the two opposite approaches is selected or whether the species will exhibit an intermediate pattern that incorporates elements of both? The answer is not always clear, and in some situations, it may not matter which approaches is followed; in many cases, probably either would serve. However, various environmental situations are most successfully taken advantage of by one or the other of the two approaches.

***r* Adaptation**

Let us first consider the case of organisms most of whose developmental stages, such as spores, seeds, eggs, or larvae, are eaten by predators or infected by disease organisms. Many such species reproduce only once a year and have adapted to predation and disease by producing more spores, seeds, eggs, larvae, or hatchlings than the existing predators or disease organisms can eat or infect at one time. Such an approach is termed *r* adaptation or *r* selection because it depends on a rapid, exponential increase in numbers at a rate close to the organism's intrinsic rate of increase (*r*). Such high productivity at one time has the double effect of increasing the chances that some offspring will survive to maturity and of discouraging predators from specializing on a type of prey that is briefly available only once a year. However, because parasites and other disease organisms also can evolve, many have adapted their own life cycles to those of their hosts, multiplying and even becoming dormant in concert with the hosts. An elaboration on the once-a-year reproductive pattern is the *mast crop*, characteristic of many tree species in temperate zones.

In these species, 2 to several years may pass with little or no production of seeds or fruits, followed by a single season in which all such trees produce a great abundance of seeds. The populations of seed-eating animals, such as birds squirrels, and insects, that depend heavily on these particular seeds or fruits may decline in the area during the trees nonreproductive years. Then, suddenly, the reduced predator population is given far more food than it can eat; the next. Year's population of predators, now larger; faces the beginning of another cycle of several lean years. As might be expected, a selection pressure exists for various species of mast-crop-producing trees that inhabit the same environments to synchronize their mast crop production. Moreover, there would be less selection pressure for a tree to produce a mast crop if similar species in the area produced their mast crops in alternate years to its crop years.

For several tree species to synchronize their mast crops and thereby discourage predation of all of them permits a much higher population density of such trees and thus a greater overall efficiency in the capture of energy and the production of biomass. The trees do not need to communicate with each other in any way. Natural selection works against those that fail to synchronize their crops. Selection pressure for *r* adaptation is also exerted by environments that place a premium on the ability to produce large numbers offspring within a short time.

Two classic examples are the annual plants and the insects that live in deserts and those that live in the Arctic. These organisms develop from eggs, seeds, or other dormant stages whose development is triggered by the arrival of favourable conditions. They begin development grow to maturity, reproduce, and set seed or lay eggs all within the space of a few weeks each year. Thus, the *r* adaptation tends to be selected for by an environment with a brief growing season followed by conditions in which a high degree of mortality is independent of population density, such as occurs with a sudden, severe onset of desert aridity or an Arctic winter. Another situation that selects for *r* adaptation is the early stage of colonizing a new island or a burned-over area. Species that are the first to colonize such an area, or pioneer species, exhibit *r* adaptation. Organisms with a high intrinsic rate of population increase ale usually small, develop rapidly to reproductive maturity, and have short life spans.

K Adaptation

What kind of a situation might select for organisms that produce only one or a few offspring at a time? Most known examples occur in situations that exert strong negative-feedback controls on population growth. Under relatively constant year-round conditions in which an environment's carrying capacity for a particular species is often reached, competition for limited resources among species with overlapping needs tends to be high; in such circumstances *K* adaptation, or *K* selection, a term derived from the symbol for environmental resistance, tends to be favoured. Such are the conditions in tropical rain forests, where birds produce relatively small clutches of eggs. But other factors can also contribute to *K* adaptation. For example, larger, longer-lived organisms, such as the blue whale and the emperor penguin, are usually *K* adapted. A longer life span permits *iteroparity*, or repeated reproduction, which compensates for raising only one offspring at a time.

Population Regulation

Simple ecosystems represented by cultures of bacteria and :protozoa have proven useful in analyzing factors that regulate population size. Among the reasons that cells begin to die in thriving cultures of microorganisms is starvation. This, eventually results from depletion of essential nutrients and is an example of a *density-dependent factor* affecting population size. A factor, such as a disease epidemic, that exerts its effect on a population irrespective of its density is a *density-independent factor*. Density-dependent factors, such as nutrient supply depletion, constitute negative feedback controls over a population's absolute size and account for an increasing of mortality as density increases. The total amount of a key nutrient present in an environment may be quite unimportant when a population is small; it becomes increasingly important and eventually limiting as the population increases in density.

In a bacterial culture that has exhausted one or more of its essential nutrients, when the starved cells die, they disintegrate releasing their nutrient components into the medium. This new supply of nutrients permits some of the other starving cells to resume growth and division. Therefore, at this stage of the culture's existence, the number of new cells tends to equal the number of dying cells, resulting in the upper plateau of the sigmoid growth curve. If a new culture is begun by introducing into fresh broth medium several cells from an old culture that had reached maximum stationary phase, a lag phase at first occurs—that is, little or no growth takes place for a time—followed by a period of exponential growth and, as the nutrients become used up, culminating once more in a maximum stationary phase.

The maximum stationary phase of a bacterial culture may last for hours, days, or weeks. But eventually, if nothing is added or removed from the culture, this phase ends as the number of deaths begins to exceed the number of cell divisions. Thus begins the death, or negative growth, phase. What happens then depends on several factors that vary under different conditions and from one species to another. Some species exhibit a log death phase that can be as steep as or steeper than the long growth rate. In other cases, every cell dies in only a few days. In yet other instances, the number of cells drops to a new low level, but the culture continues to survive, its numbers oscillating irregularly at the lower level. Such a culture may persist in this condition for months or even years.

Density-Independent Factors

As a general rule, density-independent factors usually reduce a population by a certain constant percentage no matter what the size of the population is. One type of factor that, despite adequate nutrients, can slow, stop, or reverse the growth of a culture of microorganisms is the accumulation of toxic wastes. This factor can be much more serious for the ultimate survival of the culture can starvation. In old cultures, enough nutrients may be released from the disintegration of dead cells to support growth and reproduction of other cells indefinitely. But the concentration of inhibitory substances that can neither diffuse out of the culture nor be decomposed or detoxified by the organisms in the culture, although initially having no appreciable effect, may finally reach a level incompatible with the life of any of the cells. A toxic material that has reached a concentration that kills, most of the members of a culture is usually a density-independent factor for population growth.

The resulting irruptive growth curve is strikingly different from the sigmoid curve; it is said to be J shaped. In nature, examples of density-independent factors include droughts and cold spells which by arresting growth, may bring about a nearly total population collapse. Although a sudden drop in population size results in such cases, usually not all individuals die, and, when favourable conditions resume, a second period of log growth usually occurs. Massive death from brief disease epidemics, unfavourable climatic conditions, or temporary periods of starvation also produces J-shaped growth patterns. Not all growth curves are readily classified as either sigmoid or J shaped. Nor are all factors affecting population growth .easily categorized. Both starvation and predation, which usually function as density-dependent factors, can sometimes cause population collapses resembling the downslope of a J-shaped growth curve.

Intraspecific Relationships and Carrying Capacity

From its first enunciation, some biologists criticized the concept of logistic growth as simplistic. Yet it is certainly true that natural populations maintain relatively stable levels year after year. A controversy also exists over whether density-dependent intraspecific competition or density-independent extrinsic factors are of primary importance in the regulation of population size. Careful analysis of stable populations, both in natural situations and under laboratory conditions, has revealed a wide variety of factors, both density-dependent and density-independent, that affect a population's size once it reaches a certain level.

Some of the density-dependent factors have also been revealed by laboratory experiments such as the, following on beetles, mice, and rats. In separate experiments, adults of several species of flour beetles were placed in sifted, humidified flour mixed with years. The beetles burrowed into the flour at once and laid eggs in the tunnels they made. The larvae, or meal worms, which hatched 5 to 10 days after egg deposition, also burrowed into the flour. Both the adults and the larvae ate any eggs and pupae they encountered. Male and female adult encounters, however, often resulted in copulation and the laying of more eggs. As the population increased, inhibitory substances released by the beetles accumulated. As the amount of these substances increased, females produced fewer eggs, larvae developed more slowly and adults grew to a smaller size.

A steady population levei was finally reached, characteristic of the particular species of flour beetle being tested. The carrying capacity of that environment for that species had therefore been reached. A typical result was a steady increase of population to a stable level characterized by a fixed ratio of adults to larvae and a 90 per cent or higher mortality rate of eggs through cannibalism. But some species never equilibrated; instead, their population levels oscillated repeatedly in a pattern reminiscent of a Malthusian prediction. Such oscillations can result when large cohorts, groups of individuals of the same age, reduce the size of succeeding cohorts by cannibalism. The smaller cohort is, incapable of having the same effect on the large cohort that follows it. Such an effect can persist indefinitely.

Although other factors are also involved, the beetle experiment is primarily a case of population regulation by density-dependent intraspecific competition. In another series of experiments, separate mouse populations, with unlimited food and water were established in six 2-by-8- meter pens. In three of the pens, the food, water, and nesting boxes were scattered throughout the pens; in the other three pens, all the food and water was placed at one end, and all then nesting boxes were crowded together at the other. The maximum number to which each population rose differed from 25 to 138 among the six pens. Although all were adversely affected by the crowding, the particular level that was reached in each case seemed related more to individual temperamental differences in the animals than the immediate effects of the physical, conditions. As the population grew, so did the fighting among adults.

One population showed a decline in birth rate and a high mortality in the litters. In three of the six populations, the primary limiting factor was an incrcased mortality of newborn. Fighting often was followed by trampling of nests, which led to desertion or cannibalism of the litters by the females. Although food was abundant, food intake declined as the population increased. In some populations, food intake was so reduced as to lower normal female fecundity, the ability to produce offspring. But even in populations in which enough food was eaten to maintain fecundity, birth rates dropped off. This appeared to be due to less frequent copulation under the crowded conditions, increased spontaneous abortions (miscarriages), or both. Levels of female sex hormones were shown to be affected by increased odors resulting from crowding, with the result that ovulation ceased and sexual behaviour was inhibited. Similar studies were performed on rats in crowded conditions with unlimited food and water. In these experiments, population size proved to be regulated principally through deaths from fighting.

Most deaths were due to wound infections or other aftermaths of fighting. The mouse and rat experiments are cases of density-dependent population size regulation by behavioural and physiological means. Such studies of intraspecific relationships and carrying capacity have revealed some important principles:

1. Even under controlled conditions, it is sometimes impossible to predict
   (a) whether a given population of any reach a stable level;
   (b) whether it will oscillate slightly about a certain level; or
   (c) whether it will exhibit a Malthusian pattern of' catastrophic declines alternating with periods of exponential growth.
2. The natural methods by which populations are limited vary greatly from species to species and sometimes from population to population within a species. Besides such factors as limited food supplies and predation controls that limit population size in nature appear to include
   (a) cannibalism of the young;
   (b) deaths of adults from increased fighting; and
   (c) reduced fecundity, birth rates, and rates of development as a result of the accumulation of toxic substances or other effects of crowding. Nature's way of regulating population sizes obviously includes methods that are not regarded as desirable or acceptable ways of limiting human populations.

Interspecific Relationships of Populations

We have seen how some insight into population dynamics can be gained by studying experimental ecosystems containing single species of organisms. An even better understanding of natural ,controls on populations can be obtained by studying cultures of more than one species. Two rather different examples are

1. cultures in which two similar species compete for the same resource and
2. cultures made up of a predator and its prey.

Competition between similar species

In ecological terms, competition is the interaction of two organisms seeking the same things, such as food, shelter, space, or mates. The keenest competition most often occurs between members of the same species or two similar species. A 1934 experiment performed on two species of ciliate protozoans by the Russian biologist G. F. Gause nicely demonstrated some of the effects of competition, on population dynamics. Gause found that separate cultures of *Paramecium aurelia* and *Paramecium caudatum* reached a high maximum stationary phase when raised on bacteria. When he grew both species in the same culture vessel, they competed for the available food.

As long as food was abundant, both populations grew exponentially; but as food became depleted, the growth of both populations dropped to a slower than normal rate. Later, the population of *P* caudatum severally declined, while that of *P aurelia* continued to increase. Eventually, the population of *P aurelia* reached about the same size it had achieved when grown alone. Gause obtained similar results when he grew other species of competing organisms. The experiments and various field studies led to the development of what came to be known as Gause's rule or the *competitive exclusion principle*, which states that in any one geographical area, an ecological niche cannot be fully and simultaneously occupied by stabilized populations of more than one species. This means that if two very similar species with very habits are ever found in the same locality—a rare occurrence—it can be presumed that the situation is unstable and that one of the species may soon disappear from the area.

Occasionally, seeming exceptions to this rule have been observed on closer examination, however, they have usually proved to be cases in which the niche of the two species did not overlap completely. For example, if one of two competing species of birds occasionally leaves its preferred food to its competitors and feeds on another food, the two do not occupy the same niche. In such a case, the two species may be able to coexist indefinitely in the same area. The longer two such species live in the same area, the less overlap they exhibit in appearance and behaviour, a phenomenon called *character displacement*. The less similar they become the less they will compete with each other. Such realizations led to the development of a corollary to the exclusion principle, namely, the *coexistence principle*, which states that different species that coexist indefinitely in the same ecosystem usually have different ecological niches. Exceptions to the competitive exclusion principle occur in at least two types of situations:

1. when a predator regularly feeds on two competing species, thereby preventing the potentially dominant one from excluding the other; and
2. when regular environmental disturbances, including those of human origin, favour the survival of the competing species that would have been excluded in the absence of the disturbance.

Predator-prey relationships

Other experiments performed by Gause shed light on predation as one of the checks on population size and density. Gause cultured *Paramecium caudatum* along with one of its most voracious predators, *Didinium nasutum*, another ciliate. First, Gause prepared several cultures and introduced five paramecia into each. (In the absence of predators, paramecia double in number as often as every 6 hours.) Two days later, he added three didinia to each culture. Within a few days, the didinia had eaten all the paramecia, and a few days after that, all the didinia had died of starvation Gause repeated this experiment but placed sediments at the bottom of the culture vessels to provide hiding places for the paramecia.

As before, the population of paramecia increased: when he added the didinia, they fed on the paramecia, drastically reducing their numbers. But in these cultures, some paramecia escaped by hiding. The didinia, which need a steady diet, eventually all starved and, in the total absence of didinia, the surviving paramecia increased exponentially and achieved the same dense concentration as they usually did in the predator's absence. Gause then repeated the first experiment, with no hiding places provided, but this time, he added one paramecium and one didinium to the culture every third day. This procedure was intended to simulate the occasional immigration of individuals from other area. The result was that both species, survived indefinitely in sizable numbers, although a series of population oscillations of both occurred. One principle derived from these and similar experiments is: that it is not beneficial for a predator species to eat all of its prey individuals.

Field studies have shown repeatedly that, in nature, most predators feed on several prey species, but they almost always feed most heavily on whatever prey species is, most abundant at any given time. When the numbers of a particular type of prey decline, the predator usually switches to the prey species that is then most abundant. This behaviour pattern among some predators apparently enables both prey and predator populations to survive indefinitely. It probably accounts for some of the oscillations in population size so often observed in nature. Population oscillations are particularly evident when the numbers of all of a predator's prey species are so used as to be unable to support a large population of that predator; then the predator usually starves in massive numbers. But it seldom becomes extinct in the area because some individuals usually find enough food to survive.

The population oscillations that commonly occur in many prey and predator species are especially evident in Arctic and sub-Arctic regions, which are subject to extreme annual climatic changes and in which the animal communities are composed of relatively few species. Among the factors that help both predator and prey to survive are hiding places for the prey, a variety of prey species on which the predator can feed occasional declines in predator populations, and occasional immigration of both prey and predator from other areas.

Some oscillation of prey species also occurs in the absence of predation, indicating that predation is only one, and perhaps not even the most important, of the factors that keep natural populations in check. Oscillations in food availability for herbivores, for' example, can have dramatic effects, Nevertheless, an important key to the survival of both a predator and its prey in a particular area seems to be the safeguarding of the prey from predation when its populations are at their lowest level.

Predation of canadian lynxes on snowshoe hares

Although careful field studies of fluctuations in prey and predator population sizes have been made only in the last half century, data useful for such a study had been collected since the early nineteenth century across an extensive area of northern Canada. This fortunate circumstances came to light when it occurred to D. A. MacLulich to examine the old sales and purchase records of Canada's largest fur dealer, the Hudson Bay Company. He discovered that two of the animals long trapped for their fur were the Canadian lynx and its most important prey, the snowshoe hare. The number of pelts of each that were bought and sold in a given year represented trapped samples of these two populations in northern Canada. Pelt sales probably reflected the oscillations in the populations of the two species.

The fur company records suggested that the size of the lynx population was directly related to the size of the hare population. When there were many hares, the lynxes also thrived. Then, perhaps the lynxes ate the hares at a faster rate than the hares reproduced, many lynx began to starve, and their numbers felt this would have enabled the here population to increase. As the graph based on pelt sales indicates, the lynx population peaks always followed the here population peaks by about a year. Here population peaks occurred only after a decline in lynx numbers. The important principle illustrated by this example is that predation by a carnivore is probably an important immediate check on the population size of its prey species. However, the principle is not as applicable to predation on prey that is larger and more powerful than the predator. There is a growing amount of evidence to indicate that predation by carnivores on such animals as moose, mountain sheep, and deer is largely limited to the old, the sick, and the young. Any substantial limitation of the prey population by predation in such instances is apparently achieved primarily by reducing the number of young individuals.

The Effects of Predation

In the past half century, there has been a widespread eradication of many predatory animals in various parts of the world. This has been especially true where the predator is believed, rightly or wrongly, to be a threat to the lives or safety of humans or their livestock. In many cases, the belief that a particular "varmint" is a major threat to livestock has been altogether erroneous. As a consequence of, such fears, wolves, bears, coyotes, eagles, and many other-predators have been eliminated from large areas of the United States, Asia, and Africa.

Timber Wolves and White-Tailed Deer

After timber wolves, the natural predator of white-tailed deer deliberately eliminated from northern Wisconsin by unrestricted hunting, the deer population rose to unprecedented levels. Shortly after World War II, the available vegetation was no longer adequate to support such large numbers, and thousands of deer starved to death despite the fact that emergency feeding stations were provided by the state government. The deer population is now kept down by controlling the length of the hunting season and allowing both bucks and does to be shot in some years. Wolves are also being reintroduced in some areas.

The Dingo

Another example of a campaign to eliminate a predator is that waged against the dingo, the yellow dog of Australia. This wild dog was apparently introduced 8,000 to 10,000 years age by human immigrants from Asia. It is hated by ranchers, in part for preying on beef cattle, an important export commodity in Australia. One early attempt at dingo control following World War II involved dropping poisoned meat onto cattle ranges from airplanes. Wary adult dogs were seldom poisoned by this technique, although young ones were.

A more detailed study of the dingo's eating habits then showed that they subsisted mostly on small animals such as rabbits. It now appears that only during times of severe drought on the range do the dingoes, naturally solitary hunters, form packs to prey on cattle, and then they usually kill only calves. When interviewed, some ranchers noted that the killing of calves during a long drought improves the chances of a herd's survival until the drought is over. Most ranchers, however, continue to war against the dingo.

Starfishes Versus Barnacles and Mussels

Among the benefits of predation is that it sometimes prevents local elimination of one or more prey species, a fact already noted as an exception to the exclusion principle. This unanticipated effect on species diversity was revealed by a study of the populations of barnacles and mussels in seashore areas from which all sea stars (starfish) had been removed. The starfish normally feed on both barnacles and mussels. When the starfish were removed, the numbers of barnacles and mussels increased greatly. As they competed with each other for the available spaces on rocks in the intertidal zone, one species of mussel preempted the space, reducing the number of species of marine invertebrates in these areas from fifteen to eight. Because the predator had prevented competitive exclusion from running its course, it thereby enhanced species diversity.

Benefits of predation

Paradoxically, despite limiting its size predation usually tends to be beneficial to a prey species population. When the old, weak, diseased, and crippled are eliminated by predation, the surviving population tends to consist of the healthiest and most vigorous individuals. Eliminating large numbers of young during times of population, population expansion also reduces competition among the remaining individuals for existing food and other resources. The elimination of a predator from a region has sometimes been followed by a dramatic increase in the prey population, often beyond the carrying capacity of the area.

Different rules for small predators

Although large carnivores that prey on species that con injure them may usually limit their attacks to the old, sick, crippled, and young, this is not the case with small predators, especially when the prey is relatively defenseless. A snake is likely to eat any frog, an owl any mouse, and a spider a wide variety of insects. Moreover, Some small predators feed only on a single prey species, a situation that is quite unusual. Being limited to a single prey species means that the predator depends on 'he continued existence of that prey species for its own existence. If the predator is able to attack members of the prey species of all ages with impunity, both predators and prey could be liable to extinction in areas in which both are found.

The same principle also applies to species-specific parasites and to any animal that depends on a single plant species for food. In most situations, a predator species that enters an area in which its only prey species exists may virtually eliminate that species from that area. Only chance migrants escape to establish colonies lese where. The predator then also tends to be eliminated from the area by starvation, with perhaps a few individuals surviving by encountering remaining prey or by migrating to other areas in which the prey is found. The result of such a relationship is a wide scattering of both the prey species and its predator, and both are difficult to find in anyone area. Occasionally, local population explosions of the prey species may occur, to be followed, sooner or later, by a corresponding increase in the predator species after some happen into the area. Both may then largely disappear again from the area for a time.

SYMBIOSIS

Some organisms live much of the it lives in a very special ecological relationship an important part of their environment is a member of another species. In the traditional definition, a relationship in which members of two different species living in intimate relationship to each other, is termed *symbiosis* (from the Greek *sym*, "together," and *bios*, "life"). Symbiotic relation ships, which are always beneficial to at least one member of the ,association, may vary from associations of extreme intimacy and interdependency to relatively casual associations and from very temporary relationships to those involving the entire life span of the symbionts. Some relationships are so intimate that one lives on or in the other organism is termed the *symbiont*, while the one on or in which the symbiont lives is called the host. Such highly intimate symbioses can be classified as commensalism, mutualism, or parasitism.

Commensalism

Commensals (from the Latin *com*, "together," and mensa, *"table"*) may share the same food, and the symbiont may gain shelter by living on or in the host. But even if the symbiont shares the host's food, it causes no particular harm. For example, the bacterium *Escherichia coli* lives in the human colon it benefits from the nutrients, warmth, and shelter found there, but causes no disease or even any discomfort. Similarly, the protozoan *Opalina ranarum* lives harmlessly in the colon of the frog. In both eases the symbiont depends on the host for its supply of food.

Mutualism

In *mutualism*, some reciprocal benefit accrues to both partners. The flagellate protozoans that digest the cellulose of wood particles while living in the gut of termites are good examples. The termite needs the protozoans to digest the wood it eats the protozoans need the termite for the shelter and the nutrients it provides. When termites are experimentally rid of their protozoans, they soon begin to slarve, even when provided with all the wood they can eat. Many highly evolved mutualistic associations occur in nature, some of them very interesting. For instance, lichens are associations between certain kinds of algae and fungi. Other classic examples are provided by several species of tropical acacia trees that are hosts to particular species of ants.

The trees provide the ants not only with a safe home (their sharp thorns, which the ants hollow out) but with two types of food as well (special leaf products rich in proteins, fats, and vitamins, and a special nectar rich in sugar). The ants, for their part, destroy story injurious insects that attack their tree host. Moreover, ant patrols destroy other plants, including other acacia trees, in the immediate vicinity, thereby ensuring their host tree access to adequate sunlight, soil nutrients, and moisture. Mutualism combined with parasitism exists between some fresh water calms and fish.

Most species of freshwater clams have a larval stage, the glochidium, which is a parasite of freshwater fish. The gloehidia are discharged from the female clam and usually settle to the bottom, where, upon encountering a fish of the right species, they attach to its

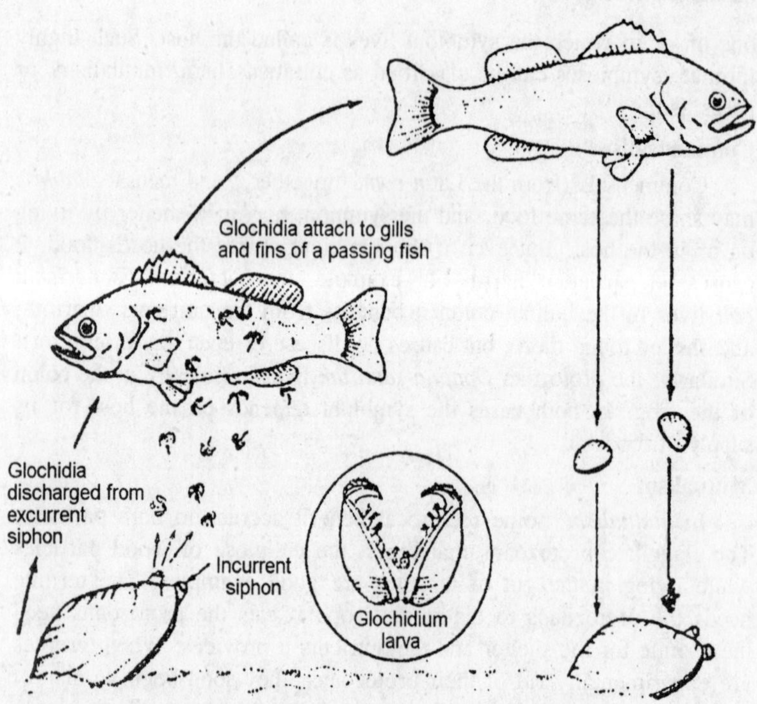

Fig. 13.8. The freshwater clam spends its larval stage, the glochidium as a parasite on the gills and fins of a fish. At metamorphosis, it leaves the fish to take up its adult life.

gills or other parts of its body. After 10 to 30 days as a parasite, they metamorphose and leave the fish as juvenile clams. The parasitic infection does not greatly harm adult fish, and the lesions soon heal after the glochidia drop off.

With some species of freshwater clam and fishes, such a parasitic relationship also involves mutualism. The siphons of the female clams become conspicuously coloured when the glochidia are ready to be expelled. The colouration attracts females of certain species of fishes when their fertilized eggs are ready for deposition. The fish inserts its ovipositor, a tube-like organ used in egg laying, into the clam's incurrent siphon and releases its eggs, which triggers the expulsion of glochidia by the clam. The fish's eggs are thereby provided a safe, well-aerated place in which to develop until hatching. The glochidia not only derive nutrients from their host but are also given a rapid and wide dispersal not otherwise available to a show-moving clam, thereby greatly increasing the species chances of survival.

Parasitism

Parasites receive benefit from their hosts while harming them. For example, the protozoan *Plasmodium*, the causative agent of malaria parasitizes both mosquitoes and vertebrates, each at a different stage in its life cycle. Another parasitic protozoan, *Trypanosoma*, produces African sleeping sickness; it is transmitted by the bite of the tsetse fly. The usual host of a particular species of parasite can be expected to exhibit a certain amount of resistance to the parasite that is it limits the degree of infection and is usually not quickly killed by the parasite. Parasites that always kill their hosts soon after invading them tend to become extinct.

14

Symbiotic Relationships

One of the most striking features of marine organisms is the large number of close association that can be observed among many different species. These are a series of association that are not predator-prey or herbivore-plant relationships in which one member consumes the other. These close relationships between unlike species generally seem to be either unharmful to either member or, more likely, to be beneficial to one or both. *Symbiosis* is the name given to such associations and means an interrelationship between two different species Symbiotic relationships are found in the terrestrial environment as well as in the aquatic, but appear to be disproportionately common, extensive, and well developed in the marine environment, such that they merit separate coverage in a book such as this.

Definitions and Coverage

Symbiotic relationships cover a broad spectrum of associations from random, casual, or facultative associations through more and more obligatory groupings that benefit one or both members to finally those that are parasitic. Such differences in the degree of association have led to the subdivision of symbiosis into more narrowly defined groupings. The term *commensalism* is often used to refer to an association that is clearly to the advantage of one member while not harming the other member. *Inquilinism* is a special subdivision of commensalism, in which an animal lives in the home of another, or in its digestive tract, without being parasitic. *Mutualism* is that form of symbiosis in which two species associate together for their mutual benefit. In a mutualism relationship the partners are often called *symbionts*. In a commensal or inquiline relationship, however, the partner gaining advantage is called the *commensal* and other the *host*. Parasitism generally refers to an association in which one species lives in or upon another and draws nourishment from that species at the expense of, or to the detriment of, the other.

In other words, it is an association in which the advantage is solely to one member at the expense of the other Parasitism is an association that seems to be extremely common, in both marine and terrestrial environments. It will not be discussed in this chapter, since it contains few features unique to the marine environment. Basically, there are two broad groupings of nonparasitic symbiotic associations in the sea: those between algal cells and various invertebrate animals and those between various animals, both vertebrate and invertebrate. We shall consider these in that order.

Symbiosis of Algae and Animals

All symbiotic relationships in the sea between plants and animals are between unicellular algae or their parts and a wide variety of marine invertebrate animals. These symbiotic relationships have been shown to be most common in tropical waters, but they also are prevalent in temperate oceans. They do, however, appear to be absent, or virtually so, from polar waters. Furthermore, the associations are, for obvious reasons, restricted to very shallow subtidal or intertidal areas or to the uppermost layers of the pelagic realm where sufficient light is present.

Types and Composition of the Associations

There are basically two types of symbiotic association between algae and invertebrates. The more common is to have the entire functioning algal cell associated with the invertebrate animal. The second is to have only the functioning chloroplasts from the algal cell incorporated into the tissues of the invertebrate body. The algal cell symbionts have been typically classified into groups on the basis of their colour. *Zooxanthellae* is the name given to those cells appearing brown, golden or brownish yellow in colour, *Zoochlorellae* to those that appear green. A third, smaller group appears blue or bluish green and has been called *Cyanellae*. These colour groups, however, do not distinguish among the actual algal species involved. Zooxanthellae, the most common in the seas of the world, are primarily species of dinoflagellates but include a few diatoms and cryptomonads. As Schoenberg and Trench have noted, the most common zooxanthellae species is the dinoflagellate *Symbiodinium microadriaticum*.

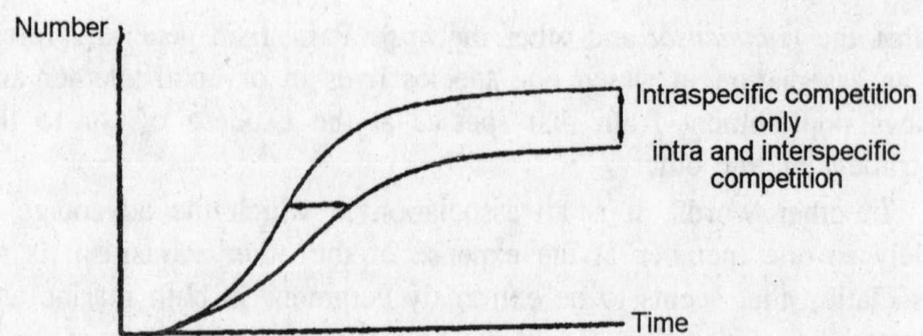

Fig. 14.1. Effect of introducing an element of interspecific competition into the population.

The zoochlorellae are much less common in the sea, but dominate symbiotic associations in fresh water. Among the marine associations, zoochlorellae include the alga *Platymonas convolutae* and unknown species of Chlorophyceae. Cyanellae are all blue-green algae, but the present, the identity of the species involved remain unknown. They are most common as symbionts in sponges and in planktonic diatoms. The algae all occur inside the bodies of the animals and most are found either within vacuoles inside individual tissue cells or else in various body spaces between or within tissue layers. Generally, the algal cells are restricted to certain tissues or areas of the host, and they grow and reproduce within the invertebrate without undergoing digestion by the host.

Table 14.1. Summary of the types of associations between algae and marine invertebrates

Algae symbiont group	*Algal taxon*	*Invertebrate animal host taxa*
Zooxanthellae	Dinoflagellata	Protozoa, Porifera Cnidaria, Platyhelminthes Mollusca
	Bacillariophyceae	Platyhelminthes *(Convoluta convoluta)*
	Cryptomonadida	Protozoa
Zoochlorellae	Pyramimonadales	Platyhelminthes (*Convoluta rescoffensis*)
	(?) Chlorophyceae	Cnidaria *(Anthopleura sp.)*
Cyanellae	Cyanophyceae	Porifera, Protozoa
Chloroplasts	Chlorophyceae	Mollusca (Sacoglossa)

Among those associations that involve only the chloroplasts and the invertebrate body, the chloroplasts are usually derived from the cells of larger green algae, which are first ingested by the animal in feeding and subsequently are transferred from the digestive tract of

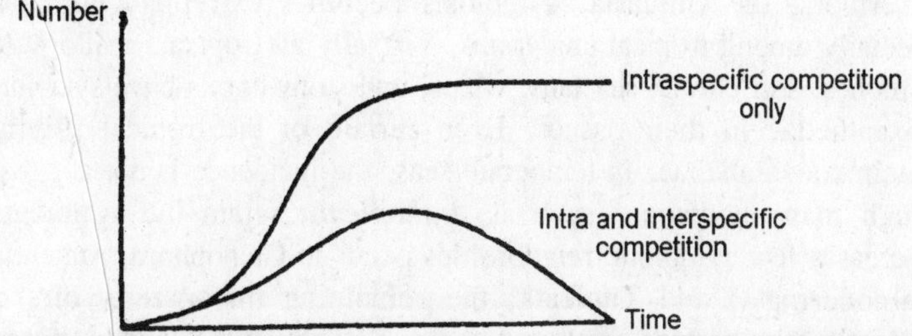

Fig. 14.2. As fig. 14.1, but strength of competition has increased such as to cause extinction.

the animal to other tissues. Algal symbiosis, either cellular or chloroplast, is widespread among various invertebrate groups. Symbiotic associations occur in Protozoa, Porifera, Cnidaria, Platythelminthes, Mollusca, and Echiura. Within these phyla, McLaughlin and Zahl report about 130 genera having algal symbionts. It is most common among Protozoa and Cnidaria.

Origin of the Association

Although we will probably never know the exact origin of algal-invertebrate symbiotic relationships, it is possible to outline some suggestions. In all the phyla mentioned above that contain symbiotic algae or chloroplasts, the final phase of digestion is intracellular, where individual cells take up particles from the stomach. It is thus possible to conceive of the origin of this association through the ingestion by cells of the digestive tract of either intact algal cells or chloroplasts, taken in by the animal during feeding. If, then, the algal cells were resistant to digestive action or the animal lacked enzymes to digest plant cellulose, the basis for a new association would be laid. This association could then evolve into a more obligatory relationship, provided that the new association conveyed an enhanced survival value to both symbionts. Since these associations, in fact, are common, they must have some selective value.

Distribution of Algae-Invertebrate Associations

Although there is not space or time to list and discuss all the species in which algal symbionts occur, it is of use to discuss briefly the various groups in which this association is prevalent. Beginning with Protozoa, symbiotic relationships are found in all of the epipelagic planktonic Radiolaria where the zooxanthellae occur in the outermost forthy layer. They also are found in a number of planktonic Foraminifera and even in marine ciliates. Symbiosis is not common among sponges (Porifera), but has been recorded in a few species, zooxanthellae in *Cliona* and cyanellae in *Demospongia*.

Among the Cnidaria, symbiosis becomes extremely common, especially among tropical cnidarians. Virtually all tropical, shallo water anemones, soft corals, sea fans, whips, and stony corals have symbiotic zooxanthellae in their tissues. Even certain of the tropical jellyfish contain zooxanthellae. In temperate seas, the incidence is not as great, though many anemones such as *Anthopleura* retain the symbionts. Whereas a few symbiotic relationships occur in Ctenophora, Annelida, Echinodermata, and Tunicata, the remaining major reservoirs of symbiotic relationships are found in the Platyhelminthes and Mollusca.

Among marine flatworms, the classic case is that of the genus *Convoluta,* which has been extensively investigated. Symbiotic relationships in the Mollusca are generally restricted to certain marine gastropods of the order Sacoglossa, where the animals retain only the chloroplasts, and to the bivalve family Tridacnidae.

Modifications Resulting from the Association

The symbiotic association between the algae and the invertebrates is generally a very close one which has resulted in rather significant anatomical and physiological changes in both the algal cells and the various invertebrate hosts. Perhaps the most profound changes that occur as a result of the association are found in the algal cells. Most marine symbiotic algae are dinoflagellates. The symbionts, however, have lost their locomotory flagellae and the characteristic grooves around the body. Furthermore, they have cell walls that are much reduced in thickness.

The zoochlorellae in the flatworm Convoluta have lost even more, in that the cell wall disappears, as does the light-sensitive stigmata. These cells become little more than bags containing chloroplasts. If the above algal cells are removed from the host and grown outside the animal in culture, they will develop the characteristic flagellae, cell walls, and other organs of a typical free-living form. It is thus apparent that all the changes observed are a direct result of the symbiotic association. In the case of the animals, the changes vary with the type of organism and with the degree of interdependence established between the symbionts.

The single universal modification is that all of these invertebrates live in very shallow water, where they can obtain adequate light so that the algae can carry on photosynthesis. Perhaps the least amount of modification occurs in the lower invertebrates, such as Protozoa and Porifera, where the algae occur as simple inclusions in the cytoplasm of the animal. No special anatomical modifications are observed.

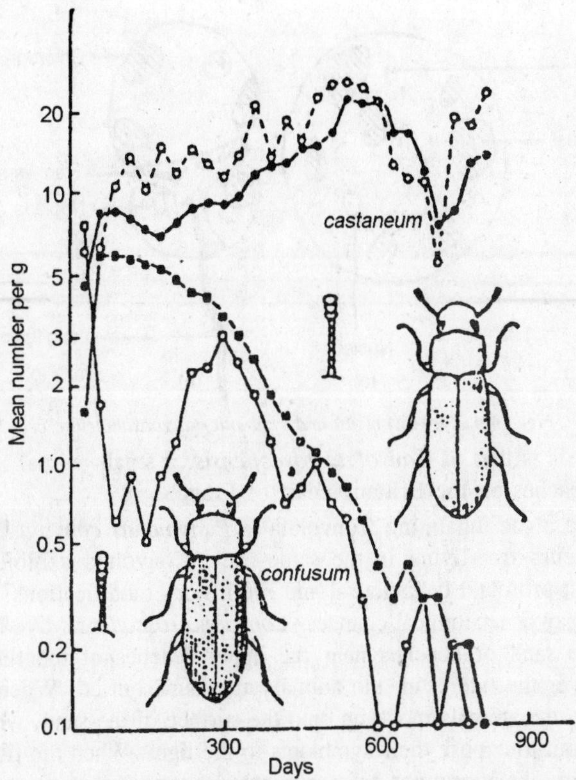

Fig. 14.3. Growth of populations of the flour beetles, Tribolium castaneum and T. confusum in separate and in mixed culture.

Among the Cnidaria, definite modifications begin to become apparent. The algae occur in marine Cnidaria in the innermost of the two cell layers, the gastrodermis. The numbers of zooxanthellae in different corals vary, and those that have the most seem to have reduced tentacle size, indicating that they are less able to capture zooplankton food.

Among certain soft corals of the family Xeniidae the digestive regions of the animal are reduced and the animals are not responsive to animal food. In the jellyfish *Cassiopeia*, a striking modification is behavioural. These animals, rather than swimming in open waters as do most jellyfishes, lie upside down on the bottom in shallow tropical waters, exposing their oral arms to the light in order to illuminate their algae. The oral arms are also much enlarged and expanded to provide more area of habitation by the algal cells. Truly a strange way of life for a jellyfish! Among flatworms, the most studied case of

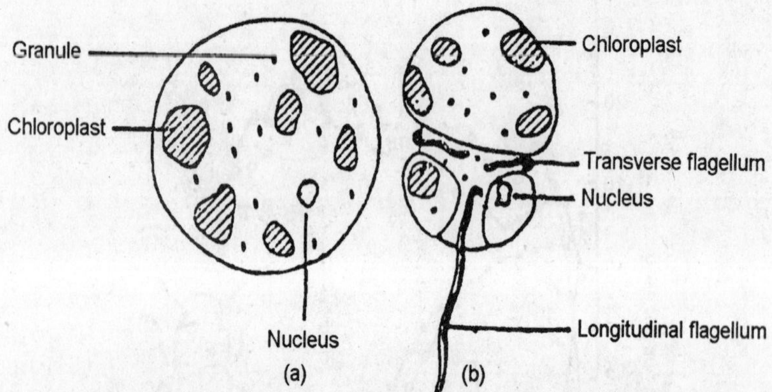

Fig. 14.4. Free-living (b) and symbiotic (a) zooxanthellae cells.

symbiosis is that of *Convoluta roscoffensis*, a small animal inhabiting sand beaches on the Brittany coast of France.

The algae inhabiting Convoluta is *Platymonas convolulae*, which also occurs free-living in the same area. Convoluta exhibit perhaps the most profound behavioural and life history modifications, but little in the way of anatomical changes. *Convoluto roscoffensis* live burrowed into the sand of beaches near the upper reaches of the tidal zone. Whenever the tide is in, the animals are found buried. When the tide recedes, the animals move up onto the surface of the sand, where they spread out to expose their symbionts to the light. When the tide begins to return, the vibrations trigger a reburrowing response so that the animals are safely under the sand again before the rising tidal waters are able to sweep them out.

Even more significantly, as Smith notes, these animals apparently do not feed as adults. Furthermore, if the young worms do not ingest the algal symbionts upon hatching, they will not complete development and will die even if they feed. It is among the mollusks, however, that we find the most dramatic changes of all resulting from the association. Although there are thousands of species of mollusks in the seas of the world, symbiotic associations with algal cells have developed in only seven species. All are bivalve mollusks in a single family, the Tridacnide, which include the giant clams. Six of the species are in the genus *Tridacna* itself and the remaining one in the genus *Hippopus*. All tridacnid clams are distributed only in the Old World tropics of the Indo-Pacific. They also include the largest bivalve mollusk in existence : *Tridacna gigas*, which has been recorded by Rosewater to reach 4ft in length and 580 lb in weight.

The other species are more modest in size, but all are still large in comparison with most other bivalve species. These tridacnid clams are inhabitants of coral reef areas, where they are found abundantly in the shallow, sunlit water. They usually have a very uncharacteristic position for a bivalve mollusk, in that they lie either on the surface of the bottom or bored into coral or coral rock with the opening between the valves facing up toward the surface of the water. The valves usually gape widely, and within this opening can be seen an extensive, brightly coloured tissue layer. It is in this tissue layer that the symbiotic zooxanthellae are found. The brightly coloured tissue exposed within the gape is siphonal tissue; the colour results from the interaction of various pigments deposited there. The reason for the bright colors is that they act to protect the tissues of the clam from the damaging effects of sunlight, while passing enough light to allow the zooxathellae photosynthesize. What is truly remarkable about tridacnids, however, is the tremendous change that their bodies have undergone from that of a typical clam in order to accommodate this symbiotic association.

Bivalves generally rest with their foot either embedded in the substrate or held against it. In this position, a normal clam has its hinge uppermost. Obviously, this would not do for a tridacnid, since such a position would not permit the tissue and zooxanthellae to be illuminated. The result, as Yonge has described, is that the entire tridacnid has undergone a tremendous rotation with respect to the foot, so that the hinge comes to lie on the underside next to the foot, and the opening of the shell faces upward. At the same time, the siphons and siphonal tissue underwent an expansion and grew and extended themselves, covering the length of the upward-facing opening and providing the expanded area for occupation by the zooxanthellae. As a result of this rotation and expansion, one of the tridacnid's shell-closing muscles was lost. Because of these profound anatomical changes, the tridacnids are markedly different in body orientation from any other bivalves. These changes can only be attributed to the association with the symbiotic algae.

Symbiotic relationships with chloroplasts have been reported by Greene in the marine gastropods of the order Sacoglossa. Animals of this order regularly consume the contents of algal cells, and some species have evolved the ability to retain the chloroplasts in a functioning position on the dorsal surface. The only modifications so far observed are that they do not feed very often and the tropical forms provide a screen of lightabsorbing material above the chloroplasts to cut down the light intensity.

Value of the Association

Since symbiotic associations with algae are so common invertebrates, and since they often result in profound anatomicals or behavioural changes in the partners, it seems only natural to assume that they must have some positive value to each of the partners. What might be the value of such associations? We have already partially answered this question with respect to corals where we saw that the corals obtained food materials from the zooxanthellae and also that the zooxanthellae enhanced the ability of corals to lay down calcium carbonate.

In turn, the zooxanthellae in corals received nutrients in the form of nitrates and phosphates produced in the metabolic process of the coral, but rare in external waters. Among most invertebrates that have a symbiotic association with algal cells, the cells retain their integrity and are not digested by the animal to obtain nutrients. Thus, the nutrients that pass are in the form of chemical compounds. Energy containing molecules, such as glycerol, produced by the zooxanthellae in photosynthesis, pass to the animal, and nitrates and phosphates, needed nutrients for the algae, pass from the invertebrate to the algal cell. A similar situation also occurs between the sacoglossan mollusks and the symbiotic chloroplasts where Green has shown translocation of organic material from chloroplast to animal.

However, in those sacoglossans thus far investigated, the duration of life of the chloroplasts appears to be much shorter and the animals must periodically supply by ingesting the contents of algal cells. Among the giant clams, however, Yong has suggested that, in addition to reciprocal transference of nutrients, the animals may also actively digest the zooxanthellae cells. This would seem to be counter productive, especially considering the great anatomical changes that these animals have undergone in order to properly provide for their algal guests. Apparently, however, the clams can discern between healthy zooxanthellae cells and those that are degenerating or senile. As a result, the calm transports only the senile or degenerating cells from the outer mantle blood spaces, where they photosynthesize, into the deeper tissues, where they are consumed by the blood cells.

In this manner, then, the clams retain the symbiotic relationship, and at the same time, they cull the unfit algal cells to obtain additional nutrients otherwise lost when the cell dies. Other values may result from the association. As we have noted, the most of these associations are tropical; these waters are classically low in plant nutrients. It may well be that the algae benefit by the association through access to a larger and more reliable source of nutrients in the form of the metabolic products of the animal (NO_2, PG_4, CO_2) than they would obtain from the open water. Tropical waters also are lower in oxygen than temperate waters, because water at higher temperatures holds less oxygen. Since the photosynthetic process produces oxygen, it may be that the animals, especially in the crowded conditions of the shallow waters of the tropics, gain through additional amounts of oxygen produced by the symbiotic algae.

Establishment and Transmission of Zooxanthellae

Whereas each generation of sacoglossans must actually feed on algae to obtain their supply of symbiotic chloroplasts, perpetuation of the symbiotic association with zooxanthellae from generation to generation of hosts is accomplished differently among the different groups. Basically, two major routes are available. Either the zooxanthellae cells must be passed on directly from the parent to the eggs or larvae, or each new generation must reinfect itself a new from algal cells in the surrounding environment. Passage of algal symbionts directly to the next generation via the egg of larvae seems to be the method used by most Cnidaria.

In most corals, for example, the zooxanthellae enter the eggs at some time prior to their release from the parent. It is also assumed, though without much evidence, that the giant clams also transmit the zooxanthellae through the eggs. It has also been suggested the clams receive them from the surrounding corals. For other hosts, the perpetuation of a symbiotic relationship depends upon the reinfection of each generation. In such cases, the algal symbionts must be obtained a new from the environment. Among Protozoa and Porifera, this is the common means of transmission. Each new generation obtains the algal cells through ingestion of the free-living form of the symbiont. Surprisingly, this is also the case in *Convoluta roscoffensis*, where the association is highly dependent one. In this case, the algae-free larvae ingest the free-living *Platymonas convolutae*.

In order to ensure that the larvae will be infected, however, there has evolved a chemical that attracts the alga to the egg cases of Canidaria so that when the young hatch and begin to feed, the algae are present. Under such conditions, the new generation is virtually assured of reinfection. A final possibility for transmission is the ingestion by the potential host of food organisms themselves with symbiotic algae. Although this has not as yet been shown to be important, it remains an area requiring study.

Symbiosis Among Animals

We are concerned here with those special associations in which members of different species are regularly associated with each other in nonparasitic relationships. Such symbiotic associations are widespread in the sea, primarily in the crowded reaches of the epipelagic and shallow subtidal zones. As with algal symbiosis, such relationships appear to be somewhat more common or spectacular in the tropics, but are certainly also common in temperate seas.

Types of Associations

Symbiotic relationships among marine animals cover a broader spectrum that the strictly mutualistic associations we have seen among plants and chloroplasts and marine invertebrates. The simplest type of associations are commensal, whereon the "guest" lives on another organism "host," or in or upon some construction of that organism, such as a tube or burrow. Such associations are similar to epiphytic relationships seen among terrestial plants, wherein the epiphyte simply uses the other plant as a substrate without actually taking sustenance from it. In these cases, the commensal usually gains in some measurable way from the association and the host is not seriously inconvenienced. Marine commensals that live on or upon other invertebrates are called *epizoites*. Those that live inside other animals but are not parasites called *endozoites*.

Epizoites are extremely abundant in marine waters and many are probably not true commensals; the relationship in the result of organisms that normally settle on the substrate settling at random on the outside of a slow-moving or sessile invertebrate. These will not be considered further here. Other epizoites are highly specific, and a symbiotic relationship is certainly the case, with the "guest" somehow seeking out the correct "host." These symbiotic epizoites are the most abundant group of commensals and are spread among the phyla Protozoa, Cnidaria, Entoprocta, Annelida, Arthropoda, and Mollusca. Many invertebrates harbour specialized ciliate protozoans, either on their external surfaces, internally in the digestive tract, or among the gills. For example, the vorticelled ciliate *Ellobiophyra donacis* is restricted to the gills of the clam *Donax vittatus*, while the collared ciliate *Lobochona prorates* is found on the telson of the isopod crustacean *Limnoria tripunctata*. Many other examples are also known.

Endozoic commensal ciliates are also common, especially in the digestive tracts of larger animals. Thus, for example, Beers found that the large green sea urchin of the cold temperate waters of North

America, *Strogylocentrotus drobachiensis*, has as many as seven species of ciliate protozoans in its gut, which are found nowhere else and die if removed. Among the Cnidaria, the class Hydrozoa furnishes most of the cases of epizoites. The best studied is the case of two species of *Probisodactyla*, two tentacled hydroids, which always occur on the rim of the tubes of polychaete worms of the genus *pseudopotamilla* in the North Pacific Ocean. Similarly, along the Pacific and, Gulf coasts, the hydroid genus *Clytia* occurs on the clams of the genus *Donax*. Other hydroids are epizoic on gorgonians pennatulids, and ascidians, but the closeness of the association is not always known.

Turbellarian flatworms are often endozoic in the digestive tracts of larger marine invertebrates and in the mantle cavity of various mollusks. Thus, the polyclad *Notoplana ovalis* occurs in the mantle cavity of the limpet *Patella oculis*, while the triclad *Nexillis epichitonius* occurs in the mantle cavity of the common Pacific coast chiton, *Mopalia hindsii*. The rhabdocoels *Syndesmis dendrastorum* and *Syndisyrinx franciscanus* are endozoites in the guts of the sand dollar, *Dendraster excentricus* and the sea urchins *Strongylocentrotus franciscanus* and *S. purpuratus*, respectively. The phylum Entoproeta is a small, little-known group, most of which, as Nielsen has documented, are epizoic on other marine invertebrates, particularly polychaete worms, where they inhabit respiratory current areas.

Among the Annelida, a number of polychaete worms are epizoites with definite host specificity. The best studied is the hesinoid *Ophiodromus pugettensis*, which occurs on the ambulacral grooves of the bat star, *Patiria miniata* on the Pacific coast. However, the whole group of so-called "scale worms" are all very common epizoites on various other echinoderms and mollusks. Among the class Crustacea, the epizoic forms occur mainly within three groups: copepods, amphipods, and decapod crabs.

Copepods associated with other marine animals are usually highly evolved and modified parasites, but there are a few that are simply commensal, scampering over the outer surfaces of various invertebrates. Such an example are various species of the genus. *Hemicyclops*, which move on the surface of the other, larger crustaceans. Still others, like *Paranthessius*, are found in the mantle cavity of various bivalves. One of the most specialized epizoic associations is that among several species of barnacles and various large whales. Two families of barnacles are found only on whales and several are species-specific. Thus *Cryptolepas rachianecti* occurs only on the California gray whales.

Eschrichtius gibbosus, and *Conchoderma auritum* attaches to *Coronula diadema*, which, in turn, is attached to the skin of whales.

A very curious form of epizoic relationship occurs between the amphipods of the suborder Hyperiida and the larger jellyfish. Hyperids are almost always epizoic or endozoic on jellyfish. Also epizoic or jellyfish are certain juveniles of various benthic crab species. Most mollusk epizoites are found among certain groups of clams. Generally, these are small clams that attach via a foot or byssus to the outside of various other invertebrates. Thus, *Montacuta ferruginosa* is found on the heart urchin, *Echinocardium cordatum*; *Orobitella rugifera*, attached via a byssus on the underside of the burrowing shrimp *Upogebia stellata*; and *Mysella pedroana* on the legs of the sand crab *Blepharipoda occidentalis*.

Among the gastropods found on other invertebrates the genus *Crepidula* is one of the most common. The above is certainly not an exhaustive list of the various epizoic and endozoic symbiotic associations, but gives an idea of the extent and diversity of such associations. Many, many more could be listed, but would not serve the purpose of this text. A second type of symbiotic association is that of organisms associated with animals forming a tube or burrow. In this case, the commensal or guest occupies the tube or burrow constructed by the host, but is not necessarily in close association with the body of the host. These commensals use the tube or burrow as a refuge from predation. They may or may not also tap the food source of the host. This type of symbiotic relationship ranges from one simply fortuitous, as in the case of an animal diving into any convenient burrow to avoid a predator, to examples of obligatory associations, where the commensals are not found other than in such tubes or burrows and are dependent upon them for continued existence.

In the former category are certain small gobiid fishes such as *Clevelandia ios*, which take refuge in any available tube or burrow but emerge to feed. Obligate tube dwellers include many small crabs of the family Pinnotheridae, known as "pea crabs." Many species of this family are known throughout the world, where they inhabit the tubes and burrows of polychaete annelids and other invertebrates, sometimes in species-specific associations. One of the best known examples of symbiosis with tube dwellers is that of the echiurid worm *Urechis caupo* on the Pacific coast of North America. This animal creates a permanent U-shaped burrow in which can be found four or more commensals. At the upper end is the fish *Clevelandia ios*, which merely uses the tube as a refuge. Further in the tube and closely associated with the worm are the scale worm *Hesperonoe adventar* and the pinnotherid crab *Scleroplax granulata*. *Hesperronoe adventor* seems restricted to *Urechis* burrows and lives against the worm, snatching food from it. Although the crabs are restricted to tubes, they may be found in the burrows of other animals as well.

Another commensal is the small clam *Cryptomya californica*, which has very short siphons and inserts them into the burrow of *Urechis*, thus enabling it to live lower in the substrate than the length of its siphons would normally permit. Occasionally the shrimp *Betaeus longidactylus* may occur in the burrow. A similar situation prevails with the tube-building polychaete *Chaetopterus*. This genus is virtually cosmopolitan, building tough parchment tubes. These tubes are also inhabited by various species of pinnotherid crabs and scale worms, the species differing depending on the geographical locality. Another type of association concerns those invertebrates that live in the mantle cavities of various mollusks. This appears to be a very common site for commensals to occupy. Thus, the nemertine *Malacobdella grossi* lives in the mantle cavity of various clams on the Pacific coast. Pinnotherid crabs occupy the mantle cavity of numerous bivalves, including oysters, where they are considered a pest.

Polychaete worms inhabit mantle cavities of chitons and limpets. In the tropics, shrimps of the family Palaemonidae are often found in mantle cavities of large bivalves such as *Pinna*, *Atrina*, and *Tridacna*. The remaining types of associations are those that appear to be mainly mutualistic rather than strictly commensal, as the aforementioned. The organisms display varying levels of behavioural and physiological modifications for the association. One of the most obvious but not well understood associations is that which occurs between various crabs and sea anemones.

Many hermit crabs all over the world have anemones attached to their shells, and in some genera, such as *Dardanus*, the presence of anemones seems to be universal. Still other crabs, such as *Hepatus*, *Munidopagurus macrocheles*, and *Stenocionops furcata*, are found with anemones attached directly to their backs. The ultimate symbiotic relationship appears to be that evolved among a few crabs, which carry anemones on their chelae and actively use them for defence and/or food capture. Interestingly, most of the anemones involved in these associations are of three genera, *Calliactis*, *Paracalliactis*, and *Adamsia*, species of which are rarely found elsewhere than with crabs.

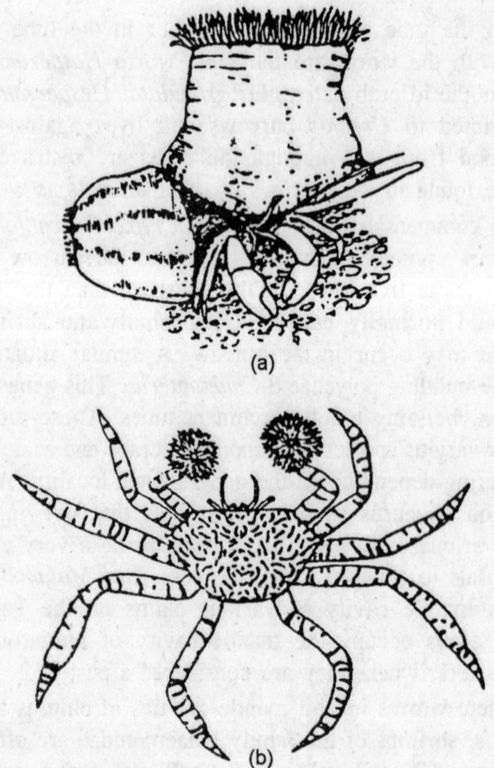

Fig. 14.5. Symbiotic relationships among crabs and anemones.

These associations must be mutualistic, because whenever the crab changes shells or molts its carapace, the anemone is usually also transferred. This is effected by the manipulatory movements of the crab, but requires a degree of cooperation on the part of the anemone in loosening its grip on the shell or carapace and allowing itself to be physically moved. Another type of association is that between other Cnidaria and fishes. Throughout the Indo-West Pacific in the coral reef areas, there is a striking association of small fishes of the genera *Amphiprion*, *Premnas*, and *Dascyllus* with various large anemones. These fishes are able to live nestled among the tentacles of the anemones by preventing discharge of the formidable nematocysts of the anemone tentacles. Other fishes are not able to prevent this discharge and the anemones can kill and eat fishes of a similar size.

A similar association seems to exist between the dangerous siphonophore, the Portuguese man-of war, and the little fish *Nomeus gronovii*, which swims among the tentacles bearing the powerful nematocysts. Still other juvenile fishes often congregate under the bells of the large scyphozoan jellyfish, where they presumably obtain protection from predation. A similar situation prevails between fishes and the long spined tropical sea urchin, *Diadema*. These urchins are circumtropical in shallow water and have enormously long, thin spines, which easily penetrate flesh. At least two tropical fishes, *Aeoliscus strigatus* and *Diademichthys deversor*, have adapted themselves to live among these spines, presumable for protection. They also look like sea urchin spines and so are camouflaged.

There is also the association of fish with other fishes, usually large, predaceous fishes. Pilot fishes and remoras are always found with other, larger fishes, other marine vertebrates such as turtles, or even any moving, inanimate objects. The final type of association is that known as "cleaning behaviour." As noted this is an association in which various species of fishes and shrimps actively attract large fishes to themselves for the purpose of cleaning them of various ectoparasites.

Origin of the Association

With such a diverse array of relationships as we have noted, it does not seem likely that there is any single cause for the origin of all the different types of associations. However, one factor common to all associations is that they have arisen in those areas of the ocean that are the most crowded with life. It would, therefore, appear reasonable that epizoites in particular could have become established due to the great competition for space among the settling invertebrate larvae. As a result, many settled upon other invertebrates and this then later evolved, in some cases, into more obligatory relationships in which those individuals that were commensal gained some advantage over the free-living forms. This is borne out by noting that many epizoites or burrow-inhabiting species are not strongly attracted to any given host species, while others are highly specific.

Similarly, many of the tube and burrow inhabitants, and those found in the mantle cavity of various mollusks, also probably had their origin in the search for space. After all, tubes, burrows, and mantle cavities really represent a certain amount of potentially occupiable, semienclosed space. The initial association of crabs and anemones may also well have begun when anemones began to take up residence on shells that were already occupied by hermit crabs.

Alternatively, the association could have arisen out of the habit of various crabs of picking up materials from the bottom and applying them to their carapaces to aid in camouflage. Many of the commensals that are epizoites or burrow-or tube-dwellers, especially the polychaeta worms, tend to be from groups that show strong *thigmotaxis*; that is, they respond positively to stimulation by contact with other, solid bodies. Such a behavioural response would naturally preadapt them to move into tubes or burrows or into various cavities in the bodies of larger animals. Once so established selection would favour the development of chemosensory clues enabling these animals to find such burrows, tubes, or cavities, and more specific associations would follow. Among the fishes and Cnidarians, it seems that the various types of symbiotic relationships have their origin in the protection from larger predators that such relationships confer.

It does not appear that this relationship stems from crowding, especially between epipelagic fishes and Cnidarians provided the symbionts could avoid having the nematocysts discharged, the tentacles of anemones and jellyfishes with their nematocysts represent a secure, formidable barrier to a larger predator. However, in the case of large sharks, pilot fishes, and remoras, the association probably originated in the search for food in the epipelagic. Pilot fishes and remoras probably were originally attracted by the amounts of food that were dropped in the feeding by the larger fishes, and they could have managed to establish themselves either by judiciously avoiding falling prey themselves to the sharks, or by being small enough to not be recognized as food by the larger fishes. Cleaning behaviour is a very complex relationship involving many fish species, and its origin and ecological functions have been confounded by various alternate hypotheses and factual inaccuracies as Gorlick et al. have discussed.

Modifications Due to the Association

As with the algal-animal symbiosis, the animal-animal relationships can result in anatomical, physiological, and behavioural modifications to one or both of the partners. The fewest modifications occur among the epizoites, especially those that are facultative epizoites and are found also in nonsymbiotic situations. Oftentimes, these animals have no anatomical modifications attributable to the symbiotic association, but those that are found with specific hosts usually have developed the means for *recognition* of the host from a distance. Such ability to recognize hosts is even more highly developed among obligate symbionts of various types. What is the basis for such recognition? Investigators such as Davenport have shown that it is more likely a chemical unique to the host. What is released into the water.

The commensal has, in turn, developed chemical sensory receptors, which detect the chemical and enable it to "home" in on its host. If such symbionts are tested in the laboratory in an apparatus designed to give them a "choice," they show a remarkable ability to choose the "host" on which they are found from among other closely related forms. Such attraction by chemical means has been demonstrated for several scale worms commensal with various starfish species; for the peculiar fishes, *Carapus*, which live in sea cucumbers; and for the polynoid *Hesperone adventor* and its host, *Urechis caupo*. Yet another modification of commensals and one that also aids in host recognition, is the ability of the larvae or young to select the appropriate substrate.

In many epizoite species that settle onto their hosts from the plankton, metamorphosis and subsequent development do not occur unless the larva is in contact with the appropriate host. The substrate discrimination abilities of some of these symbionts are truly remarkable. Some can apparently recognize substrate surface texture at the molecular level! It is not surprising, then that whale barnacles can settle not only on whale skin, but only on certain species of whales. However, in this case, it is likely that a chemical unique to the host is also present. Similarly, the clam *Modiolaria* is able to recognize the structure of tunicin, the major chemical component of the external layer of sea squirts with which it lives. A basic modifications of tube dwelling symbionts is that they are often smaller or thinner and flatter than their free-living relatives. Thus, the pinnotherid, or "pea" crabs, are among the smallest of the various marine crabs.

The shrimps that inhabit the mantle cavities of bivalve mollusks are also among the smallest of the shrimps. Similarly, the scale worms are often very flattened in comparison with other polychaetes. The fishes, such as *Clevelandia ios*, that are commensal in tubes or in the internal spaces of clams and sea cucumbers, *Apogonichthys* and *Carapus*, respectively, are very thin in order that they may fit into these restricted areas. The pearl fishes, *Carapus*, show perhaps the greatest modifications in that they have lost their scales and pelvic fins and have shifted the anal opening far forward under the head. This latter change apparently is to ensure that defecation will occur outside the body of the sea cucumber. Furthermore, as Arnold has shown, these fishes show behavioural modification in that they enter into the cucumber tail first. Such extensive modifications suggest a long history of adaptation.

Among those anemones found associated with crabs, a number of behavioural modifications occur. In the first place, the anemones must respond positively to manipulations by the crab when it rushes to transfer the anemone to another shell or from its shed carapace to a new one. This it does by releasing its grip on the old shell so that the crab can remove it with its claws. This requires a higher degree of nervous integration than we have come to believe could be accomplished by the extremely primitive, brainless nervous system of the anemone! We are thus currently ignorant of an understanding of this process among those crabs that carry anemones in their chelae.

The anemones again show a behavioural adaptation in that they allow themselves to be carried in the chelae and do not close up. In these cases, however, the crab also shows changes in that the chelae are modified to carry the anemones rather than gather food. Behaviourally, the crab extends the anemones when threatened or when gathering food; it uses its second pair of legs to transport food to the mouth, rather than the chelae, since the latter carry the anemones. Perhaps the most drastic changes undergone by the anemone member of these crab-anemone systems is that exhibited by *Adamsia palliata*, which lives with the hermit crab, *Eupagurus prideauxi*. Here, the anemone steadily expands its basal disc to completely encircle the vulnerable abdomen of the hermit crab, thus eliminating the need for the crab to change its shell as it grows. The anemone then expands to allow for the growth of the crab, meanwhile orienting itself so that the oral disc and mouth are ventral to the crab, thus enabling it to obtain food from the crab. The greatest morphological changes among fishes associated with other fishes occur in the remoras, where the animals have the dorsal fin modified to form a large sucker, which enables them to remain attached to their hosts.

Pilot fishes and the anemone fishes have no special morphological modifications. Among anemone fishes, however, there are definite behavioural and biochemical modifications. Studies have shown that anemone fishes of the genus *Amphiprion* must undergo a period of "acclimatization" to an anemone before they can, with impunity, dive into its tentacles. During this period, the fish is not capable of preventing nematocyst discharge by the anemone. During the acclimatization, Schlichter has demonstrated that the fish goes through a series of first brief and then longer encounters with the anemone, during which time, the fish coats itself with anemone mucus, thus tricking the anemone to discern the fish as itself. Hence, there is no nematocyst discharge from the anemone. When this coating is complete, the acclimatization period is over and the fish is able to dive in among the tentacles without causing nematocyst discharge. In addition to this fascinating coating process, anemone fishes have additional behavioural modifications.

In contrast to most fishes, which swim away when approached by humans or other large potential predators, anemone fishes swim out of their anemones toward the potential predator! This serves to attract the predator to them, at which time, they do an abrupt about-face and dive into the anemone. Presumably, the behaviour would tend to make an unwary potential predator chase them and subsequently be caught and consumed by the anemone. In order to further attract attention to themselves, anemone fish also have vivid color patterns, contrasting markedly with their background. In contrast to the anemone fishes, which develop immunity to the nematocyst of their host, the fishes associated with the Portuguese man-of-war do not acclimate, nor do they ever obtain complete immunity from the nematocyst.

They instead appear to spend their liver playing a perpetual game of Russian roulette with the tentacles of the siphonophore, constantly moving so as to avoid contact with the lethal nematocyst batteries! The same is true for the juveniles of various fishes that shelter below the umbrellas of the large jellyfish. Among cleaner fishes, certain morphological modifications can be noted, such as pointed, narrow snouts and forcepslike teeth, adaptations for picking up the small ectoparasites. Both cleaning shrimps and fishes also tend to have very bright and contrasting colors, which make them stand out against the background and advertise their presence.

Certain cleaning shrimps, notably the Pederson shrimp, *Periclimenes pedersoni*, have extraordinarily long antennae, which they wave about to further advertise their presence to the fishes. Cleaning behaviour relationships also impose behavioural modifications on the fishes cleaned. These animals must seek out and present themselves at the "cleaning station" and then remain motionless while the cleaner moves over their bodies. They must also open their mouths and spread their gills and opercula to allow the cleaner to enter the mouth and gill area. Of course, at the same time, the larger fish being cleaned must refrain from sudden inhalations, which might accidentally ingest the smaller cleaner!

Value of the Association

These symbiotic associations have obvious value to one or both of the members, and the relative value of the relationship to each varies depending upon the association. The major value of the symbiotic relationship to the epizoites would seem to be with the epizoite rather than the host organisms. For some epizoites, the value of the association lies in the fact that the host represents the only available suitable substrate in an otherwise unsuitable area. For example, the hydroid *Clytia bakeri*, which is epizoic on the clam *Donax gouldi* in sand areas in southern California, could not live in the area unless it had the clam to settle upon, as it cannot exist on sand. There is, however, no obvious value to the clam.

Other epizoites presumably fix themselves on the host to take advantage of feeding or respiratory currents produced by the host. This seems to be particularly common. The two-tentacled hydroids of the genus *Proboscidactyla* are fixed on the rims of the tubes of their annelid hosts to take advantage of the feeding tentacles and to remove food particles from the host. Similarly, the various epizoitic entoprocts are usually found where there are respiratory currents. In this latter case, they do not steal food directly from the host, however, but filter out the food from the moving current. Perhaps most epizoites benefit by the protection afforded them by the larger host. Thus, the various flatworms occurring in the guts and mantle cavities of their larger hosts are protected from predation and, in the case of intertidal forms, from possible desiccation. Surely, the hyperiid amphipods found on the large scyphozoan jellyfish would be much more subject to predation if free living in the plankton.

Finally, the epizoite may also gain the advantage of movement. This advantage may be of particular significance if the epizoite is itself sessile and the host freely moving. Commensals that inhabit tubes gain primarily protection from predation. This, for example, is the main value to the goby *Clevelandia ios* in the tubes of *Urechis caupo*. Other commensals, however, such as the scale worms gain not only protection but also food as they snatch morsels from their tube-dwelling hosts. Commensals in the mantle cavities of bivalve mollusks benefit from the protection of the bivalve shells, but also are able to employ the feeding and respiratory currents of the clam for their own needs.

In a few cases, these symbionts may tend toward parasitism in that they may elect to consume pieces of the gills or mantle of their bivalve host. Among the fishes associated with large jellyfishes, the Portuguese man-of-war, and the large sharks, it seems apparent that the whole value of the association is to the fishes in the form of protection from predation in an area notorious for great predation pressure. The pilot fish and remoras are also protected from predation by their association with the large predaceous sharks, but they may also obtain food when the shark is feeding. The fishes associated with sea-urchins and living in the respiratory trees of sea cucumbers are protected by their association, but their presence presumably has no correlative advantage to their host. Among the mutualistic associations, the values of the association are more pronounced and, as the term applies, extend to the host as well.

In the various crab and anemone associations, the crabs, as hosts, gain the advantage of the nematocyst batteries of the anemones, either as protection against predators or else as an offensive weapon to capture food. The anemones, on the other hand, while still sessile, gain the advantage of movement; some, such as those on hermit crabs, may also be able to obtain food fragments lost during the feeding process of the crab. A similar situation prevails between anemone fish and their anemones. The fishes gain mainly protection, but they benefit their partners by attracting prey to within reach of the tentacles and also by picking up pieces of food outside the reach of the anemone and depositing them in the anemone's mouth. In the case of "cleaning behaviour," the clear organisms gain is in the form of food, whereas the large fishes cleaned presumably enjoy an increased measure of health due to the removal of the various ectoparasites.

Luminescent Bacteria

A final category of symbiosis remains that does not fit well with the others and hence, its consideration here. That is the curious relationship that has developed between various marine animals and luminescent bacteria. In these relationships, the bacteria are usually confined to a cavity in the body of the larger animal near the outer surface. This cavity is usually connected to the outside through an opening of some sort. Such symbiotic relationships are confined to two groups of marine animals, fishes and squids, and are not common. The relationship is most common among fishes ad squids inhabiting the mesopelagic zone, but is still less common among these groups than its bioluminescence produced intrinsically without bacteria. The relationship of bacteria and the fishes or squids is a mutualistic one in which the bacteria obtain food from the larger animal.

In turn, the fishes or squid use the light produced by the bacteria for various defensive and/or offensive purposes, as discussed in a Chapter. The light produced by bacteria is usually continuous, and as a result, the fishes and squids often develop rather elaborate modifications to control the light. These anatomical developments can include the production of a reflecting layer behind the cavity to reflect the light outward, often lenses to focus and concentrate the light, and most often, a screen or shade on the outside which can be raised or lowered to either "turn on" or "turn off" the light. The bacteria show no corresponding changes in behaviour or morphology as a result of the association. The origin of these elaborate associations is not well understood, and the bacteria apparently are not passed from one generation of fishes or squid through the eggs. It would appear that each generation somehow must reinfect its light organs from the external environment.

INDEX